CURRENT

EVOLVING OURSELVES

Juan Enriquez and Steve Gullans, Ph.D., are cofounders of Excel
Venture Management, which builds start-ups in synthetic biology,
big data, and new genetic technologies. Enriquez is a bestselling
author and a global authority on the economic and political impacts
of the life sciences. He is a TED "all-star," lectures around the world,
serves on the Discovery Council at Harvard Medical School, and
was the founding director of Harvard Business School's Life Science
Project. Gullans was a professor at Harvard Medical School for eighteen years, applying breakthrough technologies to diseases such as
cancer, ALS, Parkinson's, and Alzheimer's. He has published more
than 130 scientific papers in leading journals. He was elected a Fellow of the American Association for the Advancement of Science
(AAAS) in 1998. The authors live in Boston.

Praise for
Evolving Ourselves

"This book provokes terror and inspiration in equal measure. Terror because humankind has become a potent evolutionary force affecting all life on the planet, including itself. Inspiration because with awareness comes the capacity to direct humanity's—and the planet's—evolutionary trajectory. Read this book and you will never think the same way about evolution ever again."
—Paul Saffo, technology forecaster

"Distribute free copies of *Evolving* to every high school in the country, and kids will be storming the gates of college biology programs, clamoring to become scientists."
—Genetic Literacy Project

"Juan Enriquez and Steve Gullans convincingly argue that Darwin's driving force, natural selection, no longer holds true in a world where little remains natural and free of the hand of man. Their book raises important questions that we all should consider deeply as individuals, as nations, and as a global community as we venture forth into this next phase of evolution."
—Donald Ingber, M.D., Ph.D.; founding director, Wyss Institute for Biologically Inspired Engineering, Harvard University

"Science continues to accelerate, speeding past science fiction, and the public's understanding continues to lag. In *Evolving Ourselves*, Enriquez and Gullans give us a whiplash-worthy account of the state of science in a fearless, balanced, and thoughtful way. A must-read whether you're contemplating why we're here, planning the future of your company, or wondering what your children's world will be like."
—Joichi Ito, director, MIT Media Lab

"Thoughtful and thought-provoking, *Evolving Ourselves* is rich in the fascinating details of how we work, how we got there, and how we are changing. Enriquez and Gullans take us down a path of observations and logic that leads to either your grandest hopes or your darkest nightmares. Fasten your seat belts—this rocket is about to take off."
—Martin Blaser, author of *Missing Microbes*

"A fast-paced, conversational collection of dispatches from biology's frontiers."
—*Harvard Magazine*

"Enriquez and Gullans take off all the blinders and explore the myriad, astounding ways we humans are rapidly influencing and shaping, both intentionally and not, our future selves. With mounting knowledge of what makes our species tick, they describe how we'll have the power to direct our evolutionary path."   —Linda Avey, cofounder, 23andMe

"Reading the book—it's a lot of short chapters, and every chapter is like, oh my God, I had no idea."
          —David Kirkpatrick, author of *The Facebook Effect* and CEO of Techonomy

"A provocative and sobering vision of the evolutionary forces at work in our modern post-Darwin world and the control that we humans exert over our genetic destinies. Essential reading for anyone who retains hope that humanity, through sheer application of ingenuity, will persevere, survive, and surmount the inestimable challenges that lie ahead."
          —George Daley, M.D., Ph.D.; professor, Harvard Medical School

"Timely and exceptionally rich, this book is guaranteed to stimulate reflection on what the new biology means for our species."
          —Dr. Erling Norrby, former permanent secretary,
          Royal Swedish Academy of Sciences

"Enriquez and Gullans brilliantly describe a future of extreme possibility and extreme responsibility. Thinking deeply about the implications of their work improves our chances of creating a desirable version of that future."   —Tim Brown, CEO, IDEO; author of *Change by Design*

"In a rapidly changing world, considerations of human evolution are rarely taken into account as a critical factor that shapes the immediate future, but Enriquez and Gullans will convince you otherwise. They will make you rethink what constitutes a victory (antibiotics! air travel!) over our natural biological constraints and its unintended consequences."
          —Hidde Ploegh, professor of biology, MIT; member,
          Whitehead Institute for Biomedical Research

"If this sounds fantastic, it is. Except that many of the technologies that the authors describe have a high probability of actually happening in people in the next 50–100 years."   —*The Daily Beast*

EVOLVING OURSELVES

Redesigning the Future of Humanity— One Gene at a Time

Juan Enriquez and Steve Gullans

CURRENT

CURRENT
An imprint of Penguin Random House LLC
375 Hudson Street
New York, New York 10014
penguin.com

First published in the United States of America by Current 2015
This revised paperback edition published 2016

ISBN 9781617230202 (hc.)
ISBN 9780143108344 (pbk.)

Printed in the United States of America
1   3   5   7   9   10   8   6   4   2

Set in Janson Text
Designed by Spring Hoteling

Mary and Stenie,
Diana and Nico,
Graham, Emilie, Alexei, Esme, and Dax
Our enduring love and extreme gratitude

CONTENTS

·····················:✳:·····················

EVOLVING OURSELVES

What Would Darwin Write Today?

.... :✳: ....

In the classic movie *The Graduate*, there is an iconic scene where an older guy walks a recent college grad out by the pool, puts his arm around his shoulders, and says, "One word. Just one word: 'Plastics.'" Unfortunately, the advice was wrong. The word should have been "Silicon." See, by 1967, when the movie was released, Fairchild was already selling semiconductors, and the following year, Intel was founded. Silicon Valley was about to experience the greatest single burst of creative disruption, wealth, and jobs in world history.

So now, well into the twenty-first century, what advice should a recent grad hear? Our answer is two words, just two words: *Life Code*. To get a sense of how fast and how radically humans are adapting to the modern world, try a thought experiment. Imagine we somehow plucked Charles Darwin out of his nineteenth-century home and placed him in the middle of modern-day Trafalgar Square, London. What would he make of it?

After the initial brutal disorientation, this dissector of human strengths and foibles would start to observe: How can everything be so clean and orderly? Where is the soot, horse dung, squalor, gruel, stench, ailments? No fleas? And where are all the Oliver Twist–like urchins? Everyone seems so well fed—in fact, way overfed. People look familiar but also very different; where did all these tall folks come from, and why are so many large but weak? Why are there so few kids and so many old people—and why do the white-haired folk appear so healthy? He would marvel at the variety of foods, but he would also wonder why there are so many signs that read PLEASE INFORM YOUR SERVER OF ANY ALLERGIES.[1] Then he would focus closer: Everyone looks clean and has teeth. Children can read and have free time to play . . . but why are adults running through Trafalgar Square wearing short pants and neon shoes? Why don't some of the old people lack wrinkles? Why are many kids and elderly drawing deep breaths from inhalers?

In just 150 years, the human species has changed. We have rede-signed our world and our bodies while at the same time becoming an ever more domesticated and smarter species, one that lives far lon-ger. Symptoms of rapid adaptation, perhaps even early signs of evolu-tion,* also include explosions in autism, allergies, obesity, and a host of other not so positive changes.

As espoused by Darwin in his most famous book, *On the Origin of Species by Means of Natural Selection*, two key forces determined, over the past four billion odd years, what lived and died on this planet: *natural selection* and *random mutation*.[2]

To recap Bio 101: Rule #1—to survive generation after genera-tion, a species* must adapt to a particular environment ("survival of the fittest").[3] Rule #2—the core genetic code, the DNA that underlies biological traits, randomly varies generation to generation. Usually these small changes are benign and unnoticeable. But natural life code is a casino: Occasionally significant changes help individuals reproduce and survive better than their ancestors, though sometimes changes in gene code can lead to extinction. Either existing species adapt to new environments, predators, pathogens, and opportunities, or they disappear.

---

* A few important terms marked by asterisks, as well as many other terms without asterisks throughout this book, are defined in the Glossary.

But over the past century, some of the rules changed. Our species propagated by billions, concentrated in cities, smartened, and domesticated itself and its surroundings. Alongside nature, *we* became the fundamental driver of what lives and dies. This change is so radical that were Darwin alive today, he would likely revise a significant part of his great works. There is now a second, parallel, large-scale determinant of evolution, one driven by two new rules:

Unnatural selection*

Nonrandom mutation*

Darwinian logic and nature continue to define and drive the evolution of all life in places where humans do not yet impose their will—in those spaces untouched by cities, farms, parklands, and vacation homes. But these once vast tracts are now rare. Half the landmass on Earth is now covered by what humans want, not by what would naturally grow without the intervention of our species. Oceans, rivers, and lakes are depleted. In just a few centuries, we have terraformed, fertilized, fenced, seeded, and irrigated enormous sections of what was once forest, savannah, desert, and tundra to accommodate our plants, our animals, our wishes. A cornfield is a perfect example of *unnatural selection*, something that you would never see if you were walking through a pristine forest or a fallow field.

Over the past few decades, the pace of evolution ramped up fast as humans invented ways to deliberately redesign the genetic code of living organisms. We developed powerful, cheap, and rapid ways to read, copy, and edit the life code of bacteria, plants, animals, and humans. When we engineer cancer-prone mice and long-lived worms, we alter the essential nature of the species. By introducing *nonrandom mutations*, we subvert random, slow evolution with rapid, deliberate, sometimes intelligent life changes.

What thrives on Earth now depends on a new evolutionary seesaw. On one side sits the full weight of nature, of the traditional forces of evolution, natural selection and random mutation, leading to extraordinary diversity, continuous speciation, and extinction. On the other side sit the wishes of a single species, *H. sapiens*. Although Darwin wrote extensively on what happens when plants and animals are domesticated and redesigned through breeding by humans, he did not make the logical leap that human influence could eventually match and even exceed the forces of natural selection and random mutation.[4]

And while many people instinctively associate the word "unnatural" only with bad stuff, our transition away from "Nature" toward a more gentle and weakened "nature" has been spectacularly successful and beneficial to humans. Life expectancy in the UK in 1856 was 40.4 years.[5] Now, because we are the only species to have put a damper on natural selection, we live almost twice as long and, in the developed world, more than 99 percent of our babies and children survive.[6] In domesticating and de-wilding plants, animals, environments, and ourselves we increased not just survival but also our quality of life.

Along this journey, consciously or not, we acquired an awesome responsibility; as we select and design what lives and dies on this planet, we drive evolution. It is ever faster and cheaper to redesign flowers, develop exotic foods, build bacteria to manufacture therapeutics, and design animals that serve and entertain.

The first part of this book showcases various symptoms of rapid human-driven adaptation and evolution. The second section explains the various ways we alter life. In the third section we cover what happens as we begin to edit life forms on a grand scale. Then we discuss how we might choose to evolve ourselves and some of the ethical implications of these choices. Finally we explore using our newfound powers to revive and restore the extinct, speciate ourselves, design synthetic life forms, and perhaps even leave the planet. A new epilogue discusses how fast life code is advancing and how complex the ethical choices are as we begin to choose to evolve ourselves.

As we increasingly bend nature to our own desires, many of the sick or weak are no longer relentlessly culled by natural selection. This transition away from nature operating and guiding life forms to humans doing so may just be the single greatest achievement and challenge, so far, for our species. As a result we can begin to answer questions like: How do we want to design life? What do we want humans to look like in a few hundred years? Do we want other hominin species walking around? What should we do with all the synthetic life forms we are creating? While there are certainly many wrong and perilous answers to each of these questions, getting it right potentially means continuing to improve the overall human condition, leading to better health, longer life span, and greater control over our daily lives. There is already much discovery to be proud of, and the adventure of controlling and guiding life has just begun.

SYMPTOMS OF REAL-TIME EVOLUTION

• • • • • • • • • • • • • • • • • • • •

Is Autism a Harbinger of Our Changing Brains?

.... :✳: ....

The *Mortality and Morbidity Weekly Report* (MMWR) is in some ways a medical version of the *Kelley Blue Book*, the publication that provides the value ranges of used cars. The MMWR provides, in mind-numbing detail, just how many people got sick or died last week. It's not exactly beach reading, and it's usually as exciting as watching paint dry. But within the endless columns and statistics of the MMWR, the patient and persistent can spot long-term trends and occasionally find serious short-term discontinuities.

Physicians and epidemiologists get excited by short-term discontinuities; a sudden increase in an extremely rare tumor, like Kaposi's sarcoma, can be a harbinger of a massive infectious disease epidemic with a long incubation period, AIDS. The dozens of patients entering the hospital with this rare tumor in 1982 grew into 75,457 full-fledged AIDS cases in the United States by 1992.[1]

Conditions and diseases develop and spread at different rates. A rapid spike in airborne or waterborne infectious diseases like the flu or cholera is tragic but normal. A rapid spike in what was thought to be a genetic condition, like autism, is abnormal; when you see the latter, it is reasonable to think something has really changed, and not for the better.

Usually changes in the incidence of a genetically driven disease take place slowly, across generations.[2] Diseases such as cystic fibrosis or sickle cell anemia result from well-characterized DNA mutations in single genes, and the inheritance pattern is well understood: If parents carry the gene and pass it to a child, the child will be affected. Cystic fibrosis occurs in 1 of 3,700 newborns in the United States each year with no significant change in incidence over many years.[3] Similarly, sickle cell anemia is a genetic disease. One of every 500 African Americans acquires the errant gene from both parents, and we can predict the incidence of sickle cell anemia with some regularity.[4] You cannot "catch" these kinds of conditions by sharing a room with someone; you inherit them. If your sibling has cystic fibrosis or sickle cell anemia, then you have a 1 in 4 chance of also being sick.

We know there can be a strong genetic component to autism—so much so that until recently autism was thought to be a primarily genetic disease: It was diagnosed in 1 percent of individuals in Asia, Europe, and North America, and 2.6 percent of South Koreans.[5] If one identical twin has autism, the probability that the other is also affected is around 70 percent. Until recently, the sibling of an autistic child, even though sharing many of the same parental genes and overall home environment, had only a 1 in 20 probability of being afflicted. Meanwhile, the neighbor's child, genetically unrelated, has only a 0.6 percent probability.[6] But even though millions of dollars have been spent trying to identify "the genes" for autism, so far the picture is still murky. The hundreds of gene mutations identified in the past decade do not explain the majority of today's cases.[7] And while we searched for genes, a big epidemic was brewing.

In 2008, when the MMWR reported a 78 percent increase in autism—a noncontagious condition—occurring in fewer than eight years, alarm bells began to go off in the medical community.[8] By 2010 the Centers for Disease Control and Prevention (CDC) was reporting a further 30 percent rise in autism in just two years.[9] This is not the way traditional genetic diseases are supposed to act. This rate of change in autism was so shocking and unexpected that the first reaction of many MDs was that it wasn't really that serious. Many argued, and some continue to argue, that we simply got better at diagnosing (and overdiagnosing) what was already there.[10] But as case after case accumulates and overwhelms parents, school districts, and health-care systems, there is a growing sense that something is going horribly wrong, and no one really knows why.

What we do know, because of a May 2014 study that looked at more than 2 million children, is that environmental factors are driving more and more autism cases. Whereas autism used to be 80 to 90 percent explained/predicted by genetics, now genetics is only 50 percent predictive.[11] We have taken a disease we mostly inherited and rapidly turned it into a disease we can trigger. Now the chance of a brother or sister of an autistic child developing autism is 1 in 8 instead of 1 in 20.

The rapid pace of today's human-driven environmental changes may be leading to maladaptive changes in humanity as our biology misinterprets how to acclimate to this new unnatural environment. Autism may be just one harbinger, one symptom, of our radically changing world. Almost every aspect of human life has changed— moving from rural to urban; living in an antiseptic environment; eating very different sugars, fats, and preservatives; experiencing novel man-made stimuli; ingesting large quantities of medicines and chemicals; being sedentary; having children later; and living indoors. Given so many transformations, it would be surprising if our bodies and brains did not change as a result.

The DarWa Theory Revisited

·············· :✳: ··············

Charles Robert Darwin was a man of extraordinary courage and integrity. As a teen he intended to be a preacher, yet what he observed in nature was increasingly at odds with his intended profession. Except for his grandfather, much of his family opposed his theory of evolution.[1] His wife was especially concerned he would be condemned to hell. Darwin himself wrestled with serious doubts; he never even used the word "evolution" until forty-two years after he started writing about biology.[2]

Over decades Darwin accumulated examples, wrestled with his faith, and refrained from publishing, despite evidence that creatures evolve.[3] And then came an extraordinary crisis. Darwin received a letter from an obscure specimen collector working in the far reaches of present-day Indonesia. Wracked by malaria, Alfred Russel Wallace had nevertheless crystallized his observations, after decades of traveling the Amazon, Malaysia, and Indonesia, into a single letter—one that described in detail what would become the theory of evolution.

Wallace was not a "gentleman," in a class-conscious era when only gentlemen were supposed to become scientists, gain entry into learned societies, and get published. So Wallace shyly asked Darwin whether he thought his ideas and theories were any good and, if so, might Darwin be kind enough to forward them for publication. Darwin received this single correspondence and immediately concluded there could not have been a better short abstract of his own grand theory.[4]

Darwin could have simply ignored the letter, dealt with a growing and serious family crisis, and immediately published what he'd been developing on his own for decades. After all, the correspondence came from a relatively unknown individual, on the other side of the planet, during a period when ships sank. And Darwin had devoted years of work and thought to his unpublished theory, having sailed to the Galápagos Islands more than 25 years earlier on the *Beagle* and thereafter continued minutely recording, detailing, and strengthening his arguments. Instead, after a night of extreme anguish, Darwin forwarded the letter to his scientific colleagues, advocating rapid publication with full credit to Wallace. Darwin realized that this action would likely preempt his getting credit for the theory of evolution.

While Darwin had the strength of character to encourage the primacy of Wallace's work, his friends had other ideas. Through years of discussion and correspondence, they were well aware of Darwin's developing theory. So, unbeknownst to Darwin, they chose to place some of Darwin's own unpublished work in the same journal, alongside the Wallace letter.

On the first of July, 1858, a small group at the Linnean Society of London, which thought it was merely attending a sleepy memorial meeting for one of its deceased presidents, first heard the theory of evolution. Almost no one realized they had just witnessed history. Wallace, still in Asia, wouldn't find out for weeks his theory had been aired. Neither did Darwin attend the meeting; he was burying his baby son.[5]

Even though Wallace and Darwin co-discovered and espoused a similar theory of evolution (as we're calling it, "DarWa 1.0"), their view of the driving mechanisms differed. Darwin primarily stressed the importance of the individual; Wallace, the environment. Their combined work built and strengthened the foundation for what most today refer to as the theory of evolution. (Darwinism as we know it synthesizes a lot of people's work, and has been augmented, edited, and refined by many; for instance, Herbert Spencer first coined the famous phrase that we associate with Darwin's theories: "survival of the fittest.")[6]

While Darwin and Wallace got a great deal right about evolution, they remained puzzled by two mysteries: What was the mechanism by which adaptations happened, and why did the fossil record seem to show that sometimes, rather than proceeding in slow, small, incremental steps, evolution happened quite rapidly? Unfortunately they never met or corresponded with the third critical actor-creator of the theory of evolution. Augustinian friar Gregor Mendel was too modest and dedicated to his monastic life to focus on fame. So his key article, on the genetics of peas, the one that founded modern genetic science, remained obscure and ignored for decades.

The discovery of units of heredity, later called genes, which began with Mendel in 1866, offered a solution to the first of Darwin's mysteries: the rules of how traits get passed on.[7] But even without modern genetics included, DarWa 1.0 fundamentally transformed our view of life and its future. Throughout the early 1800s, we thought the Earth was only 10,000 years old; there was no need for evolution, plate tectonics, or other grand schemes.[8] Only in the last two centuries—a minuscule fraction of the time humans have been alive—have we realized that continents and mountains move, and species emerge, disappear, and change. By the time Darwin died, the generally agreed-upon age of the Earth, among scientists, was 100 million years. Today there is overwhelming data, from chemistry, biology, geology, astronomy, and anatomy, evidencing a 4.54-billion-year-old Earth.[9]

DarWa 1.0 still provides an accurate and detailed framework for understanding life forms and their development, successes, and failures. But it is now impossible to overlook or even underestimate the effect humans have had on the planet in just the past few centuries. A very few animals and crops, genetically selected to fulfill our needs and desires, now dominate large swaths of the world's landmass. An area equivalent to the whole of South America is under cultivation just so we can feed ourselves and our animals. A further 8 billion acres are used for livestock. More than 19 billion chickens and 1.4 billion cows live in highly urbanized animal environments we call "farms."[10] Despite most of us regarding these settings as rural, the animal population density often exceeds even the most crowded of our cities. The wild is becoming rural, the rural urban. Nature is not "selecting" what lives and dies in these environments. Humans are. And the species that humans have unnaturally selected, consciously or unconsciously, are increasingly dominating the planet.

Under DarWa 1.0 one would never see evolution occurring the way it's depicted on modern T-shirts; remember those cartoons where a creature crawling out of a primordial ooze eventually begets a small mammal, then an ape, then a hunched Neanderthal, and finally a handsome human? (Who then de-evolves into a poor, hunched schlub at a computer desk.) This depiction of evolution is anathema to natural selection and random mutation because it depicts a linear, orderly, and logical progression from one model to the next. The real history and fossil record of evolution, of natural selection, looks like a complex, overlapping, messy bush—an astonishingly promiscuous, interesting, tangled, semi-chaotic web of life.[11] Yes, there may be a trunk to this tree, a common ancestor, but we also observe many, many subtypes, varieties, subspecies, and new species rising and falling as environments change. There is no overall plan and logic; rather, ancestors throw the dice. Some offspring will come up sevens or elevens and be winners. Most will not. The cycle is neither predictable nor foreordained. The dice were not, until very recently, loaded.

That is, until 1972, when Paul Berg began combining genes from different organisms.[12] This DNA came together not because the momma creature and daddy creature did the nasty, or because of a random mutation in a given offspring. Instead, it was the first instance of a group of scientists deliberately taking genes from one organism and inserting them into another species.[13] It was not nature randomly engineering but rather humans beginning to discover and apply intelligent design to life. Thus it is almost the exact opposite logic of DarWa 1.0. And it all started with domestication.

Twenty Generations to Domesticate Humans

······················ :✳: ······················

Exiled by Stalin, scientist Dmitry Belyaev ended up far from a lab or university, in the endlessness of Siberia. An experimentalist at heart, he made do with what he had and began breeding wild foxes. But he did not do this at random; each year he graded each fox as particularly friendly and tame or nasty. Only those at the extremes were chosen to breed. One-fifth of the nicest, as well as one-fifth of the most violent, were segregated and bred, generation after generation.

The offspring of the nice foxes rapidly evolved floppy ears, short tails, lighter-colored fur, a less pungent odor, and bigger heads—quite weird, given that the original nice-enough-to-breed animals were picked for behavioral, not physical, traits.[1] One potential explanation is that as the adrenaline levels in tamer animals dropped, melanin fell as well, which in turn altered and lightened fur color. Looks may, at least in foxes, reflect temperament, and there may be some genetic basis to stereotypes like floppy ears.

Eventually Belyaev's "de-wilding" experiment was so successful in modifying a wild species that a subset of these creatures began acting like Labrador retrievers, so much so that they got shipped off to the United States and sold as "gentle house pets, perfect for children." It appears you can rather quickly breed out aggression (or, as pit bull owners know, breed it in).[2]

Even relatively minor environmental adjustments can lead to extreme diversity in birds, plants, and animals—even within the same geographical area. If you take wild sheep and corral them for a few generations on a farm, you observe massive change and domestication. And when you move animal, plant, or bacterial species from mostly rural environments to 70 percent urban environments in just five generations, you can expect to see extremely rapid mutations—or extinctions—as a consequence. Some species of pigeons, such as everyday city pigeons (that is, the feral pigeon, which descended from the rock dove), went from shy cliff-dwelling creatures to the dive-bombing pests of Trafalgar Square and Piazza San Marco. Meanwhile, billions of passenger pigeons did not adapt, remaining too trusting and vulnerable to hunters, and went from literally blackening the skies to extinction in about a century. Now over a couple of centuries, we as a species have changed just about every factor that would promote rapid evolution in any other species.

Darwin understood the consequences and implications of human-driven unnatural selection as it was applied to the domestication of plants and animals; he called this "artificial selection."[3] While studying fancy pigeon breeding, Darwin chronicled rapid species changes. Some even argue that pigeons were more important, and more conclusive, to his theory of evolution than his iconic finches. He realized that if human breeders could manipulate a single species to such an extent in captivity, then perhaps nature could also manipulate, and change, all species in the wild. Darwin just did not follow through on the logical consequences of this thought, and did not foretell that humans would soon wield enormous power over the planet, most other species, and themselves. If he had, he would have understood unnatural selection as a key driver of global change.[4]

As humans have evolved further and further from being an "all-natural" species, we too have quickly and dramatically de-wilded ourselves. The concrete urban environments that now house the majority of the human species tend to be cleaner, better lit, and safer than the jungle and rural landscapes we abandoned. We now think it normal and natural to have a tap nearby that delivers clean water and a toilet to whisk away waste. Never mind that none of this was true for 99 percent of our history. We are now so used to our completely unnatural, human-designed environments that we simply take them for granted; they have become "human rights."

Our domestication occurred very rapidly, mostly in the course of just over ten generations. The wild used to be a few yards from where we slept, as did the source of most food. The natural size of our ancestral tribes was around 150 individuals. When groups became much larger than that, strife ensued and subgroups splintered off, moving a little ways away, colonizing their own little swath of savannah and beginning a new chapter of hunt-and-gather. Survival was far from a given.[5]

About 7,500 generations ago, our type, *Homo sapiens*, began to build, create, and pillage small villages. What we refer to as "civilization" began about 500 generations ago, with the advent of agriculture. As late as 2000 BCE, the total world population was in the tens of millions, most of them broadly dispersed.[6] Eventually substantial cities began to emerge in the Fertile Crescent, China, India, the Americas, and even in Europe, but cities were uncommon. In the year 1300, less than 5 percent of England was urban; throughout the Industrial Revolution, rural was the norm. Even in 1910 only 2 out of 10 people lived in cities.[7] Yet by 2007, a majority of the world population had urbanized. That means a massive, global urban migration took fewer than one hundred years—about five generations. To put this in a historical context, there have been at least 125,000 generations since the first hominins began to walk around. (By the way, hominins are humans plus their extinct ancestors, whereas hominids are hominins plus great apes and their ancestors.)

Globally, city populations will likely double by 2030. Within the next twelve years, China alone intends to move 250 million people from rural environments into cities.[8] Accommodating these new residents will require building the collective equivalent of Tokyo,

Mexico City, Seoul, New York, Mumbai, Jakarta, São Paulo, Delhi, Osaka, Shanghai, Manila, and a few dozen other major cities. Consider that in the 1980s, 8 out of 10 Chinese lived in the countryside. By 2025 only 3 out of 10 will be rural residents.

For most of our history, days were filled with malnutrition, disease, and violence.[9] Unnatural urban environments have been very good for humans; as we domesticated ourselves and our environments, we gradually removed the obstacles to a long life span.

Predators of all kinds were far more common until our massive and deliberate kill-off modified our environment to such an extent that we must now search carefully to find any of the once-common big animal predators. Grizzly bears no longer pose a threat in most of the United States. Saber-toothed cats are mere fossils. We have largely taken ourselves out of the food chain, and we tend to die in our beds and in hospitals.[10] We still worry a little about sharks, but have unfortunately managed to turn most of the species into shark-fin soup. While the rare shark attack triggers global news coverage,[11] deer crashing into cars now kill eleven times more people every year than do sharks. It's exceedingly rare to encounter poisonous snakes in most cities, other than a few politicians.[12] Although Paleo diets and lifestyles are a growing fad, sensible people might ask why anyone would want to go back to a period when only 10 percent of the population made it beyond age forty.[13]

By far the most dangerous and common predator remains other humans; on average we are 11,000 times more likely be killed in war today than by all the sharks in all the oceans.[14] But even in this arena, despite daily mayhem and bloodshed, there is a broad and deep trend toward domestication. The number of wars and violence in general has gradually decreased almost everywhere.[15] Yes, there are still horrendous incidents and ghastly regional conflicts, but the world is a far more peaceful place than it used to be, and the chances of an eighteen-year-old being required to fight in a war in his or her lifetime are at an all-time low. Even terrorism, post-2001, has all but disappeared from the United States and much of Europe.[16] Which is not to say that we aren't prone to irrational fears, extreme miscalculations of risk, and manipulation; between 2007 and 2012 about 4.6 Americans died annually from terrorist attacks on U.S. soil. About 100 times more drowned in their bathtubs. But the fact is, Americans spend $400,000,000 per year on homeland security and defense per victim, while we spend $9,000 per cancer victim and $80 per victim of stroke and heart disease.[17]

The nature and scale of state and religious retribution against individuals also changed dramatically. Priests no longer burn people at stakes in town squares or torture them in towers. While the United States still incarcerates a disproportionate number of people vis-à-vis other developed countries, the chances of an individual suffering torture, murder, or execution while imprisoned has dropped precipitously. Europe is even further ahead. People used to be routinely and publicly burned at the stake, beheaded, drawn and quartered, boiled, crushed, mutilated, and hanged as a matter of grand public spectacle and entertainment. Henry VIII alone reputedly executed 72,000 during his reign. In eighteenth-century Britain, 222 offenses led to the death penalty.[18] In contrast, no one has suffered the death penalty in England since 1964.

With human domestication, individual muggings and murders also dropped precipitously. When Thomas Hobbes described life as nasty, brutish, and short in 1651, 1 in 1,000 Europeans was a victim of murder.[19] Today's murder rate in the United States is twentyfold less, at fewer than 5 in 100,000.[20]

On the other hand, not everywhere is safe and trending safer. In 2014, Nigerian terrorist group Boko Haram kidnapped hundreds of girls in raids. This was horrifying, but on a very different order of magnitude from when slave traders would capture and kill or enslave millions. Some leaders of African nations have voted themselves immune from human-rights prosecutions, even by their own African Court on Human and Peoples' Rights.[21] Although this move angered human-rights watchers, and was a clear step backward, it was also a signal that someone is beginning to hold formerly all-powerful and sovereign dictators accountable—and that global norms and expectations are effective enough that tyrants now fear the consequences and want to insulate themselves.

As we tamed ourselves and our leaders, we also began to tame our environment. Exposure to the elements used to be a major killer. Although we still incessantly comment on and complain about the weather, with rare exceptions, its consequences are far less deadly. Tornadoes and hurricanes, floods and droughts, and even "snowpocalypses" are described in detail and followed by the media. We now think it normal to have advance warning, but for much of human history, weather was one more brutal and constant driver of natural selection.

Beyond extreme weather, perhaps the biggest change we've seen has been in overall exposure and temperature flux. For most land-based species, coping with extreme temperature and weather differentials is a fact of daily life. In the space of just about one century, however, our bodies, which had evolved to deal with this flux, have come to live, eat, and sleep in unchanging climates. We are coddled by temperature-controlled buildings, houses, cars, and offices, living our lives between 68 and 75 degrees Fahrenheit and sheltered from rain, wind, and snow. This lifestyle is completely unnatural and abnormal, and quite pleasant.

Until very recently, even in advanced countries, finding, gathering, and consuming enough calories was a constant and daily preoccupation. Now famine is unusual and far from the massive killer it once was.

For most humans, access to food is no longer a life-or-death issue, unless it is an excess of calories. Yes, too many people still go to bed hungry or malnourished, but starvation is far from the scourge it once was. The percentage of undernourished people in developing countries fell from 24 percent in 1990 to 15 percent in 2010.[22] In 1900, in the oldest children's hospital in the UK, almost 1 in 5 infants died from malnutrition and gastrointestinal diseases.[23]

It would be surprising if such massive changes, in every aspect of humans' daily lives, did not lead to rapid adaptation and ultimately to speciation. Today's humans interact, coexist, compete, and cooperate with thousands of others on a regular basis. Had we not domesticated ourselves, had we not reduced the historic levels of violence, had we not somewhat educated most children, global urbanization would have been a cross between *Lord of the Flies* and *Mad Max*. So when you wonder where, seemingly all of a sudden, the extraordinary increases in life span, intelligence, height, and other positive traits we now just take for granted, come from . . . you may wish to remember how profoundly unnatural and nonrandom we have made our environments. We have de facto domesticated ourselves just as thoroughly and completely as we have our own cats and dogs, with massive evolutionary consequences.[24]

Side Effects of Nonviolence

·····································:✳:···································

We are in the midst of an epidemic of nonviolence, which has the effect of casting aside one of the greatest drivers of natural selection. As peace and prosperity spread, as well as human rights and women's rights, our chances of passing on our genes skyrocket. Within the United States, if one parses the fifty categories in which one might inflict violence on children and youths, not a single one increased between 2003 and 2011.[1]

Given the daily news and general mayhem reported, this trend is hard to see, but we are—with notable exceptions—a far more peaceful and domesticated species than we once were. Italian homicide rates have steadily declined from 73 per 100,000 in 1450 to 2 per 100,000 in 2010.[2] The same is true in country after country. Violence used to be a core and unremitting part of most humans' experience. Now, much of Europe, which over the past 32 centuries has been the most war-loving region on the planet, has given up investing in functioning, real armies.[3]

And we are also choosing far more diverse mates. Tolerance of interracial couples as well as of people with different beliefs and religions is on the rise. This is a really recent phenomenon; until June 12,

1967, in seventeen U.S. states it was illegal to marry someone from a different race. It took a Supreme Court decision, in the most appropriately named *Loving v. Virginia* case, to invalidate the previous jurisprudence, including a ruling by a Virginia judge who stated, "Almighty God created the races of White, Black, Yellow, Malay, and Red, and He placed them on separate continents . . . The fact that He separated the races shows that he did not intend for the races to mix."

Only in 1991 did the majority of Americans agree with the statement that interracial marriages are OK.[4] And race is far from the only barrier. Even at the beginning of the 2010s, more than 86 percent of U.S. relationships are between people of the same faith. Mormons, conservative Christians, Muslims, and Sikhs are especially wary of relationships with those of other faiths.[5]

But the overall trend toward more openness and acceptance of others, especially among the young and educated, is overwhelming. Travel and study-abroad programs abound and social media has exploded, providing opportunities to interact with far more people than our grandparents ever dreamt of. Among the many benefits of multiculturalism, however, there may also be losses. For example, in the 1900s, about one-half of Americans were blue-eyed. By 1950, only one-third. By the end of the twentieth century, fewer than 1 in 6. Blue eyes are a recessive trait, and whereas 80 percent of people used to marry within their ethnic group, now they tend to marry based on where and how long they go to school.[6] Imagine saying someday, "Once upon a time there were people with blue eyes."

Eye color may not be the only trait we are losing in an effort to domesticate ourselves and our behavior. As we cram together in ever-expanding cities, it's worth asking whether we would want to *chemically* tame or untame ourselves. We can alter trends in both directions. Want to make men more macho? Between 2000 and 2011, in 37 out of 41 countries surveyed, monthly sales of testosterone increased, and increased, and increased.[7] (Walmart sells testosterone boosters for $8.98.) What does testosterone do to the body? Supposedly a lot of good things—more muscle mass, increased sex drive, greater confidence, and perhaps . . . more aggression? The underlying cause of why someone blows up in a violent rage or bites off part of their own tongue is, as baffled scientists like to say, "multifactorial."[8] In essence, it is hard to sort out causality, because there are so many variables: Was it the testosterone, steroids, booze, or a sad country song that set off the sudden rampage? Certain chemical use is correlated with violence. In peaceful Sweden, whenever police jailed someone and then sent urine samples to labs, 33.5 percent of those samples came back positive for steroids.[9]

Famine, disease, and war drove natural selection. While ugly and heartless, such events also drove human genetic diversity in many closed and isolated communities. Genomes record the history of exactly who had sex with whom and when, as well as where one's ancestors came from. As gene sequencing gets ever cheaper, we are seeing, in granular detail, the genetic makeup of people from each country, region, and tribe. For example, today's average Maya gene admixture includes: Colombian, Pima, Karitiâna, Spanish, Surui, Uzbek, Irish, Japanese, and Yoruba.[10] And you can get far more personal: In 1998, Dr. Eugene Foster greatly upset Southern gentility, and possibly even the Daughters of the American Revolution, tracing DNA from descendants of Thomas Jefferson and his black slave Sally Hemings to prove they had several children together.[11]

Race blending wasn't uncommon in an era of slavery, exploration, and conquering. But what happens now as peace prevails and some human niches isolate while maintaining their traditions? Consider the area around the Arabian Peninsula: The cultural mandate that women not stray far from the family circle when seeking a husband, as a way to ensure their honor, has resulted in an extraordinary number of close-relative intermarriages. Studies estimating consanguinity in Egypt range from 20 percent to 42 percent. In other Arab countries and regions, rates of consanguinity can reach 60 percent.[12]

People once mostly lived in closed, isolated, rural enclaves and tribes. The normal state of affairs, for much of the peninsula's history, was that you could and did marry close relatives. Many of these tight familial clusters were periodically disrupted by outsiders and invaders. But now that the region has enjoyed a period of relative peace for much of the past century, stability—plus extreme cultural strictures—may have accentuated three trends affecting the genetic makeup of the population. First, if you continue to marry first cousins and second cousins, and have few interactions with outsiders, you tend to get less and less gene variation. If a bad gene enters that reduced pool, it takes longer to breed out. (The converse is also true: that a rare beneficial mutation can propagate more quickly through a small inbred population, perhaps enhancing survival or reproduction.)

A secondary effect of the lack of violence, chaos, and disruption on a population—as well as more modern medical care—is that a lot of children with birth defects, who likely wouldn't have survived harsh nomadic conditions, now do survive and have kids themselves. Because of humane choices and policies, traits that might have been bred out under natural selection are reinforced and passed on.

And a tertiary effect of nonviolence: Many people manage to get wealthier. The temptation to keep growing wealth and status within the family, or at least within the tribe, may further reinforce the incentive not to marry outside that group, much less a commoner or a foreigner.

Consanguinity doubles and triples the risk of birth defects.[13] These days 8 percent of all kids born in Saudi Arabia have severe genetic defects (about three times the rate in European countries).[14] Type 2 diabetes affects 32 percent of adults. Hypertension hits 33 percent.[15] In the United Arab Emirates, congenital anomalies account for 40.3 percent of infant mortality. In Qatar, high rates of consanguinity have led to a higher incidence of rare genetic diseases, as well as increases in "common adult diseases like cancer, mental disorders, heart diseases, gastrointestinal disorders, hypertension, and hearing deficit."[16]

Even extremely rare genetic defects can be passed on and spread quickly within a specific undisrupted societal subgroup, which is precisely the problem that afflicted the royal families of Europe in the nineteenth and twentieth centuries; hemophilia became known as the "Royal Disease" after Queen Victoria passed it on to two of her daughters, her son, and the royal houses of Spain, Germany, and Russia. (Blessed be Kate Middleton, commoner.) An extreme example of long-term effects is the inbreeding that took place for six centuries among the Hapsburgs, who eventually developed "Hapsburg lips"—extreme malformations that prevented one of them from chewing food.[17] Hapsburgism may now be occurring across whole nations. So, as we engineer massive, positive change in the form of nonviolence, we may also want to revisit culture and its long-held norms.

Allergies: Another Harbinger of Our Evolving Bodies?

·······················:✳:······················

Think back to some of your earliest childhood memories. Perhaps some of them had to do with food. Coffee brewing, Mom singing in the kitchen, preparing something you loved: cookies, bread, pudding, a cake, or a roast. You would spy on dinner parties, hiding behind doors and staircases, just to glimpse the guests, hear snippets of conversations. You overheard a lot, but can you ever recall your grandparents or parents asking every guest whether they had any food allergies?

At a recent dinner of very prominent scientists, one answer to Juan's question "Any allergies?" was, "I usually never eat at anyone's house, because I am so allergic to almost everything. The only thing I could conceivably eat is a piece of meat or chicken, cooked in a saucepan, with no butter, oil, spices, or condiments of any kind. I'll also have a glass of water or two."

Hospital visits related to food allergies tripled between 1998 and 2006. More than 17 million Americans suffer from food allergies. A host not only has to be courteous in asking every guest about allergies, but he or she also has to be mindful of the serious consequences of not getting it right. So too does every restaurant, school cafeteria, and food cart. In kids, reported food allergies increased by 50 percent in just over a decade, from 3.4 percent (1997–99) to 5.1 percent (2009–11).[1] One in ten preschoolers is now allergic to at least one thing.[2] The overwhelming prevalence of allergies, plus fear of lawsuits, led Dunkin' Donuts to simply post the following message: "Dear Valued Customers: Please be advised that any of our products may contain allergens."[3]

It would be one thing if all of these allergies were triggered by radically new chemicals, food colorants, or preservatives. But for the most part, it's not artificial flavorings and additives that people react to so violently. Oddly enough, seven common foods that we have been consuming for millennia account for 90 percent of food allergies: milk, eggs, nuts, fish, shellfish, soy, and wheat.[4] And speaking of going nuts . . . one survey shows the incidence of adult nut allergies has remained stable from 1997 through 2008. But for kids under eighteen, allergic reactions have increased 3.5 times, from 0.6 percent to 2.1 percent of the population.[5] How did kids become so sensitive to foods that have been around for ages in just one or two generations? Are kids simply exaggerating? Could hives, itching, and closed airways be simply figments of our collective imagination? Is this "new normal" a sign of something far more profound?

Many of the world's best scientists have been racking their brains to figure out the puzzle of allergy prevalence.[6] Among today's multitude of alternative causal hypotheses, one of the most popular explanations is the "hygiene hypothesis."[7] Its supporters argue that we have unnaturally designed and domesticated our surroundings and environments to such extremes that they are now too clean, protected, and isolated. Advocates of this hypothesis point out how kids raised on farms near animals, dirt, and microbes rarely get allergies.[8] And kids living in homes with an income twice that of the poverty line are more likely to suffer food and respiratory allergies than poor kids.[9]

The overall explanation, says the hygiene hypothesis medical subtribe, is that our immune system, which developed over millennia to cope with eating raw meat off of dirt floors, has ever less to attack in our overwashed, scrubbed, sanitized, de-bacterialized environments.

So our own defense mechanisms, having ever less to focus on, become ever more sensitive to the smallest perturbations. This makes sense. But before we revert to the "good old days" we should remember where we came from and what led to this new body state. . . .

We are a species that evolved consuming raw meat, uncooked and unwashed veggies, and dirty water. Even today, when the Hadza people in Tanzania catch an impala, they field dress the animal while "washing" their hands using the digested grass inside the animal's stomach; they then invite their guests to join the feast as they "slice the stomach into pieces, and toss [pieces] into their mouths like popcorn."[10] All too often, when one travels in many countries one sees mothers taking water from severely contaminated streams or babies bathing in ponds alongside animal waste. Living with a lot of dirt and microbes has been, and for billions still remains, normal, natural, and dangerous.[11]

Many of those who yearn for a lost "all-natural" food system often forget that "organic" doesn't necessarily mean safe. For much of history death came from "natural" foods and unprocessed water. It still does in parts of the developing world. Diarrhea kills close to 1.9 million kids under the age of five worldwide every year, accounting for about 19 percent of all childhood deaths.[12] According to the United Nations, dirty water still kills more people than all wars.[13] Only very recently have the majority of humans on Earth gained some access to clean water, sewage collection, and shrink-wrapped, preserved, and safer foods. Unnatural sanitizers and cleaning products, modern soaps, disinfectants, and chlorination have saved billions of lives.

We may have tipped the balance a little too far. Some of the very changes that make our lives cleaner, safer, and better—changes that have increased our very survival rate as a species—bring along unintended consequences. So we sneeze, itch, and wheeze ever more. Fortunately, we can readily reach for yet another unnatural, but often effective, solution—Zyrtec, Claritin, Benadryl, and other assorted antihistamines. But as we seek pharmaceutical assistance, we may also wish to consider the latest allergy theory: They are caused because we wiped out part of our symbiotic microbes.

Our Unnatural "All-Natural" World

······················· :✳: ·······················

If you really are what you eat, then we are already quite a different species. Bodies that for hundreds of thousands of years ate "all-natural" have been challenged to adapt fast to tidal waves of nachos and pizza.

Dental plaque provides a small window through which to view this massive evolutionary upheaval.[1] Anyone who has been to the dentist knows how tough it is to remove plaque. Bad for you, but good for science. Its toughness makes plaque a great reservoir of data for bioanthropologists. Diet affects plaque, and by comparing the plaque in ancient and modern human teeth, scientists can infer what kinds of things we ate and what lived in our mouths. In the pre-Twinkie era, both early humans and our close relatives had mouths that were quite healthy. There are almost no examples of Neanderthal cavities. Paleolithic and Mesolithic human skulls are almost devoid of cavities.

As human diets began to modernize, as we began cooking and cleaning more of our daily foodstuffs, a strange thing happened: The bacterial colonies in our mouths became far less diverse. Hunter-gatherers from seven thousand years ago had far more microbial diversity in their mouths than did Stone Age agriculturalists.[2] Bacteria that had coexisted and coevolved with our bodies were crowded out by a new environment, and our mouths became colonized by nastier bacteria. We further repopulated our mouths with the ever more processed sugars. The incidence of cavities exploded. We began to suffer chronic oral disease, something that became most bothersome, and sometimes even deadly, in the pre-antibiotic, pre-brushing, pre-dentist era.

These days we need to do things no self-respecting Neanderthal would have considered: brush our teeth three times a day, floss, drink fluoridated water, fill cavities, and use dentures. (These practices are far from common, or necessary, in any wild animals.)

Our changing diet wasn't just hard on our mouths. Average male height during the ninth to eleventh centuries was just below that of modern men. But the transition into the Middle Ages, the Enlightenment, and the Industrial Revolution was brutal. Disease, wars, serfdom, and filthy towns changed the morphology of men; by the 1700s, the average Northern European was 2.5 inches shorter than before and did not recover until the twentieth century.[3]

Evolving diets redesign our bodies in other ways. The average U.S. male increased his weight from 166 pounds in 1962 to 191 pounds by 2002. By 2011 we had gained, on average, a further five pounds.[4] If we had observed this kind of generalized transformation in the bodies of a wild species, we would be shocked.

In altering our desires and diets we did not just alter our own bodies, we also guided the evolution of broad swaths of nature. Think about it: Just what exactly is an "all-natural" food? Consider the humble potato. When conjecturing about the remote origin of his French fry, a teenager might blurt out Ireland or Idaho.[5] Both would be wrong. It was the old Inca Empire.[6]

There's good reason for humans to tinker with foodstuffs. Pre-Tiwanaku (the ancestors of the Inca), only about two hundred all-natural potato species existed, some of which were poisonous. Potatoes are related to deadly nightshade, which is why Indians would dunk tubers in a water-and-clay mixture; the clay would bind with and absorb the poisonous solanine and tomatine.[7] In pre-Inca times, when food was quite scarce, eating potentially poisonous food was worth the risk; few crops grew on those steep, terraced hillsides, particularly edible plants that survived tropical and subtropical pests and insects.

Gradually the Peruvians' ancestors learned how to pick, clean, cook, and de-poison potatoes. By the time the Spaniards showed up to rape and pillage, the original two hundred varieties had proliferated into more than three thousand varieties. Many of these you can still sample in Cuzco; its municipal markets are full of bright-yellow, purple, pink, dried, bright-white, starchy, fibrous, and crimson potatoes to snack on—a far cry from the dirty-looking brown, dowdy tubers we tend to see in Western supermarkets.[8]

And just what's natural about having potatoes grow in Europe, anyway? They are in no way a native species.[9] But boy, have they been useful. The Irish and Germans avoided famine for centuries thanks to this Incan import. However, they reverted from the cornucopia of Incan breed varieties toward a few standardized seeds a costly error known as monoculturing. If you monoculture, and end up depending on the "Irish Lumper," a single blight can wipe out a lot of plants and a lot of people, which is precisely what happened during the Gorta Mór, the Great Hunger, in Ireland.[10] Lack of potato-disease resistance, and lack of alternative species, cost the country almost a quarter of its population; a million starved and almost a million emigrated. An equivalent disaster in the United States today would mean losing 80 million citizens. (Not that we have all learned this lesson . . . Just in case you don't get bulletins from the Dairy Cattle Reproduction Council, it turns out that inbreeding of Holsteins, Jerseys, and Brown Swiss is at an all-time high.[11] And Florida orange growers are also relearning the perils of monoculture as their crop withers away, a victim of a citrus plague.[12])

A fun and ironic field trip involves going to "all-natural, organic" farmers' markets and counting the number of completely unnatural, human-designed varieties of produce. Those wonderful, multicolored, oddly shaped, different-size heirloom tomatoes? Why, surely they're "all-natural." Well, not exactly. Tomatoes used to be small, green, and slightly poisonous. Summer farmers pride themselves on the variety of shapes, colors, sizes, and flavors they can coax into being. But this process is human-driven, not natural selection at work. Extreme tomato engineering begat some great specimens and some horrid options as well. Next time you bite into a gorgeously red but utterly tasteless big-box-store tomato slice, you can blame the lack of SIGLK2, the genetic switch that generates a tomato's naturally sweet taste. It's been turned off through breeding in most commercial varieties, in favor of that alluring bright-red skin.[13]

And while we're on the topic of "all-natural" plants . . . How often do you ship your loved one true wildflowers? Odds are you tend to send weird-looking, colorful, fragrant, newly bred, long-lasting flowers. As you continue wandering through the "organic" market, how about those artisanal cheeses? Might these be the result of centuries of unnatural human tinkering too? When was the last time you found "wild cheese" in the forest? Even the animals that produce the basic ingredients for cheese are unnatural; should you truly seek a partly all-natural Brie or Camembert, it best be made with aurochs' milk (which might be a little hard since the last aurochs died in Poland in the early 1600s).[14] Or how about those beautiful "sheep's milk" cheeses? Make sure they're from a mouflon or Orkney, as all current domesticated varieties of sheep are the product of our unnatural meddling.[15] We propagate hundreds of examples of useful nonnative species; otherwise there would be no corn in Europe, no horses in America. Britain alone hosts 1,800 nonnative plant species, which provide many staples at "all-natural" markets. Since 1930, scientists have created and spread over 3,200 mutagenic plant varietals, whose seeds have been modified using X-rays, gamma rays, and various chemicals. Think about that next time you consume grains, pears, peppermint, peas, grapefruits, sesame, and a wide variety of other "non-GMO" plants.

Many nonnative species hybridized with native species, enormously increasing variety and number. We have so modified, bred, molded, shaped, and distorted the "all-natural" to our own tastes and purposes that we forget, or take for granted, that much of what we live with, eat, and desire has evolved into something quite different from all-natural. In the words of ecologist Chris Thomas, "Nothing is entirely natural any longer."[16]

Our eco-engineering and unnatural practices also lead to unnatural selection throughout the world's oceans. If you put on a scuba tank in January and take a brisk dive off Monterey, California, it's clear, cold, and breathtakingly beautiful; forests of kelp sway as seals whiz by. You can pick up an octopus and watch birds dart past seeking fish thirty feet under. Countless starfish and anemones wave at large fish. But you don't see any large sardine-bait balls. Which is strange for two reasons: This area is one of the most protected, and earliest-protected, eco-coastlines on the planet. Also, this was once the "sardine capital of the world." (That's what fueled John Steinbeck's novel *Cannery Row*.)

There are two parallel evolutionary systems at work off Monterey. A natural one in which humans, the predators, decimate their prey—in this case fish. And an unnatural one in which humans use what they have learned to sometimes consciously redesign an ecosystem, to preserve and protect. A lot of species come back from the brink not because they adapt and adopt but because we will it; often we find them cute and they have a celebrity advocate.

Our attempts at eco-engineering and unnatural selection can also beget very rapid adaptive and heritable changes in a species.[17] In an attempt to prevent massive factory boats from simply netting all things in the water regardless of their usefulness, various countries have put minimum size requirements in place for fishing. The theory: Let creatures grow till they can reproduce, then harvest them. Good idea, except that when you drive evolution in such a cut-and-dried way, in a matter of just a few generations the most successful survivors and reproducers are the runts.[18] As more and more runts survive and reproduce, dwarf fish become the norm. This in turn changes not just the ecosystem of the specific semi-protected species but also those of its predators, symbionts, and co-dwellers.

We dominate so much that we can push unnatural selection (and rapid propagation) in some really unlikely environments. If you sail by the cold, foggy islands of Maine, in addition to fog, hidden rocks, and strong tides, you also have to cope with the extreme proliferation of colorful lobster buoys. Each year more appear. One would have thought that not a lobster would be left, but the opposite's true. The past few years have seen an 80 percent increase in catch, and a halving of price. A combination of global warming, abundant food in traps, size limits, depletion of cod and other ground-feeding fish, and fisheries management has exploded the lobster population.[19]

One glimpse of how unnatural the world (and how important human-directed evolution) has become can be found while examining the patterns of species extinction. Humans are truly handy at wiping things out. Within a few thousand years of arriving in Australia, we killed every land animal, reptile, and bird weighing more than 220 pounds: fifty-five species, erased.[20] The current rate of extinction is estimated to be 1,000 to 10,000 times higher than the normal rate of 1 to 5 species per year.[21] Even the poor frog named in Darwin's honor is near kaput.[22] But again, while many species collapse because of our actions, we are also creating/driving/begetting an unprecedented number of new species to satisfy our hunger for food, beauty, and companionship.

Just over a decade ago some scientists predicted one-third to one-half of all land animals would go extinct due to human climate tinkering.[23] Now some of these same folks witness and document a surge of new species entering the niches left by disappearing creatures.[24] So while we are in the throes of a great extinction, human design and accident have also led to an enormous explosion in variety and varietals.[25] Cal State's Madhusudan Katti has documented how urban encroachment sometimes leads to the extinction of many native species, but he also found rapid adaptation and the flourishing of various species in their new urban environments; one-fifth of known bird species now also inhabit urban areas.[26] Many birds develop a different song in order to cope with background urban noise; San Francisco's white-crowned sparrows are now far noisier. Migration patterns also change. European blackcap warblers now just stay in southern England rather than flying off to Africa. Eating habits, weight, and appearance alter as birds adopt new behaviors and food becomes constantly available in outdoor cafés. Mammals also adapt; Drs. Emilie Snell-Rood and Naomi Wick took skulls from ten mammals and minutely measured the differences in brain size between those that lived in the country and those that lived in the Twin Cities. Two of the city species, white-footed mice and meadow voles, which have short reproductive cycles, grew bigger brains than their rural cousins.[27] Apparently, at least for small mammals, it takes extra brains, guts, and street smarts to survive in the city.

Consider dogs. The closer a dog is to a coyote, or a wolf, the more "natural" it is.[28] Curiously, one of the closest things to a wolf, in people's homes, is a pet Afghan.[29] These big, thin, hairy, energetic animals, originally bred to hunt tigers, are now shampooed, combed, ribboned, and pampered. As we get further and further away from all-natural, we deliberately cultivate a crazy quilt of breeds, each designed for our own pleasure and purposes. Dog genetics is big business. We buy our pets for very specific purposes: to hunt, guard, look adorable in a Chanel bag, to make us look more macho. Buyer's choice. Successful human-driven pet designs tend to get reinforced and reaffirmed rapidly, through rebreeding. Breeds or mixes that do not please us rarely re-wild, much less survive and thrive; instead, they either gradually die out, mongrelize, or end up euthanized in shelters. Often what we like and desire is completely unnatural; let a dozen Chihuahuas or Lhasa Apsos loose on the African plain and watch what happens. Likely not many, if any, would survive. So exactly what about today's average dog is natural selection?[30]

Dogs are a perfect example of the two new core drivers of modern evolution: unnatural selection and nonrandom mutation. Man's best friend is a reflection of our genetic whims and desires. But of course it's not just dogs we are changing, and it's not just veggies, farms, and pets we alter; we are also fundamentally changing ourselves.

Fat Humans, Fat Animals: Another Symptom?

There's rumored to be another great big trend out there, perhaps one you may have pondered as you sat in the middle seat on your last flight. Just a few folks are getting just a little fatter, fast. Global obesity nearly doubled between 1980 and 2014.[1] One in three Americans is now clinically obese (the definition of which is a body mass index, or BMI, over 30).[2]

Obesity kills more people than does malnutrition.[3] The number of really fat folks in the developing world exceeds those living in rich and developed lands; obese Mexicans are now a greater percentage of the population of Mexico than obese Americans are of the United States.[4] A quick-and-dirty explanation has been offhandedly attributed to supersized soft drinks and fast food. Certainly that's a significant part of the story, given that a drink in a movie theater can be equal to a quart of syrup water, with the equivalent of twenty-seven sugar cubes dissolved inside.[5] But it is not just food that has changed.

Obesity also results from too few steps.[6] The substitution of machines for muscles is a trend that started a long time ago with water works, accelerated with the Industrial Revolution, and became dominant in the last half century. In 1960, at least half of all U.S. private-sector jobs required at least moderate physical activity. By 2010, fewer than 1 out of 5 jobs required physical exertion. Very few of us have mules outside to help us do our work, and most of us are nowhere near as muscle-bound as our ancestors were a century ago. In rural areas, where hard manual labor was always the norm, the rapid adoption of tractors, combines, mowers, pickup trucks, and other substitutes for backbreaking work has often led to acute obesity, even when farmers are compared with quite stout urban dwellers.[7] The average U.S. male now burns 142 fewer calories per day, during work hours, than his grandfather did. This factor alone predicts a 16 percent greater body weight for adult males today compared to 1960.[8]

As a species we have been conducting a large-scale experiment, over a very few decades, as to what happens to a human body when it massively increases its average weight—a trend that may have long-term evolutionary consequences. Fat mothers tend to have fatter babies.[9] Older mothers tend to have fatter babies; every five-year increase in Mom's age increases the chances that her newborn will become obese by 14 percent. Some medical practices are now seeing 16 percent obesity in infants.[10] Problem is, kids born obese, or who gain weight early in life, are also prone to adult obesity and far more chronic illnesses.[11] And they, in turn, are likely to have fatter babies.

For millennia, a chubby baby was a good sign, but too much of a good thing tips the scales (heavy-handed pun intended). A study of 25 million children across twenty-eight countries timed how long it takes them to run one mile; on average they were a minute and a half slower than their parents were at the same age.[12] Every decade, average global physical fitness falls 5 percent. (Yikes!) We are, without a doubt, eating too much and working out too little. Got it. But what if there were other human and environmental factors that might also be driving the extreme obesity epidemic?

One clue lies within recent insurance claims . . . for dogs and cats. During 2007, one provider's health-care claims related to obesity rose 19 percent.[13] Six years later, the percentage of overweight dogs had increased by an additional 37 percent. Overweight cats? More than a 90 percent increase in seven years.[14] Most of these animals are typically not eating or drinking the same fast food we are. Many are on quite controlled and standardized diets. So why is obesity exploding in this population as well? Perhaps we simply coddle and overfeed our furry domestic friends.[15] But if that's the case, then why does a broad-scale animal obesity study show that in both males and females, across twelve species, everything, everywhere, is getting fatter?

One might understand why city rats, Dumpster-diving for fast food and sugary drink, might begin to look like Porky Pig. But why would the same phenomenon be occurring in lab mice and rats? One of the key rules in science research is to reduce the variables, keep things standardized so you can really compare results. If lab animals are consistently treated and fed the same way, why would the odds of a male lab macaque being obese increase by 86 percent between 1971 and 2006? Obesity odds in female lab macaques increased by 144 percent.[16]

Were one tempted to dismiss these findings as simply the result of changes in lab food and practices, then one might ask why the obesity index for pasture-fed horses has increased 19 percent. Is the grass in these fields really different, or is there something globally systemic going on?[17]

Antibiotics may be one culprit. Many farmers mix antibiotics into animal feed. The objective is not to cure sick animals but to induce weight gain. These sub-therapeutic doses of antibiotics kill off some of the gut bacteria and allow more nutrients per feeding to get into the animal's body. While the amount dosed per animal is small, the overall systemic effect is enormous. Industrial farms used 30 million pounds of antibiotics in 2011 alone—about four times the amount sick humans used.[18] Some of these chemicals leach into the environment.

Trace amounts of chemicals in effluents, left over in animal waste, water, and feed, constantly contaminate human food sources and environments.[19] In China, where antibiotics are underregulated and overused, these chemicals can concentrate wherever water is recycled.

Some parks, where agricultural and human wastewater is reused, harbor 99 to 8,655 times more antibiotic-resistant genes than parks using fresh water.[20] Spreading low-dose antibiotics so broadly and consistently increases drug-resistant bacteria, because many bacteria are weakened but not killed, so they have a chance to survive and adapt. This makes each individual antibiotic less and less effective in combating human diseases as well. So might the widespread prevalence of low-dose antibiotics in our environment, just enough to promote weight gain in animals, be one factor in the generalized weight gain of wild animals and perhaps humans?

Aside from antibiotics, many other compounds, known by the cuddly title of "obesogens," may activate fat cells. Because some of these substances are nontoxic and noncarcinogenic, their effects tend to be understudied and underappreciated. And yet they commonly appear in our food chain; in 2009, some 56,231 pounds of triflumizole—a common fungicide and possible obesogen—were sprayed all over leafy green vegetables, fruits, and flowers.[21] And while we consumed our "fresh" salads, we may also have been drinking water from hard plastic bottles and thus consuming bisphenol A (BPA), another possible obesogen.[22]

Overall stress and sleeplessness can also affect human obesity. And a tendency toward obesity can be passed on generationally; constant maternal stress alters children's immune systems as well as their behaviors. How does this work? Studies have shown that pregnant rodents exposed to stress have a lot of hormones, such as cortisol, running through their blood, and their offspring in turn display anxiety behavior. What is interesting is that if the offspring are fed a specific probiotic, then the stress-driven changes in hormones and brain biology, and the resultant anxiety behavior, are significantly reduced.[23] This implies that diet, the environment, and chemicals all interact with the bacteria that live within our guts and are connected to our immune and endocrine systems. (The gut is the second "smartest" organ in our bodies with regard to nerve density.) Human studies indicate that irritable bowel syndrome (IBS) often occurs in individuals with psychiatric disorders, and is often exacerbated by stress.[24] As we change our lifestyles, diets, and stress levels, as well as those of the animals that live around us, we likely also influence obesity, even in utero.

Furthermore, in a purely Darwinian world, nature's cruelty might have taken care of extraordinarily obese babies, at least in the short term; many women would have been physically unable to deliver these ever-larger babies. But in yet another instance of unnatural selection, cesarean sections, removed this natural constraint. Throughout history, before sterile procedures and anesthetics, cesarean sections were almost always death sentences for the mother. Now they are commonly chosen, even for cosmetic or convenience purposes. More than one-third of U.S. babies and more than 40 percent of Chinese babies come into the world in a way that is very different from the historical norm.[25] In our human-driven evolutionary world there are no more natural vetoes on size at birth.

An Evolutionary Anomaly: More Sex, Less Reproduction

.... :✳: ....

In terms of evolution, sex is the bottom line.[1] No sex, no reproduction; no future genes, no evolution. Period.

Normally any species that has food, space, health, and peace tends to have lots and lots of sex, and the population expands accordingly. Antibiotic-resistant bacteria spread, lilies take over ponds, and rabbits overwhelm. It's normal and natural, in times of plenty, to have a lot of Boogie Nights. Beating the odds of natural selection requires that your specific gene code survive, thrive, and spread.

And while there are some very weird customs and habits out there in the animal world, even within the context of really unusual behaviors, such as mating plugs that glue a squirrel's vagina closed, Indian stick insects that copulate for ten weeks straight, honeybees' exploding testicles, and fig wasps that decapitate their lovers, no species except humans systematically practices birth control, abstinence, or childlessness when surrounded by abundance.[2] *That* is truly unusual and kinky.

Widespread birth control and new attitudes toward sex seriously bend Darwin's rules of evolution toward our wishes. So too does domestication/urbanization. Not that you'd want to really chat about this with your kids, or parents for that matter, but think back to when you were sixteen. Perhaps you might have had just a little curiosity about, interest in, and obsession with sex? Likely you indulged some of that curiosity, desire, and the occasional fantasy? Contrast this with today's Japanese teenage sex habits.

According to the Japan Family Planning Association, 59 percent of women aged sixteen to nineteen have no interest in sex.[3] Perhaps far more shocking is that reportedly 36 percent of teen guys, those grow- ing bundles of acne and testosterone, have no interest in sex either. This is why the Japanese population is collapsing at such a rapid rate. The demographics are already so unbalanced that each month the police arrest more elderly shoplifters than teenage ones. By 2060, the Japanese population will likely be one-third of what it is today. (And speaking of strange imbalances, within the Japanese porn industry there are more than 10,000 female actresses and fewer than 100 males, leading one overworked male star to argue that his kind are now rarer than Bengal tigers.)[4]

It's not just Japan. Globally, there are fewer kids born. In many countries, next generations grow ever smaller. In the United States in the 1970s, 1 out of 10 U.S. women was childless. Now it's 2 in 10.[5] Only 1 in 100 women in 1970 had her first child after age thirty-five; today, 8 in 100 do. This trend quickly cascades upward, generating ever-larger gaps between generations; in 1990, about 90 percent of women ages sixty to sixty-four had at least one grandkid. Today, fewer than 75 percent in this age range are grandmothers; soon it will be less than 50 percent.[6]

Prior to the mid-twentieth century, humans attempted to pro- tect and pass on their genes in some most creative ways. They imple- mented grand societal plans, attempting to codify rules and norms for all. They individually carried out various experiments in dark bedrooms and closets. Many wished for long-term, stable love, but they also experimented, officially and unofficially, with arrange- ments between men and women, men and men, women and women, lovers, mistresses, friends with benefits, communes, grandfathering, co-ops, male violence, coercion, polygamy, matriarchy ... As Bernard Chapais comments in the journal *Evolutionary Anthropology*, "The human mating system is extremely flexible."[7] Indeed, it is. The sheer variety and breadth of options is breathtaking. Only 17 percent of human cultures are supposedly strictly monogamous (one partner for life, period). All other societies allow some within their borders to operate under very different, overlapping, sometimes quite contradictory laws, beliefs, and morals.[8] But there have always been kids, lots and lots of kids, running around. That's how we got to 7 billion souls on Earth today.

Sometimes populations decline because humans impose wars, privation, and extreme oppression on other humans. Russia is no stranger to depopulation. It occurred from 1917 to 1923 as the empire became the "People's Paradise." At least a further 2 percent of the entire country's population was lost just from 1933 to 1934 as Stalin collectivized agriculture.[9] World War II and its aftermath killed off a further 13 million. Ironically the greatest driver of Russian depopulation hasn't been death and despair. It's been relative democracy and peace. No period of Russian depopulation has been as long and lasting as that which began with the dissolution of the USSR in 1992. In the final sixteen years of the Communist era, births in Russia exceeded deaths in Russia by 11.4 million. But in the first sixteen years after glasnost and perestroika, deaths exceeded births by 12.4 million.[10]

Meanwhile, China, even as it reformed its economy and acquired wealth, enforced a one-child-per-couple law, resulting in an inverted population pyramid: eight great-grandparents relying for retirement on four grandparents who depend on two parents who had one child. But 8:4:2:1 is not a normal-looking population structure, and neither is it a structure that bodes well for future evolution. While initially one child per couple was a rational policy that cut down on explosive population growth and allowed decades of increasing productivity as the population rapidly matured, by 2012 these policies started to harm the labor force. For the first time, the absolute number of workers aged fifteen to sixty declined 0.6 percent. And it will continue to do so, year after year, for the foreseeable future.[11]

China recently slightly loosened its regulations—if both parents had themselves been only children, they could have two kids. But this change may be too little too late; young couples feel it is normal to be an only child and fear the coming burden of caring for aging parents. So China is beginning to follow the patterns of South Korea and Japan; in these countries it's not legal restrictions on childbearing but rather expensive real estate, brutal education policies, lack of child-support networks, a desire for more personal freedom, and changing social norms that are rapidly shrinking overall populations. There may be a monetary or political logic to these choices as well, but this trend is not a natural Darwinian pattern. It's unnatural selection that as nations get richer, more comfortable, more relatively open, less violent, and better at educating women (female literacy is a good predictor of birth rate), they also choose a massive decline in their current population.[12]

Entire nations have decided, through the choices made sometimes by governments, sometimes by individuals, not to grow, not to reproduce, not to create more kids. We think this is normal. In the traditional terms of evolutionary biology, however, it's just plain weird. And it is only one of the really strange trends going on in sex and mating that will change how many of us there are and whom we mate with, thus determining the genes of our descendants.

Perhaps the greatest irony in our recent sexual evolution is the profound change in the number of years in which we *can* reproduce, physiologically speaking. While the average age of conception has been rising and rising, the onset of puberty has been steadily falling. Since the 1750s the average age of male puberty has fallen 0.2 years per decade.[13] Between the mid-nineteenth and mid-twentieth centuries, girls in the United States and several European countries began menstruating younger and younger.[14] Every decade, the ability to reproduce comes 0.3 years earlier. In the United States, the average age of first menses dropped from seventeen to around thirteen between 1830 and

1962. The average U.S. boy begins to enter puberty at age ten.[15] As we continue to alter our environment, our chemistry, our habits, and our hormones in unnatural ways, ever more girls and boys enter puberty while their birthdays are still in single digits.[16] At some point, we may consider whether to alter the most appropriately named KISS1 gene, which influences when puberty switches on or off.[17] The point is, we have a choice. How we address this choice will influence our evolution.

Meanwhile, there is another major, and opposite, trend in the reproductive viability of men; in the United States and parts of Europe, male sperm count fell by up to 50 percent in the latter half of the twentieth century.[18] In France, male sperm count fell 1.9 percent per year from 1989 to 2005.[19] (So the rumor that men are ever more useless may be partly true?)

Today's reproductive technologies and choices would bring some crimson to Darwin's Victorian cheeks, but they would also fascinate him. As far as Darwin (and even your grandparents) knew, there was one, and only one, way to get pregnant. Today there are dozens. Start with the de-linking of sex and birth. Historically, a couple would mate, and often less than a year later, a baby arrived. Now birth control allows far more choice of when to conceive. We have decoupled sex from time, and are stretching the calendar way, way out. None of our apelike human ancestors could reproduce outside the body, keep the embryos locked away inside frozen nitrogen far from home, and then implant the embryo inside a surrogate mother. No other mammal could freeze its eggs. But we can. Decoupling reproduction from age so you can have kids years, or decades, later? Nope, not on previous menus of sex acts either. And the oddest thing? We take it all for granted. Just before the birth of the first in vitro fertilized (IVF) baby, 4 of 5 Americans thought "test tube babies" were against God's will. But within a month of Louise Brown's July 1978 birth, an abstract procedure was transmogrified into a healthy, smiling baby. Suddenly more than 60 percent of those surveyed decided they were OK with IVF.[20] Now it's just part of the "natural" order of things, something to be discussed and thought about over a Starbucks latte with one's partner.

In 2012, about 1.5 percent of all births involved in vitro fertilization, in which hormone injections stimulate a woman's ovaries to release multiple eggs, which are taken outside the body, put together with sperm from the woman's male partner or a donor, and grown into minute embryos that are then transplanted back into the woman or a surrogate.[21] The use of surrogates further allows a woman's babies to be born years after she and her partner pass away; a baby can be born a year, five years, or ten years after an egg or sperm leaves the body. Two parents or three, or four, is not the way Grandma did it.

And while we are on the topic of the elderly, chemicals have also radically changed sexual habits, and even parenthood, among this demographic. As the smart folks at *Wired* so aptly put it, "Yesterday's drugs were about need, today's are about desire."[22] While "overweight, depressed, bald, and impotent were merely descriptive terms, in today's America they're considered illnesses and are fought back with some $44 billion a year in direct medical expenses—fast approaching the $50 billion we spend battling cancer." (By the way, this article was published in 2002.) As old folks sought and found better sex through chemistry, the widespread availability of Viagra, Cialis, and other erectile-promoting drugs led to some strange outcomes: The fastest-growing population for sexually transmitted disease? Not the inner cities, but old folks' homes.[23] It also led to some rather elderly fathers and mothers; Maria del Carmen Bousada was just about sixty-seven when she gave birth to twins. An unnatural trend that again promises nonconventional evolution opportunities.

And speaking of rapid evolution, aging parents drive a more rapid change of the human genome. As men age, their sperm acquire mutations (accumulating about two per year). This means around 75 percent of the natural human rate of mutation has been driven by men.[24] The older the man, the more mutations; fathers thirty to thirty-four years old increase the chances that their kids will have neural tube defects by 20 percent compared with twenty- to twenty-five-year-old dads. Fathers aged fifty and up have a 230 percent higher likelihood of engendering these, and many other types of birth defects.[25]

Finally, there is the option of picking and choosing embryos outside the body based on the fitness of their DNA (pre-implantation diagnostics reveal chromosomal defects such as Tay-Sachs or Down syndrome). And we're introducing new genes into embryos, or blocking some genes from expressing, pre-birth. And we're operating on embryos. The list goes on and on. When a species takes such deliberate and clear technological control over its own reproduction, something very fundamental has changed. And of course we're just getting started. The unnatural selection of future humans will be greatly augmented by nonrandom mutations arising from cloning, asexual reproduction, and edited copies of ourselves born decades in the future.

HOW DOES EVOLUTION REALLY WORK NOW?

The Nature Versus Nurture Wars

······························· ⁑ ·····························

Were folks sainted in academia, Edward O. Wilson would surely be a candidate. With seemingly endless time to talk to students, especially anyone even vaguely interested in ants, Wilson built an infrastructure for entire fields of study. His focus ranges from the minute aspects of exactly what makes one species of ant different from others to how entire ecosystems work. A systemic thinker, seeking to unite religions, morals, environments, and myriad science disciplines, Wilson thinks about how whole environments interact with ingrained genetic programming, and how this whole shapes individual and community behavior.

One of Wilson's major intellectual opponents was a cantankerous Nobel laureate, James Watson. While certainly very smart, and a co-discoverer of the structure of DNA, Watson has been dogged by constant controversy. Early in his career he minimized the contribution made by chemist Rosalind Franklin to the discovery of DNA, despite having looked at her crystallography pictures of DNA without proper permission.[1] Pre-Nobel, having driven a part of the British establishment batty with his somewhat aggressive and self-assured manner, Watson decamped to Harvard, unpacked, looked around, and promptly declared his biology colleague E. O. Wilson to be a mere "stamp collector."[2] In Watson's world, only the molecular part of biology really mattered.

Wilson was not pleased. But being a Southern gentleman and a gentle soul, he held his tongue . . . for a while. After constant bludgeoning, however, Wilson's almost infinite patience ran out and he finally said Watson "was the most unpleasant man he had ever met"— Wilson's equivalent of calling him every obscenity known to man. He wasn't the only one who felt this way. Soon Harvard's campus was bitterly split by factions arguing over whether to approach biology primarily from a top-down, systemic, environmental viewpoint (Wilsonians) or strictly from a bottom-up, reductionist, molecular-biology perspective (Watsonians).

This was the continuation of a debate that started early in the twentieth century, when our understanding of nature-environment versus heredity took on a notably different tenor after the rediscovery of Gregor Mendel's work and the concept of units of heredity or genes. Then, in the 1940s and '50s, Oswald Avery, James Watson, Francis Crick, Sydney Brenner, and a multitude of other scientists began discovering and mapping how DNA, proteins, and other forms of life code operate; life is encoded in a spiral staircase, a double helix. The rungs on this staircase are four chemicals: adenine, thiamine, cytosine, and guanine, abbreviated as a four-letter code, A, T, C, and G. Just as you write a sentence using a long string of letters, or a computer program using a binary code of 1s and 0s, so too you write a genome using a long string of A's, T's, C's, and G's. Further, nearly every human cell contains twenty-three pairs of chromosomes, made of DNA combined and inherited from each parent. Within these twenty-three chromosomes lie 3.2 billion ATCGs. All of this code equals one human genome, which you could print out in a very boring 262,000-page book.[3] A strand of DNA is so thin (two molecules across) that if all of the chromosomes in a single cell were stretched end-to-end they would be six feet long. Remarkably, nature creates nearly perfect copies of your DNA in every one of your cells every time they split.

Initially DNA was a dark-horse molecule in the scientific race to find the basis of heredity. While geneticists knew that DNA existed in many life forms, at a macro level it looked chemically identical in all species, so it was assumed to be unimportant—a bystander and by-product—because surely something far more complex must determine the differences and similarities between a telemarketer and a gummy paramecium. Only after Oswald Avery, Colin MacLeod, and Maclyn McCarty performed their seminal experiment in 1944, showing the DNA in pneumonia was what transmitted different types of infections, did a few other scientists begin to focus on DNA as the driver of evolution.[4] And still this critical discovery was largely ignored by the scientific community until the 1950s (and mistakenly never recognized by the Nobel Prize Committee).

Post Watson-Crick, the world gradually began to realize that life code could be read like an instruction book. For much of the remainder of the twentieth century, molecular biologists strove to understand life code's many expressions and variants and to explain it all on a microscopic level. Traditional biologists, ecologists, and environmentalists were often ignored or ridiculed by an emerging scientific priesthood that believed it, and only it, held the key to all life. Watson shouted that every biologist who looked top-down at environments, at how animals or plants acted, was fair game and should be fired. He was interested only in biomolecules. Nature's code drove everything.

As rapid, automated DNA-sequencing machines got ever more powerful and cheaper, landmark studies of DNA uncovered the genetic mutations that cause diseases such as cystic fibrosis, many cancers, and various heritable traits, including a bent pinkie finger, tongue rolling, dimples, and muscle type. Each of these discoveries reinforced and re-proved Darwinian ideas as to how random DNA mutations can create differences among species—even among people. Entire libraries of increasingly complex genomes began to provide detailed blueprints of how various creatures were built and differed, culminating with the draft human genome in 2000.[5]

But the more we learned about genomics, the odder it seemed.

Instead of having the 100,000+ genes scientists predicted, humans only have about 19,000.[6] The difference between you and your horrid neighbor is a mere 0.1 percent of the core ATCG code.[7] Yes, genomics can sometimes predict skin color, high blood pressure, eye color, dangling earlobes, and hairy backs, but for more complex traits the predictive power of genomics and its ability to connect genes to traits drops precipitously.

Gradually, even the most fundamentalist of the "only molecules matter" crowd has had to recognize the role of the surrounding environment in genetic expression. An organism's environment can seriously influence how its DNA is expressed and how this expression is passed on to future generations. The DNA you are born with may be seriously altered by your surrounding environment. As geneticist MDs now like to say, nature loads the bullets (the genes) and nurture pulls the trigger (the expression of those genes).

As we learn to modify gene code, the difference between what nature has evolved and chosen in DNA molecules and what we choose to modify or rewrite inside a multitude of species further complicates the nature-nurture division. Even at Harvard both factions are now on speaking terms and informing each other—so much so that Watson and Wilson came together onstage in 2009 and had quite a civilized conversation. (However, during Q&A, when asked, "Can a gentle person do well in science?" seemingly in reference to Wilson, Watson's answer was wholly in character: "Jesus would not have succeeded.")[8] We owe much to James Watson; he helped map out the exact mechanisms of DNA replication, and he became an early advocate and leader of the Human Genome Project. Under his watch, Cold Spring Harbor Laboratory produced first-rate discoveries, mapping out crucial pathways in inherited diseases and cancers. But he sometimes let out his Mr. Hyde.[9] And he was so extreme in arguing how only gene sequences mattered that he killed a lot of promising careers and research.

So what happened with the nature versus nurture debate during these decades of extraordinary discovery? Biology turned out not to be deterministic. The context in which a set of genes is expressed matters—a lot. Both Watson and Wilson were right, in part. And their theses were far more accurate when brought together and integrated—something Wilson realized by 1998 when he wrote *Consilience: The Unity of Knowledge*.

How does this affect you? It makes it far harder to predict exactly what will make you sick or the consequences of taking any particular medicine or attempting to "upgrade" your body. As humanity begins writing and releasing genetic codes into our domesticated and human-design-centric environments, both nature and nurture will come together to drive evolution. As we modify, shape, and domesticate our environment, we also modify ourselves, our gene code, and that of our descendants. Our cultural norms, the "nurture" part of the equation—how we choose to act; what we do; and how we deal with microbes, plants, insects, animals, and ourselves—reflect an increasing awareness of and belief in human dominance, and in turn this shapes human evolution. A revised theory of evolution has to account for both how we will write in gene code and how we will redesign the environments this new code is expressed in. Otherwise we will, once again, end up with "missing heredity."

Missing Heredity, Mysterious Toxins

····················· :✳: ·····················

Had Central Casting been looking for the prototypical white-haired, ramrod-straight, caring, brilliant physician, Victor McKusick, MD, Ph.D., would have fit the bill. Beginning in 1966, he spent his life cataloging every known genetic disease variant into one core reference book: the Online Mendelian Inheritance in Man (OMIM).[1]

McKusick was tireless and relentless. His catalog expanded at a rapid pace, with mutated genes responsible for numerous common and rare diseases added every day, as well as genes that are indicators of earwax type and some forms of athletic ability. The OMIM describes rare diseases in excruciating detail, as well as variants hinting at predisposition to neurological conditions such as Alzheimer's, Parkinson's, and ALS. McKusick also cataloged the small variants that partially determine eye, hair, and skin color; curly or straight hair; height or body weight. In all there are over 21,565 entries connecting genes to diseases, everything from alopecia to Zellweger syndrome.[2]

OMIM provides an extraordinary guide to the many things that may happen to you because you did, or did not, happen to pick your parents wisely. Take height, for example. Many a high school basketball coach, upon seeing some young kids in the local grocery store tagging along with their super-tall parents, will strike up a conversation to get a sense of future prospects for the high school team. In most instances one can tell, by looking at parents, who is most likely to become 6'5" tall. Variations or mutations in forty different genes are known to correlate with height.[3] As long as malnutrition is not a factor, perhaps 80 percent of your height is determined by your genes, with tall and short stature running in families and being highly correlated among siblings and non-identical twins separated at birth.

And yet the forty identified genes that "predict" height still cover only 5 percent of the genetic basis of height. Despite all of McKusick's cataloging, millions of research dollars, countless hours of thought, and the most advanced genomics technologies, we are still unable to pinpoint the principal genetic causes of being tall or being short—or the genetics behind a host of other traits that seem pretty obviously distinguishable.

Wonderful as he was, McKusick was so successful, dominant, and smart that he, like James Watson, may have unwittingly and partially misguided many people in terms of the predictive power of genes alone. Early on in the study of genetics some simple disease or non-disease traits could be reduced to a single mutation in a gene, but these successes appear to have been the "low-hanging fruit." Most traits are complex. Often genes tell us very little about the why and leave large gaps in the how. But one rarely reads about this aspect of genetics because almost all publications, and subsequent breathless stories in the media, primarily focus on what geneticists *can* find and not on what they can't.

Much of the general public, therefore, has ended up with the mistaken impression that almost every imaginable disease has a clear and readable genetic basis. But very few diseases operate like a light switch or correlate in a one-to-one way whereby if you have gene X then you get disease Y. Time and again, studies fail to find the genes or genetic variants responsible for a majority of the cases of heritable diseases or traits, including for Type 2 diabetes (only predictive in 6 percent of a population), good cholesterol (5 percent), early myocardial infarction

(3 percent), and familial breast cancer (10 percent). In every case, a few genes are identified as causal, but in only a small subset of patients. This includes the recent sensational headlines that "specific gene variants" predict whether taxes deter cigarette smoking in one group versus another. You have to read the article, and the footnotes, before you realize this is true, but only for 1 to 2 percent of the population.[4]

So when you read yet another story headlined "Scientists Find a Gene for X," read the endnotes, qualifications, and caveats carefully. They may have found a statistically significant correlation, but it's usually only relevant to small subsets of a population. For the many complex diseases that run in families, including cancer, hypertension, neurodegeneration, autoimmunity, diabetes, obesity, schizophrenia, and depression, the majority of supposedly causal genes are missing. Typically, in the strongest correlations, scientists uncover genetic differences that account for only 5 to 15 percent of the cases.[5] This nasty, large, dark secret among the lab-coated is known as "missing heredity."

Not finding specific, causal, predictive genes for inherited traits, diseases, and conditions remains far more common than finding the specific genetic signatures that allow one to say, yes, a person has this, will develop that, will grow to be that. But geneticists, being very smart people who often trained in mathematics and physics before switching into life code, focus on results and success, and rarely write commentaries about why they can't find the missing genes.[6]

The occasional "missing gene" commentaries fall into a category all their own; these are among the most difficult and densest papers you will ever attempt to read. They include terms like *epistasis, allelic architecture, biological dark matter, linkage disequilibrium,* or *polymorphism,* along with an ungodly number of equations and correlations. But fancy terms and mathematics aside, when you finally drag yourself to the conclusion of any one of these papers, you basically read the same four explanations over and over again: (1) "It's complicated and related to many, many interacting genes"; (2) "We need more data (and certainly more grant money); (3) we will get the answers once we sequence the genomes of millions of people"; (4) "The genes aren't really missing, they are just hidden and our new strategy will find them." Or, if they truly found nothing, they might just adopt a serious demeanor, look over their glasses, harrumph, raise their eyebrows, and expectorate an old chestnut: "It's a combination of nature-nurture . . ."

Despite billions of dollars spent, most labs suffer from constant cases of *gene interruptus*, leading everyone to focus on and endlessly cite the few genes and codes that do clearly correlate, leaving a deafening silence on the majority of conditions, diseases, and traits that do not have clear, mathematically probable, peer-publishable results. Grant after grant gets spent on results that basically say, "My lab rats ate my genes."

Intelligence is a good example of missing heredity; a substantial part, perhaps 50 to 85 percent, is assumed to be inherited.[7] But try as they might, gene scientists just cannot find "the gene" or genes associated with this minor and non-obvious human trait. In a forced march, reminiscent of the early sieges of Everest, the Beijing Genomics Institute launched a controversial study of a mere 126,559 individuals in an attempt to find intelligence's missing heredity. After analyzing tens of thousands of small genetic differences that existed within this group, and looking for correlations, the study showed very minor effects from three variants within a few genes.[8] How can it be that we cannot find any evidence of such a supposedly predictable genetic predisposition?

Most human diseases, behaviors, and traits involve a combination of the genes we got from our parents *as well as* events we experience in our everyday lives, especially in utero and as children. You might even throw some microbes, or the lack thereof, into the predictive equations. Many of the conditions that seem to run in families, such as cancer, depression, intelligence, asthma, athletic prowess, height, addiction, happiness, autism, hypertension, musical talent, body weight, childhood aggression, longevity, altruism, heart disease, and schizophrenia, are part heredity and part environment . . . It's Wilson plus Watson.

Chemical exposure is a good example. *Sax's Dangerous Properties of Industrial Materials* catalogs 28,000 toxic substances, including more than 2,000 known teratogens—agents that cause birth defects in a fetus.[9] The list reads like a chemistry text, filled with many unpronounceable words.[10] Some of these chemicals are clearly, unequivocally bad actors and have measurable, observable, publishable bad-juju effects. There are countless examples of severe adverse events on unborn or young children in connection with these chemicals. But in any given neighborhood, even if everyone is exposed to some of these chemicals, only a few may be damaged long-term.

So why don't we just stop using chemicals right now? For one thing, this is completely unrealistic. About 96 percent of all manufactured products have "chemistry" as a basic input.[11] Second, the chemistry that surrounds us, day in and day out, usually makes our lives better and safer. If you wanted to get a quick sense of what chemicals do for us, you might go for a weeklong hike in a Florida swamp and take nothing at all chemically related with you. To make it even more amusing, let your good friends hike with you and use "chemicals," such as dried foods, a tent, efficient clothes, a flashlight, mosquito repellent, sunblock, water purifiers, mosquito nets, and antiseptics. After you get back from your jaunt, try a similar experiment in a city. Use and touch no chemically derived products at all, while your friends get on with their daily lives. You might find your urban existence is not quite so pleasant without transport, housing, roads, most foods, and myriad other unnatural essentials we take for granted.

What we can do instead is recognize that some chemicals drive evolution, and we must get better at identifying their effects. This isn't easy, even given clear hints that some chemicals affect us in specific ways; as is the case with genes, it is far more common to conclude research with ambiguous results that do not clearly demonstrate causality. One analysis of congenital birth defects indicates that while 20 to 25 percent are caused by known chromosomal or gene anomalies, only 7 percent can be attributed to environmental agents including radiation, drugs, and environmental chemicals.[12] The remaining 70 percent of birth defects have an unknown origin, which again leads right back to the type of language one so often sees when scientists haven't got a clue: "It's a multifactorial combination of genes and environment."

For almost three-quarters of birth defects, no one gene variant or chemical has been identified as the smoking gun. Which means it's hard to get funding, publish, and prove that a specific chemical is what causes or contributes to a specific condition. As a result, many suspected toxins continue to be produced, sold, and spread; often those who want to continue to market a product have quite a bit of money and power, and can sow doubts as well as present many alternative causes to be explored before taking their product off the shelf. Even the most egregious offenders can thrive for decades; after completing internal studies showing extremely negative health effects, tobacco companies cynically paid for advertisements that screamed, "More Doctors smoke CAMELS than any other cigarette," and "20,679 Physicians say LUCKIES are less irritating."[13]

In our increasingly complex world, as we encounter thousands of unnatural toxins, typically in varying combinations, every person becomes a relatively unique specimen unto him- or herself, one that cannot be replicated in a laboratory setting. The best we can do is look for instances where a large population has been exposed to an agent. So while MDs love certainty and want to provide definitive diagnoses, many areas of medicine today still rely on a wait-and-see, or maybe-this-is-causing-that, or let's-see-if-this-works-at-all approach.[14] In fact, in many cases we have no idea what is causing a particular syndrome, disease, or epidemic. We just notice that unexpectedly someone is sick for no apparent reason.

We are increasingly driving evolution, but often without knowing the real causes of many afflictions. Are we really that much different, then, from the shamans of yore, explaining and treating fevers, headaches, plagues, deformities as products of mysterious spirits? Not surprisingly, given so much hedging and uncertainty among scientists, the afflicted, desperate and afraid, sometimes turn to educated—and sometimes uneducated—guesses of what caused harm to their bodies. They implement personal preventive strategies and unconventional treatments to ward off or treat a variety of conditions. The popularity of back-to-nature living, various niche organic foods, homeopathies, vitamin therapies, detoxifiers, and (ouch!) colonics grows year after year, even when scientific studies fail to support them.[15]

Chemicals continue to spread. In 2013, Canadians monitoring sewage indirectly discovered yet one more way to demonstrate just how exposed we all are to multiple, accumulating unnatural substances. After finding ways to type and trace various unnatural compounds that often show up in effluents, these researchers began mapping concentrations of certain chemicals within various rivers and lakes. The resulting data told them who was dumping how much sewage where. But it also told them what was entering and leaving human bodies. What goes in must come out, so by studying the concentrations of four artificial sweeteners found in waterways—cyclamate, saccharin, sucralose, and acesulfame—they could trace what we put into our bodies, process through our gut, and expel into sewage. Artificial sweeteners are particularly easy to trace because they pass right through our bodies, into treatment plants, and then get deposited downstream in measurable quantities. Then they get recycled and concentrated even more, in our bodies, in our waters. What the Canadian team found at twenty-three sites in southern Ontario, in household water taps, was significant traces of sweeteners, the highest environmental levels of these substances reported anywhere.[16]

Sweeteners provide a cool set of markers for all kinds of useful information: One can trace and separate human and animal waste streams and pollutants, assess nutrient assimilation, determine pollution patterns and dispersal rates, and figure out where emerging contaminants are coming from. On the other hand, having these and many, many other compounds persist and spread to such an extent might have some effect on the ecology of the Great Lakes, never mind our own bodies. Chemical exposures of various types, often at irregular, low levels, for long periods of time are more than likely inducing gradual changes in our bodies. Just one more way in which we are, deliberately or not, inducing nonrandom mutations and driving our own evolution.

Transgenerational Inheritance—aka "Voodoo Biology"

·❋·

Let's consider some new science that even biologists find a little freaky. As WWII peaked, the Nazis blocked all food and fuel supplies to the Netherlands, leading to famine. Many babies born during the famine suffered long-term effects, including a higher incidence of a variety of conditions such as heart disease, obesity, glucose intolerance, and obstructed airways. Severe trauma altered the victims' gene code for life, even if the victim had yet to be born.[1] So far, geneticists had no issue with this, because one could argue that under the rules of traditional genetics, the famine had also acted directly on the genes of these babies in utero.

But here is the weird part: The effects didn't stop with a child or with a generation. Postwar and post-famine, later-born siblings were also affected. Even in periods when food was available and the war over, a genetic memory lingered. And it appears to linger a long time. In follow-up studies, the daughters of Dutch mothers who had suffered through WWII's famine while pregnant in turn had daughters with twice the average rate of schizophrenia.[2] In other words, mothers' wartime duress was passed on to their daughters, in the form of mental illness, and then on to the granddaughters: a genetic scar, inherited collectively by many individuals across at least two generations. Somehow, genes had been altered even for those who had no direct contact with the famine itself.[3]

If our gene code can change in real time because of our surrounding environment, and if these changes can be passed on, then a long-discredited biologist, Jean-Baptiste Lamarck, may not have been 100 percent wrong. In the early nineteenth century, Lamarck was run out of bio-town for daring to suggest that evolution can take place in one generation; he argued that if giraffes stretch their necks to reach the upper branches of trees, their necks will lengthen and this beneficial trait will be passed to their progeny. In other words, Lamarck was saying that evolution isn't the very slow and apparently haphazard process Darwin described. Soon thereafter, Lamarck's theory was disproven. Darwin won, to the point where an AP Biology review states, "We now know that Lamarck's theory was wrong. This is because acquired changes (changes at a 'macro' level in somatic cells) cannot be passed on to germ cells."[4] Cut and dried, case closed . . . except that the Dutch famine cases seem to contradict this assertion.

Until very recently, "transgenerational inheritance" was a concept typically banned from all polite geneticists' conversations. But then doubts began to creep in when scientists performed experiments and observed the various nifty tricks and speed with which various bacteria adapted to new environments. The experimenters realized two things: First, there was a very low likelihood that rapid adaptation was taking place due to random, beneficial mutations. Second, given how fast a trait like antibiotic resistance could spread within a species and across many species of microbes, there had to be some real-time evolutionary reset mechanism.[5] So a few brave souls revived the term "epigenetics," first coined in 1942 by Conrad H. Waddington, a British scientist.[6]

Most early epigeneticists were ignored or written off as "voodoo biologists."[7] What they preached was such a radically different discipline from core genetics that as long as their experiments were confined to bacteria, the outcomes and modes of action could be considered a fluke. But then came tomatoes, in which scientists observed and quantified transgenerational changes from mother to daughter to granddaughter tomato after exposure to drought, extreme cold, or great heat.[8] The discoveries kept piling on; in 2013, a Cornell team demonstrated that epigenetics, not gene code, was a critical factor when trying to figure out when and why a tomato ripens.[9] Similar epigenetic effects were discovered in worms, fruit flies, and rodents; a creative and slightly meanspirited experiment let mice smell sweet almonds and then shocked their feet. Soon mice were terrified of the smell of almonds. When these mice reproduced, the kids were never shocked, but they were still quite afraid of the same smell. So were the grandkids. The brains of all three generations had modified "M71 glomeruli," the specific neurons sensitive to that type of smell.[10] We do not yet know how many generations epigenetic tags can survive for, but in rats the effects can last at least four generations. In worms, disrupting epigenetic control mechanisms can have consequences persisting for seventy generations.[11]

What seems to be happening is that core/linear DNA code, the code reflected in an organism's sequenced genome, takes a long time to change. But there are various other mechanisms that change/modify/speed/slow the expression of the core gene code. It's analogous to the way a word can take on a completely different meaning—say, if one's name is spoken by one's parent in one tone or another, with a sharp or soft emphasis, slowly or very fast. The name has not been modified, nor has its spelling, but the tone can sure change the same word. And a parent who over a lifetime always pronounces a child's name with a sigh of disappointment has a very different influence from one who always says the kid's name with sunny optimism and joy.

What is interesting about epigenetics is that it argues that external and internal stressors and de-stressors can alter gene code not just in individuals but over generations; in other words, epigenetics posits that even the same exact gene code (that is, identical twins) can be reversibly modified with small chemical tags, effectively "on/off" switches that activate or silence particular genes. This implies that an environmental stimulus (for example, famine, stress, toxins, affection) can be transmitted via the nervous, endocrine, or immune systems to the DNA in each cell, which in turn sets switches that express hereditary code to silence or activate in a particular situation. Under siege by some invaders? Flip a few switches to cope. Fall harvest plentiful? Flip a few switches to store fat, procreate, and ramp up metabolism. A plague in the neighborhood? Flip a few switches to enhance resistance. Your DNA genome has "on/off" chemical switches that collectively are known as your epigenome. So your epigenome is unique and changes every time a switch is flipped.

Because your epigenome's switches are considered reversible when they are passed from parent to child, many scientists view this to be "soft evolution," i.e., not guaranteed to be enduring as occurs when a mutation arises in the core DNA genome. The epigenome can be passed on, sometimes reversed, sometimes reinforced. Unlike in classic Mendelian genetics, it is hard to predict and quantify, so you can just imagine how this variation in experimental outcomes has driven many careful, traditional scientists who believed the DNA code was the be-all and end-all of heredity completely crazy; they would try to eliminate all the variables, use genetically identical rats, and sometimes get completely different results. So it is no surprise that for decades epigenetics was ignored or pooh-poohed by funders, senior biologists, and science magazines. There was no reliable way to trace the precipitating event and no way to easily predict which individuals would be affected in future generations.[12]

So how do our epigenomes become informed about life around us, particularly the epigenome of a fetus or a yet-to-be-conceived child? Most of the science points to our neural, endocrine, and immune systems. Our brains, glands, and immune cells sense the outside world and secrete hormones, growth factors, neurotransmitters, and other biological signaling molecules to tell every organ in the body that it needs to adapt to a changing world. As we experience stress, love, aging, fear, pleasure, infection, pain, exercise, or hunger, various hormones adjust various physical responses within our bodies. Hormones surge through our blood; changes in cortisol, testosterone, estrogen, interleukin, leptin, insulin, oxytocin, thyroid hormone, growth hormone, and adrenaline make us behave and develop in different ways. And they signal to our epigenomes, "Time to flip some switches!" Genes get shut off or turned on as the world around us changes.

But exactly what are these "switches," and how do they operate? Soft evolution is analogous to an annotated book. The basic text and argument of the book remain the same. But if the text is gradually surrounded by margin notes and comments, then those who read different annotations of the exact same book may end up with very different learning, depending on who annotated the particular copy they borrowed, how they treated the original text, how the reader decided to interpret the interplay between the original printed text and the annotations, and whether some of the annotations were erased or modified by other readers. Some annotations are so important that they eventually get incorporated into future edits/editions/interpretations of the core text. (See for instance the Bible, King James edition, which is very different from far older, closer-to-the-original-source Aramaic versions.)

There are multiple ways to add in rapid, inherited epigenetic adaptations without any change in the core DNA code.[13] One basic and common mechanism is "DNA methylation." In nerd-speak, enzymes in our cells attach a methyl group (CH_3) to a cytosine (C) located next to a guanine (G) in our DNA, forming a methylated island. This tells the gene that follows next, "Shhh, do not express yourself." Recent studies revealed 6-methyl adenine (A) to be the opposing genomic "on" switch. One of the key reasons for human diversity is that about 70 percent, or roughly 14,000, of our genes have these "on/off" switches plus random mutations among them, so there are countless combinations of ways that these switches are flipped in the human population.[14]

Sperm and eggs get a nearly fresh start vis-à-vis epigenetic tags and inheritance. An estimated 90 percent of the switches are erased before conception occurs, which means most epigenetic memories are lost. But there is still a lot of recent news/data/knowledge moving from generation to generation. Because most scientists neither suspected nor believed that an unborn child could hear so much "hormonal gossip," even pre-conception, one can find naïve articles on core vehicles of heredity with titles like "Receptors in Spermatozoa: Are They Real?"[15] But those who described sperm as simple bags of DNA with a tail could never explain why sperm had so many receptors for so many hormones not directly related to reproduction, including leptin (one of the obesity genes), as well as nineteen growth factors, cytokines, and neurotransmitters. Epigenetic switches can be flipped on and off in sperm, eggs, or embryos, so your kids and grandkids can share your environmental experiences and knowledge, and be better prepared for the environment they will soon be entering. For instance, if you were a male smoker, and your brother was not, twenty-eight epigenetic signals in your sperm would be different from his.[16] So yes, sperm are listening. And they sometimes survive to tell tales not just of a man's genetics and age, but also of childhood drama and crazy teen years with wild raging hormones. Fertilized and unfertilized eggs also carry receptors that are continually informed by Mom's hormones. At conception, your grandchildren listen to distant tales, and sometimes pass them on.

Scientists are just beginning to decode the epigenetic signatures that are transmitted through sperm and eggs to future generations.[17] For example, pregnant mothers exposed to lead and other toxins pass deleterious epigenetic memories to their unexposed grandchildren. A study of paternal sperm uncovered an epigenetic signature that correlates with signs of autism in young children. A study of mice found that sperm and eggs obtained from obese parents carry epigenetic signals that persist even when the embryo is carried by a normal weight surrogate mother, resulting in overweight, diabetic offspring. So remember, prior to conception your sperm and eggs are listening, but we can't yet decode what they hear. We are all deeply influenced by what our parents and grandparents did, ate, and breathed.

Many studies now confirm the transgenerational Dutch famine studies in humans. Calorie and protein deficiencies lead to underweight babies. A child entering a world of little food benefits from being smaller and burning fewer calories (a phenomenon termed the "thrifty phenotype").[18] During the Dutch famine, babies were underweight. Though they appeared healthy once normal amounts of food became available, as adults they often experienced a higher incidence of obesity, diabetes, and cardiovascular disease.[19] Mom actively communicates with her baby using hormones that regulate metabolism.[20] The rapid rise in Type 2 diabetes among Cambodians in their late thirties, whose fetal development coincided with the famine inflicted by the Khmer Rouge in the late '70s, is likely related to parental stress.[21] Even after moving away from a war zone, immigrants to Canada, Denmark, and Sweden have children who still suffer far more mental illnesses despite being conceived and raised in a safe and supportive environment.[22] However, one has to remember our lives used to be quite a bit tougher and more stressful than they are today, so there are instances where even tragic events can have favorable effects. For example, Swedes whose grandfathers experienced starvation before puberty suffered fewer heart attacks and cases of diabetes.[23] Life reprograms itself in strange ways.

How epigenetic gene switches are set can affect brain development, memory, or whether you get cancer. Which particular switches are on or off varies from birth through old age; as your body ages, you lose many of your protective cell functions, and switches can get stuck in what seem to be wrong positions as our ancient evolutionarily acquired epigenetic programs attempt to adapt to our modern lifestyle.[24] Interestingly, recent studies of rodents showed how drugs that alter DNA methylation can block new memories or restore memory losses.[25] We are just beginning to learn the logic of which switches pass to subsequent generations; some appear to be random and others convey instructions critical to survival. The latter instructions and traits have a chance of eventually becoming embedded into one's core genome.

Given that our environment communicates with our DNA and with that of our grandkids, one has to consider the overall effect on our own evolution caused by the recent and very rapid change in where one lives, what one eats, and what one is exposed to. Epigenetics may solve part of the puzzle, as a very powerful agent of genetic change, and explain why the incidence of so many conditions, such as allergies, obesity, and autism, is exploding. A significant part of the story is not what we are exposed to but what our grandparents were exposed to.

As the rate at which we change the human condition accelerates, gene switches are being activated or deactivated all over the place, sometimes with favorable outcomes and sometimes with unfavorable outcomes as nature and nurture seesaw back and forth. Why is this so important? Because it provides a mechanism for very rapid evolution, a means by which humans—or any species—can adapt and evolve in a few generations, and is linked to current environmental events, in which traits can be passed on and reinforced within time periods that Darwin would have thought impossible. Epigenetics is a second, parallel path to random or deliberate genetic engineering; and your epigenome, which you inherit from both parents, acts like a second genome.

WWIV: Nuking Our Microbes

·············· :✳: ··············

Most history books teach our species-centric view of life, so we grow up believing that our species is pretty deft at mayhem, carnage, and death. But for all of history, at least until we got proficient at making hydrogen bombs, it turns out we were rank amateurs.[1] By far the overwhelming causes of death in wars and conflicts haven't been bullets, bombs, steel, arrows, rocks, or fire. They were bacteria, viruses, and parasites.[2] Infections killed far more soldiers and civilians than did any general's strategic acumen or dictator's deranged decrees.[3]

The military's emphasis on order, spit and polish, short hair, digging latrines, and quarantines isn't coincidental or merely aesthetic. It comes from a millennial history of dirt and infectious diseases overwhelming and wiping out troops and civilians. One of the reasons it was so easy for a few Spaniards, Brits, and French to conquer vast tracts of land and millions of people is that up to 90 percent of the First Peoples (peoples indigenous to any given geographic area) may have been decimated by smallpox, leptospirosis, yellow fever, and plague brought in by early explorers.[4]

During the Crimean War, 2,755 UK soldiers were killed in action. Wounds ended a further 2,019 soldiers' lives. And disease wiped out

16,323.[5] In the U.S. Civil War twice as many Union soldiers died of infections as from Confederate attacks.[6] It is only in the last seven decades of world history that we've radically tipped the mortality balance away from pathogens and in our favor, a great example of unnatural selection. In a real and quantifiable way this required winning the four greatest global wars humans have ever fought—not against one another, but against our microscopic foes.

Micro-WWI: Vaccines. The power of vaccines is extraordinary; they are a weapon that turns the tables on many of the microbes that formerly practiced a whole lot of "natural selection" on us and fundamentally shaped human civilizations. In modern times, as almost all humans are exposed to compounds that can stimulate their immune systems, we are driving whole classes of microbes toward unnatural extinction. For example, as late as 1988, polio was crippling more than 200,000 people in the most populous nation on Earth, India. In 2009, Indians suffered half of the world's total cases of polio. On March 27, 2014, India was declared polio-free. Wild polio now only survives in Nigeria, Afghanistan, and Pakistan.[7]

Polio is not the only victim of our deliberate selection. If you were born after 1972, you simply never acquired the round scar on your arm or inner thigh that your parents and grandparents had; you were never vaccinated against smallpox. Thus, one of the deadliest and most prevalent of all diseases became a historical footnote.[8] Vaccines, along with clean water and basic nutrition, may just be the cheapest and most effective thing humans have done to reduce infant mortality.

Micro-WWII: Antiseptics. It's hard to overstate the success of this second Great War on microbes. Before Joseph Lister's sterilization methods became standard operating procedure, going into a hospital, even for minor surgery, frequently meant a death sentence. Compound fractures, which led to immediate amputation and often to subsequent massive infection, had a death toll of about 68 percent.[9] A 1915 *Science* magazine article describes pre-Lister hospitals as "reeking with pus and emptied by death. After operating, when the roll was called reporting a mortality of 40, 50, 75, 90, and 100 percent . . . Now we have hospitals of immaculate whiteness and the roll call reveals few mortalities exceeding 10 percent, most having fallen to 5 percent." Despite accumulating evidence, many of the most prominent surgeons of the day continued to doubt the existence of "microbes"; these doubters' wards remained death chambers for decades. But eventually even the stupid understood. Now you're very unlikely to hear what was once a common phrase: "She died of hospital gangrene."[10]

Micro-WWIII: Antibiotics. We no longer see the waves of epidemics that were once common and incredibly deadly; bubonic plague alone killed about 25 million people, or one-third of Europeans, in the mid-fourteenth century.[11] The antibiotic miracle launched just as the real World War II began to turn in favor of the Allies, when bacterial infections remained an opponent far deadlier than any opposing army.[12] Discovered by accident in a poorly protected lab petri dish, a "mold juice" (aka penicillin) that was so effective at controlling and killing bacterial colonies suddenly emerged as a "miracle cure" for human infections. By the end of WWII the balance of power in the war on bacteria was beginning to tip as well. Suddenly you could stop syphilis, gonorrhea, septicemia, infections that had been felling armies. Consider how radical and how recent this change is: The first U.S. civilian ever treated with antibiotics, in March 1942, died in a New Haven hospital in 1999.[13]

Unnaturally selecting non-pathogenic microbes is essential, but it can also generate blowback. Initially cultured in bedpans and rough vats, penicillin was rare and expensive. Very few harmful microorganisms had been exposed to the substance before, and when they did encounter it almost all died. Seeing how the antibiotics were so effective, we quickly learned to scale production and began to use the substance with abandon. However in thus carpet-bombing microbial communities, we began to discover a few isolated instances of antibiotic resistance. Then, unlike what had happened with antiseptics and vaccines, three disasters occurred almost simultaneously.

Think about the typical ads you see for antibacterial soaps, mouthwashes, and cleaners. Generally they advertise that if you use them regularly, they will kill 99 percent of all bacteria. And they do. But the few bacteria left behind are tough survivors. They are resistant to that particular antibiotic or antibacterial agent and, because so many of their brethren and competitors have been wiped out, the survivors have space, food, and light—a perfect environment in which to spread and colonize. So they reproduce, fast. Places that are constantly scrubbed, disinfected, and sanitized are particularly vulnerable; humans inadvertently turned hospitals, cruise ships, spotless kitchens, and professional sports locker rooms into evolutionary accelerators for some very bad bugs, which is why you hear more and more about friends or relatives getting "hospital-acquired infections."[14]

To our peril, we also began to realize that microbial genes are very promiscuous; they hop from place to place. So after a few deadly bugs adapted to their newly toxic human environment, resistance began to increase rapidly across many species of pathogens. And because they reproduce so fast and share resistant genes, Darwin's evolutionary time scale was seriously compressed. It became a brutal race between miracle cures and resistant pathogens.

The third disaster was overuse. Sometimes humanity's combined greed and stupidity knows few bounds.[15] Initially, doctors used antibiotics when there was little choice, because a raging infection could and did kill within hours and there was no other option. Eventually we got lazy. We overprescribed. We quit finishing our prescribed doses. We began to use antibiotics not just for ourselves and our crying kids, but for animals of all sorts. We even sprayed fruit trees, filling whole environments and generating resistance on a massive scale.

Eventually, entire classes of antibiotics were overwhelmed by brutally competitive bugs. Scientists scoured the globe, digging up dirt from remote regions to find new naturally occurring antibiotics that had not yet been applied in a medical context. Even this, which led to semisynthetic compounds, was not enough. They too were overused and lost effectiveness. As a result, in the United States alone, more than 2 million people suffer an antibiotic-resistant disease every year. A really conservative estimate of the direct death toll, every year in the United States alone, exceeds 23,000. To put this number in context, it's almost twice the number of HIV/AIDS deaths.[16]

More than half of the babies born vaginally today in the United States acquire microbes at birth with tetracycline-resistant genes from their mothers.[17] Micro-WWIII is far from over. A survey of 2,039 U.S. hospitals in 2009–10 found that 20 percent of hospital-associated infections (HAIs) involved multidrug-resistant organisms.[18] Only in December 2013 did the United States begin to put some mild, voluntary guidelines in place to phase out indiscriminate use of antibiotics in animals, to be gradually implemented over three years.[19] The situation in many less-developed countries is far worse: Inexpensive antibiotics are available over the counter to anyone for any reason, leading to a terrifying rise in multidrug-resistant bacterial infections, particularly in regions such as Asia.

When engaging in these three major wars against microbes, humanity did not just dip a toe into the microbial environment to cure life-threatening conditions in a few sick people; we engaged in massive environmental engineering and unnatural selection. The assumption that we could so fundamentally alter the environment and evolution of microbes that have coexisted and coevolved with us for millennia without suffering any consequences is an interesting one. Clearly we have seen incredible benefits in terms of overall human mortality by declaring wars on microbes, but we must also ask, as we nonrandomly attempt to napalm the bugs that live on, within, and near us, what might be the results of these actions? We best consider this sooner rather than later because we are just beginning to launch another major offensive. . . .

Micro-WWIV: Antivirals. There are more viruses on Earth, by a factor of 100 million, than there are stars in the universe.[20] And they mutate and multiply faster than any other organism. While we have significantly tamed our bacterial environment, it's only within the past decade or two that we have systematically started to understand and attack complex viral pathogens as well.[21] The AIDS epidemic significantly accelerated our entry into this broad and complex conflict. In times past, when someone would come down with a virus, such as the flu, the doctor would say, "Antibiotics are useless; get some rest until you feel better." But when thousands and thousands of people began to die from AIDS, huge protests led to massive funding, reduced regulatory constraints, and a clear focus on killing HIV.

Like antibiotics, the initial strategy worked. We fought HIV to a standstill, monitoring, controlling, and reducing viral loads. Someday we may be able to eliminate the virus altogether. And we are taking on more and more viruses, including hepatitis, Ebola, and flu.

But as we continue to carry out WWIV against viruses and as we hijack/domesticate/co-opt a whole new microenvironment, we might reflect on what we learned from WWIII and the overuse of antibiotics. First and foremost lesson? It works. Far more humans are alive today and have descendants because we keep performing unnatural selection on the bacterial environment in favor of "survival of the unfit." But there can be blowback, in the form of some resistance. And viruses mutate hyper-fast, which is why we need to redesign flu vaccines each year, and even then the new vaccine sometimes doesn't work. This is why we use three or four medicines at the same time to attack the rapidly mutating HIV. It also implies that were viruses to acquire broad immunity to combination therapies, they could come back with a vengeance.

We know little about viruses, and they are so varied and complex that current antivirals only work on a narrow set of pathogens. We will likely go after SARS, MRSA, bird flu, and swine flu on a case-by-case, medicine-by-medicine basis. Broad-spectrum antivirals are under development but remain a ways away, which is a bad thing for patients with certain exotic and rare diseases but may not be horrible from an evolutionary perspective.

We may need to learn a whole lot more about what we are unnaturally selecting on a macro scale before we seriously go after the 10^{31} viruses that live on Earth. And should we begin using broad-spectrum antivirals, we might wish to remember that our gut contains tenfold more viruses than bacteria, and we are virtually ignorant about what species they are and what roles they play in our health and wellness.[22]

As we seriously begin to tinker with viral ecologies, we might want to ask three questions: (1) Are we also killing a lot of useful and symbiotic viruses? (2) What are the functions of these viruses? (3) How do we prevent engendering superviruses? Even today, after investing a whole lot of research and money, scientists discuss individual viruses, but they rarely discuss the *virome*—the way our whole virus population interacts with our bodies and cells every day. The numbers are completely overwhelming; we have about a hundredfold more viruses living on us and in us than we have total human cells. If we remain in the Dark Ages in terms of understanding our symbiosis with viruses, especially the benevolent viruses living in our bodies, our allies could become casualties of our new drug war.[23]

Billions of people are alive because we have fought and largely won major wars against microbes. But sometimes it's best to reflect on just how different and unnatural this state of affairs is. The natural norm for humans and animals was to be perpetually covered with dirt and coevolving with microbes. It is humans, not the environment, that redesigned, reformulated, and restructured entire microbial ecologies to suit our current needs. So as we reach for the Lysol, Clorox, Listerine, Dial, Secret, Mr. Clean, or Spic and Span . . . or when you get a scrape and run for the Bacitracin, Neosporin, Mercurochrome, alcohol, or iodine . . . you are unnaturally selecting your microbial environment.

The benefits of doing this can be enormous, but there can also be consequences to overindulging. No one's arguing that antiseptics, vaccines, antibiotics, or antivirals are a bad thing, or that we shouldn't have them. What we have achieved with their help is extraordinary. But there are sometimes consequences to such large-scale bioengineering. Restaurants and custodial businesses are particularly prone to severe antibiotic-resistant infections.[24] Washing your hands more than twenty times a day makes you 2.8 times likelier to contract dermatitis.[25] A general explosion of allergies and resistance may be just one more symptom of how fast we are redefining and redesigning our interactions and relationships with the microbial world. And as far as our species is concerned, our microbial and viral ecology are just as much a part of who we are as our cells. When we unnaturally select microbes, we are driving our own evolution.

The "Yucky" Stuff Inside You

.... :✳: ....

Take it as a compliment: You are a symbiont. In addition to your human cells, skin, blood, and organs, you carry around a large and diverse microbiome comprising of a community of thousands of different species of bacteria, fungus, and other microorganisms. In fact, there are as many microbes in your body as there are human cells.[1] Science is just beginning to understand how the several pounds of microbes that inhabit your body affect your everyday life.

Your ensemble of bacterial genomes is unique to you, and it changes as you age, travel, eat various diets, and take antibiotics.[2] Your microbiome responds to weather, pets, seasons, friends, lovers, disease, and many other variables in your daily life. Moreover, there's clearly an interplay between each person's genome and specific bacteria, so that people with different genetic makeups exposed to the same microbes will experience differences in microbial colonization, digestive conditions, acne, body odor, and countless other traits. The microbiome also affects obesity, heart disease, autism, immunity, memory, cancer, and aging. Collaborative communities of bacteria living in your gut and on your skin work to convert food to nutrients, heal wounds, help keep away really nasty competing microbes, make vitamins, create or break down toxins (and, of course, generate foul-smelling gases and noises at the most awkward possible times).[3] Without these microbes, you would likely not be alive, and certainly not healthy.

Each of the myriad species inhabiting your body has its own genome. Your microbiome contains far more genomes of DNA than do your "human" cells. And, because you are a symbiont, rapid changes in these microbes can also drive rapid changes in your body and evolution; think of it as your third genome (alongside your core DNA genome and your epigenome).

Darwin had no clue about any of this. Until the last decade, most of us had no clue. Because most microbes don't survive outside their native habitat, when scientists would take samples from various environments and then grow out colonies in petri dishes to see what microbes they could find, the microbial community appeared quite small. Nothing could be further from the truth; the only thing that's quite small is the subset of microbes that are amenable to being taken outside their particular habitat and then survive and reproduce inside a sterile lab. To find microbial diversity, you don't go to a lab, you go where microbes live day to day. One option is to go sailing.

J. Craig Venter hates to get bored. An adrenaline junkie who races motorcycles, collects antique cars, and takes on world governments, in 2000 he completed the first draft sequence of a human genome, racing against an international public consortium that spent far more money, adopted his sequencing methods, and sometimes deliberately discredited his work. In the end the race was declared a "tie," and President Clinton said, "We are here to celebrate the completion of the first survey of the entire human genome. Without a doubt, this is the most important, most wondrous map ever produced by humankind. . . ."[4]

In the black Suburban heading home after the event, Venter was exhilarated, exhausted, and above all somewhat relieved. The war between the public and private consortia had been escalating: James Watson was accusing Venter of being a modern-day Hitler who wanted to annex and privatize the human genome.[5] John Sulston, a wonderful humanist-scientist, who would soon win a Nobel for his work on how organs develop, was telling everyone that patenting genes was a crime against humanity.[6] And as governments threatened to change the rules on intellectual property with respect to genes, the stock prices of biotech companies were in a nose dive. Declaring a tie between the public and private gene-sequencing efforts therefore took part of the pressure off, even if both sets of genomes really were in draft form and far from finished.[7] Now all were declared heroes and a temporary truce was in place.

But after a few months, and a few scotches, having been ousted from his start-up, Celera Genomics, a restless Venter began to ask himself: What might a scientist do after sequencing the human genome? Without a clear answer, he went back to his default response when faced with difficult questions . . . It's time to go sailing.

Venter being Venter, it was not enough to simply take a hundred-foot sailboat out, cruise, and party. He also began to study what microbes live in the ocean; putting a hose over the side, gathering a couple hundred gallons of seawater, and pumping the water through micro filters is not that hard. You end up with pieces of filter paper, about the size of a dinner plate, that look slightly off-white. These filters were then folded, frozen, and shipped off to a big lab in Maryland. Whatever microorganisms lived in that specific seawater sample were then run through the same gene sequencers that had just completed sequencing a human genome.

The first samples were taken in what was until then considered the Sahara Desert of the oceans—the Sargasso Sea. Sampled and studied for centuries, there was supposedly little left to discover in this patch of ocean. The traditional method of study—gather water, take it back to the lab, grow the microbes in a petri dish—had indicated there's not much living there at all. But Venter knew most microbes won't grow in the artificial world of a lab. And now that exquisitely sensitive sequencing machines and massive computers were available to untangle genetic code, one could gather and study microbial genetic code directly.

The first ten samples of seawater from the *Sorcerer II* expedition increased the number of all known genes, from all species, by a factor of ten. Turns out there is a lot of microbial life in the relatively "lifeless" Sargasso Sea. This led to a series of follow-up questions: Are the oceans, their waves and currents, a giant Cuisinart that intermixes all microlife so that what lives in the oceans is all quite similar? Or are there very distinct microbial communities and ecosystems? And if the latter, how large and how diverse are they?

There was only one practical and fun way to get to these answers: Go sailing again. But this time sail around the world, retrace Darwin's original voyage, and collect seawater every two hundred miles.[8] Over the next three years, starting in the Bay of Fundy, coping with massive tides, through the Panama Canal, the Galápagos, the South Pacific, Australia, Africa, and the South Atlantic, the crew of *Sorcerer II* dealt with pirates, storms, and strange diseases, and gathered many, many tall tales. Finicky governments were the biggest single obstacle to their quest; for instance, the Brits, having issued sampling permits, suddenly got terribly touchy when the expedition passed near islands where the UK government had conducted nuclear tests. One morning a naval gunship appeared astern and haughty commandoes boarded *Sorcerer II*. Diplomats consulted for days, and the sailboat eventually left, escorted by a warship. (Quite a few resources expended by the nation that produced Darwin, Wallace, and many other great scientists, all to stop some basic research . . .)

In its first phase, the *Sorcerer II* expedition reported more than 6 million new genes, doubled the number of protein families, and described a hundredfold more ways to convert sunlight into energy to power an organism.[9] More than half the mass of all living creatures on this planet turned out to be microbes that are equally as important as trees in processing CO_2 and oxygen for the planet.[10] Every two hundred miles, the microbial community varies substantially; there is far more diversity within the various oceans than there is in the Amazon. This microbe-biosphere interface is essential for life as we know it.

As Venter's teams got ever more proficient at collecting, analyzing, and cataloging massive, intermingled, complex microbial communities, they then began to shift their focus from the ecology of microbial communities of the oceans toward the ecology of the human body. They began studying how microbes interact with humans on three levels: What lives on your skin? What lives in the orifices of your body, between the external and internal organs? What lives inside your body?

In 2006, microbiologists Claire Fraser, then Venter's wife, and Karen Nelson, a no-nonsense Caribbean/African American powerhouse who now heads the Venter Institute, coauthored a paper that profiled the enormous genetic diversity among the microbes doing the backstroke inside your intestines. Turns out the level of diversity you see in the seas is rivaled by what resides in the human gut.[11] And this was just the start.

One's body is a huge and diverse microbial ecosystem. The stuff that resides in your mouth is, fortunately, quite different from that which thrives in your gut.[12] As the scientific results began coming back, we began seeing our bodies in terms of different geographies, reflecting microbial ecosystems that have coevolved with us for millennia. One square centimeter of human skin can sustain 10 million microbes and host mini deserts and rain forests; your forearms are relatively fertile, home to about 44 species.[13] A less microbe-friendly niche lies behind your ears—only 19 species (Mom was wrong; wash your forearms and forget washing behind your ears). Totally unexpectedly, only 17 percent of the microbes that live on your left hand also inhabit your right.[14] Why? Well, which hand do most people use to shake, touch, and eat with?

Building up one's microbial community begins at birth. Vaginal births provide many of the early and critical base microorganisms that first colonize a baby and help it develop and digest. As we change birthing methods, we can also change our most basic selves; the gut microbiome in babies born by cesarean is less diverse and includes many bacteria that normally live on the skin.[15] Differences in a child's birth microbiome may persist for a long time; C-section babies take a year to develop the same variety and density of microflora as babies born vaginally.[16] And even then differences persist; the microbiomes of seven-year-olds can reveal how they were delivered.[17]

Microbial exposure and symbiosis do not stop at birth. A physician friend of Steve's, Emine, born in Anatolia, Turkey, was raised according to traditional customs; upon birth, she was placed on a bed of herbs, known as *üzerlik otu*, or Syrian rue, which is well-known locally for curing or preventing almost everything.[18] Old wives' tale, right? Except that a teaspoon of dirt can hold fifty billion microbes.[19] Many of Emine's cousins with even more traditional mothers were constantly wrapped in the traditional heated clay/argillaceous soil *höllük*.[20] This warm soil generates a cocoon for the baby, especially during the winter months, and also serves as a diaper that gets exchanged as it gets wet, undoubtedly home to a host of symbiotic ammonia-metabolizing microbes that are no longer a part of our lives. (Ever wonder what parents did before diapers?) Yet again a clear example of how yesterday's infant microbial world, and immune system, was strikingly different from that of today's kids.[21]

What one is born with, or acquires early, can provide some immunity. As one enjoys global travel and exotic foods, one sometimes also ingests a host of new microbes that may not be compatible with the current residents of one's gut, leading to such colorfully named afflictions as Montezuma's revenge, Delhi belly, Turkey trots, Cairo cramps, Dakar dash, Rangoon run, and so on. Even as your hosts seem completely unaffected after eating the exact same foods.

But because we have coevolved for so long, the vast majority of our little gut micro-buddies are friendly and treat us well. Altering a microbial environment can alter our development, our health, and our behavior.[22] Bacteria make bioactive chemicals such as vitamins, nutrients, and toxins, and they wage battles whenever we ingest a few new species (ever notice that kids love to eat dirt?). Our microbes break down tough-to-digest foods, nibbling on them a little and passing the rest along in an easier-to-digest form.

While most microbes mutate, adapt, and interact with our bodies in positive ways, there are plenty of bad bugs out there, which we are ever better at taming. Today's microbiomes are very different from yesterday's; a very good thing because, in most parts of the world, until very, very recently, even the act of being born was truly dangerous. In the United States in 1900, 1 in 10 kids died before age fifteen. Altering microbial communities with sanitation, vaccination, antibiotics, and clean water worked spectacularly well at combatting early mortality; by 2008 on average, 99.1 percent of U.S. kids survived to age fifteen.[23] And from 2000 to 2013, global childhood mortality fell 41 percent.[24]

Unnaturally selecting and reshaping our microbiomes, internal and external, can have real evolutionary consequences, as can changing our living conditions and eating habits. We coevolved even with "bad bugs," including tuberculosis (TB), a chronic infection that lives for years in our bodies, never completely killed by our immune systems, and that infects millions every year. Yet the human immune system seems to view TB bugs as an old friend and does not usually destroy them; in most cases these bugs lie dormant, sitting quietly in the lungs or elsewhere. Only 10 percent of the time do they grow deadly. Why?

Humans evolved large brains partly thanks to TB; large brains require lots of energy, energy production requires lots of vitamin B3 (aka nicotinamide). B3 comes from meat, but throughout human history there have been times when meat was a rare treat (for example, cold winters and periods of drought). So when meat was in short supply the brain would suffer, resulting in poor cognition, brain atrophy, reduced development in childhood, and other challenges to survival. According to Drs. Adrian Williams and Robin Dunbar, TB bugs filled the void, making vitamin B3 naturally and ensuring healthy brain development and function.[25] The mechanism Williams and Dunbar discovered was that when vitamin B3 levels fall due to dietary deficiency, TB microbes awaken and make the vitamin, but they also make more people sick as they proliferate. But when dietary B3 levels rise, TB goes dormant. In fact, vitamin B3 was once used therapeutically for TB infections. (And to ensure that many people get the evolutionary benefit of TB, the few who become sick actively cough on non-infected people to pass it to future generations.) Williams and Dunbar identified an inverse correlation between dietary meat consumption and the incidence of TB deaths in Britain over a hundred-year period from 1850 to 1950. Indeed, diet, microbes, and human evolution are coupled together.

Whereas our ancestors typically lived for generations in one region, eating the same foods, we race around the planet exchanging, importing, and exporting our own bacteria. Americans spent $867 billion on travel in 2012. And globally, just in the first half of 2013, 494 million earthlings traveled to another country.[26] As we add bacteria from Mexican tacos, Thai salads, Chinese veggies, and Patagonian seafoods to our bodies, and then to our sewage after we return home, we mix, match, and alter our niche microbial ecosystems. One might wonder about the consequences, good and bad, as we continue coevolving unnaturally fast; when scientists sequenced the microbiome of a 5,200-year-old frozen mummy and compared it to today's peoples, the microbial communities were totally different.[27]

One's microbial ancestry has real consequences. Túquerres is a little town in the semitropical highlands of Colombia, wracked by drug violence, next to an imposing volcano, hosting a minor church.[28] Tumaco is a tropical, drug-ridden, violent coastal city in that same country.[29] Neither is a tourism hot spot, but the two locations are frontiers in understanding stomach cancer and the genomic interplay between microbes and humans. Those who live in Túquerres are twenty-five times more likely to have their inflamed guts progress into gastric cancer than those from Tumaco.

More than 90 percent of the folks in both towns harbor the same species of a common bacterium, *Helicobacter pylori*, often associated with stomach ulcers. That is not uncommon. The longer one lives, the higher the chances of being infected. By age fifty, an average American has a 1 in 2 chance of being colonized by *H. pylori*.[30] The likelihood's even higher in developing countries. However, which specific strain of H. *pylori* colonizes a body can tell you a lot about human history. By tracing the variant in the gut, scientists can detect broad trends in colonization, the various patterns and waves of migration, and perhaps even how long ago your ancestors left Africa and where they wandered.[31] Certain variants make you more prone to ulcers and cancers; others have coevolved for so long that humans have essentially domesticated them. In the Colombian case, the highlands population descended from Spanish whites and native South Americans; their virulent H. *pylori* infection comes from Europe. The lowlands population is primarily of African descent, like the milder H. *pylori* that colonizes them. And the extreme differences in health outcomes can be directly traced to how the two populations coevolved with different subtypes of the same species of bug.[32]

If the findings of this study of Colombian H. *pylori* apply to other diseases—and there's increasing evidence that they do—we have to rethink how we diagnose and treat many conditions, including obesity. Particularly given that gut bugs don't just give you diarrhea, they can also moderate your metabolism.[33] Two great bacterial families battle for dominance within our stomachs, and the microbiome of obese people contains a greater abundance of Firmicutes and a lower abundance of Bacteroidetes. The prevalence of these species changes significantly in people who lose weight or undergo gastric bypass.[34] So in the case of gastric bypass it may not just be the surgery itself that affects the patient but also the resulting change in the gut microbiome, a theory further supported by an experiment that transplanted gut bacteria from skinny mice into genetically identical but overweight mice. This simple intervention led to weight loss, without any change in diet.[35] Some humans also seem susceptible to these kinds of treatments; overweight women who swallowed two probiotic capsules a day containing *Lactobacillus rhamnosus* lost ten pounds in just twelve weeks.[36] (Curiously, men on the same regimen lost no weight. As we continue to study the difference in gut biomes between males and females, we may begin to understand why.)

One way to treat the extreme and chronic diarrhea caused by a nasty little bug (with a most appropriate name, *C. difficile*) is to find a loved one with a healthy digestive tract, gather some of their poo, put it in a blender, and introduce it into the patient's intestine.[37] Sounds incredibly gross, but the procedure, whose recipe the FDA will have fun trying to regulate, can recolonize a gut and restore health almost immediately. As antibiotic resistance grows, these kinds of microbial recolonizations will likely become commonplace.

The list of how gut microbes change the most basic functions within our bodies goes on and on. Some people are more prone to heart disease when eating meat because their gut microbes convert carnitine to TMAO, which promotes atherosclerosis.[38] Certain species of gut microbes make an enzyme that breaks down bile acids, reducing fat absorption and lowering cholesterol levels.[39] Transferring microbes from a mouse with cancer to one without can cause colon cancer, and antibiotics can sometimes reduce the size and number of tumors.[40] Autoimmune diseases, such as irritable-bowel disease, rheumatoid arthritis, multiple sclerosis, and Guillain-Barré syndrome, have been tied to particular microbes.[41] The link between fast food, obesity, and insulin resistance also appears to involve our gut microbes.[42] Microengineering one's body has consequences way beyond weight loss; we know that as many as 36 percent of the natural chemicals in our blood are produced by our microbes, including neurotransmitters such as serotonin; some researchers and clinicians are linking gut microbes to neuropsychiatric disorders such as depression, anxiety, autism, cognition, and obsessive-compulsive disorder.[43] Eventually we will engineer supplements containing specific probiotic bacteria optimized to treat various digestive and other illnesses.[44]

As we edit and alter our microbial communities, we can alter our lives, our very survival, and that of our descendants. Bacterial communities can even affect how you smell and taste, who you mate with, and therefore who your descendants are. A 2010 fruit-fly experiment showed that animals on the same diet tended to mate with one another and ignore the rest. But as soon as the creatures took antibiotics . . . they went back to freely mating with everyone else.[45] Nevertheless, we still mix, till, raze, and overwhelm micro-ecosystems with singular abandon. Day by day, we move plants, animals, insects, and microbes from country to country; transport water and air between continents; and warm our planet. And in so doing we modify the invisible microbial world we live in. We create environments that favor one microorganism over another. Sometimes we do it with forethought, but more often we don't.

Is there any evidence that altering a microbiome, or even a part of a microbiome, can drive rapid evolution and speciation? Consider jewel wasps. Many years ago, one of three species of wasps branched off far enough that it could no longer produce offspring with the other two; the driver of this divergence was gut bacteria.[46] When raised in a controlled lab environment exposed to the same food and bacteria, all three types of wasps could eventually interbreed again. Speciation arose because of what had colonized their stomachs; in the wild, the new bacteria killed any offspring carrying genes from the wrong type of parent. New diet, new microbiome, new species.

Our microbial DNA acts as a third genome, a gigantic symbiotic genome, one that is an integrator of nature and nurture. What your microbes do, or don't do, affects you, and it changes your descendants. Microbes concentrate enormous power for assisting in and shaping evolution. And we have been busily selecting them, and even redesigning them, to suit our purposes.

Autism Revisited: Three Potential Drivers

.... ✳ ....

A ndrey Rzhetsky, a new breed of biologist, brought up in Russia's third largest city, Novosibirsk, is a big deal—tenured at the University of Chicago, director of the Conte Center for Computational Neuropsychiatric Genomics, and senior fellow of the Institute for Genomics & Systems Biology and Computation Institute. Never one to shy away from controversial and difficult subjects, he takes on big, complex, controversial topics like what's behind the autism epidemic.[1]

While most biologists spend hours in a lab, Rzhetsky thinks there is more than enough macro and micro data available already; he queried 100 million medical records trying to figure out the best correlations between environmental changes and autism, and concluded that autism might just be a milder form of extensive chemical poisoning.

It was already known that when certain chemicals, such as thalidomide, enter the human body, particularly during crucial periods like gestation, they can seriously deform various body parts. So Rzhetsky began searching for various extreme malformations, especially in boys, since they are far more prone to autism. These kinds of conditions, like micropenises and undescended testicles, are rare, occurring in only 0.27 percent of the male population. But wherever Rzhetsky found a cluster of these congenital malformation cases, he could also predict a massive increase in the incidence of autism in the surrounding community.[2]

In some parts of the world, the trends are more than a little worrisome. In Korea (2000–2005), the rates of undescended testes increased 348 percent. Malformations were far more common near chemical parks than near nonindustrial areas: 198 cases per 10,000 births in the petrochemical-plagued Yeocheon versus 11 cases in the city of Chuncheon.[3] Less than a year after these data were published, the first major Korean autism study found three times more cases of autism per capita than occur in the United States.[4] (And remember the United States is already in the throes of a massive autism epidemic.)

According to Rzhetsky's correlations, the closer you get to certain chemical clusters, the likelier you are to see a population with devastating birth defects, and the likelier a far larger percentage of the population will be affected by autism. Deformities increase substantially if mothers are janitors, maids, landscapers, or farmers. If Rzhetsky's correlations are right, then wherever you see an uptick in serious deformities you should see an explosion in autism cases. It might pay to keep a close eye on parts of Brazil, given the extraordinary rates of micropenises, especially in boys whose parents worked with or near pesticides.[5]

We live in a far more complex and varied chemical environment than our ancestors did. A single grad student can now synthesize hundreds of novel chemicals in a month, whereas it used to take a great chemist years to come up with a few new molecules. New chemical disruptors may be changing how the human brain develops, but because human clinical trials for such things would be unethical, and the variety and interactions of various chemicals is so extensive, it's very hard to trace exactly which chemicals, in what combinations, alter the brain. We do know that children are far more sensitive to certain chemicals and medicines than adults are. A child's brain grows to 80 percent of an adult-size brain during the first two years of life, and many of its initial connections are established during gestation and early childhood. Perhaps some of the materials we use every day—in our homes, such as clothes, foods, and utensils; or in the course of other activities—are having serious disruptive effects on the brains of young boys.

In a sense, boys seem to be acting like the proverbial canary in a coal mine, especially vulnerable to environmental insults from the pesticides, plasticizers, sex-hormone analogues, drugs, synthetic molecules, and other chemicals that surround us. Even overall sex ratios may be affected by these agents; Russian pesticide workers exposed to dioxins father fewer boys.[6] Something similar is occurring in Taiwan.[7] And in heavily industrialized Sarnia, Ontario, the reduction was even more brutal: Just one baby in three was male.[8] Overall, mothers seem less affected by most chemicals, and the same is true of daughters. (And in the United States, boys are diagnosed with autism 4.5 times more frequently than girls.) So Rzhetsky advocates measuring specific chemicals to test whether, as parents expose their bodies and their children's bodies to more and more chemicals, one of the immediate and cumulative effects is that their kids suffer autism.

Not everyone agrees with Rzhetsky. A second major research camp believes parents have so modified their bodies and breeding habits that they themselves are the primary cause of the autism explosion. In 1900, in the United States, fewer than 1 in 5 women entered the workforce. Today it's almost 4 out of 5. Working women have fewer offspring, and the children they have are born later in life. Late conception increases the likelihood and prevalence of some birth defects.[9] Many of these new mothers might not have been able to conceive naturally, so they flooded their bodies with synthetic hormones. In 2009, more than 6 percent of U.S. births came after ovulation treatments.[10] The specific chemicals used to induce childbearing past the age when your body naturally says, "It may be too late" may also have unintended side effects on the development of infants' brains.[11] Add to this a bevy of stereotypical macho men, the ones over age sixty, with the new convertibles and young wives, who are also fathering kids later—not always a smart idea, given that their sperm accumulate an average of two new mutations per year.[12] These unfortunate combinations result in kids who are far likelier to suffer autism, schizophrenia, and cognitive impairment.[13]

A third major faction in autism research is focused on gastric disorders; many autistic children seem especially prone to stomach troubles.[14] The question is, how does this relate to brain development? Well, it turns out "gut instinct" is absolutely real; one of the oddest of anatomical discoveries is the number and reach of neurons scattered throughout your digestive tract. At least 500 million neurons stretch from your throat to your . . . ahem, rear end. The existence of this "enteric nervous system" means you have the equivalent of 6.6 mouse brains lying outside your skull telling your body what to do, what to eat, and how to react.[15] Having a partially decentralized brain helps you decide how to digest, what to expel, what to crave, but the import of this enteric system goes far beyond food. The neurons in your gut generate as much dopamine as your brain; this is the key neurotransmitter that helps control the body's reward and pleasure sensors as well as emotional responses.[16] Furthermore, up to 95 percent of the serotonin in your body, the neurotransmitter that inhibits appetite, controls sexual behavior, and reduces pain also resides within your enteric system.[17] Celiac disease increases the probability of having an autistic child by 350 percent. So when the gut-neuronal system is disrupted, there can be serious consequences, including an increase in the chances of getting autism.

Various other factors that disrupt early neuronal development may also be driving autism. Mothers hospitalized for viral infections during their first trimester triple their chances of having a child with autism. But, ironically, some communities plagued with infectious diseases have *lower* incidences of autism, so part of the inflammatory epidemic may be due to our living in ever more sanitized and pristine environments where our own immune systems less frequently fight off infections and parasites.[18] And beyond these explanations are still others such as antibiotics, gene mutations, TV exposure, premature birth, assortative mating, obesity . . .

Whichever of these explanations, or combination of explanations, turns out to be right, we are still left with a mass of open questions: How are we so quickly altering our descendants' brain circuits? Is it a cumulative exposure to more and different chemicals so that what our grandparents and parents were exposed to is now affecting our children? After generations of low-level chemical exposure, have we accumulated a "memory" of this toxicity in our bodies so that now we somehow transmit a torrent of acquired nonrandom mutations to our children? Or is there one particular new class of chemicals whose use is growing fast? We don't know. We do know we have an epidemic on our hands. Our children's brains are evolving fast.

Viruses: The Roadrunners of Evolution

.... :✳: ....

O ther than the late great Carl Woese, not a lot of folks can claim to have discovered a whole new "branch of life." While biologists were happily puttering around claiming there were only two branches, prokaryotes (bacteria) and eukaryotes (most everything else), Woese began arguing that there were far older, more primitive ancestral life forms. Most of his colleagues laughed.

These ancient microbes, called *Archaea*, today are recognized as descendants of some of the oldest life forms on Earth.[1] Our true ancestors aren't just a small tree-dwelling hominin, but also far earlier, sulfuric-acid-loving creatures that lived in the equivalent of boiling battery acid. They were far better adapted to a much hotter and at times even caustic and violent Earth. Now that we know what to look for, we find them everywhere, especially in the most extreme environments—near volcanoes, in sulfur and tar pits, in geysers, in Antarctica, and in undersea vents.

As the life on Earth evolved through the epochs, from a hot hell to a frozen snowball, from an *Archaea*-friendly and warm methane atmosphere into an atmosphere full of a new poisonous substance— oxygen—life adapted and adopted. But when Woese and his colleague

Nigel Goldenfeld studied microbial evolution and compared it to the standard rate and timetable for evolutionary mutation, they came to a startling conclusion: The Mendelian-Darwinian vertical "tree" orthodoxy (one thing begets another and evolves into yet another) was partially incorrect.[2] The basis of their argument was simple mathematics: In a microbe-dominated world, there was simply not enough evolutionary time for all life to have converged on a single genetic code, on one basic genetic machinery.

As far as we know, all life is encoded in DNA, in four letters.[3] But chemically it did not have to be this way. There are other chemicals besides GATC that could have become the chemical properties for passing on heredity and coding evolution. Because there are many ways to stabilize and reproduce life code, and given the very different environments, climates, and chemistries that predated our photosynthetic, oxygen-rich world, it's likely that other basic life languages also evolved, using life-code variants besides DNA and RNA.[4] Woese's argument is that the only way for life code to have optimized, unified, and become as error-resistant as it is was not through a slow evolutionary process but through massive horizontal gene transfer—that is, a direct transfer of genetic material from one species to another. No messy sex, no worries about compatibility, just swap and go.

In Woese's model, evolution takes place not just with the birth of each new generation, after combining DNA from mother and father (the classic vertical/tree-branch type of evolution), but rather is also a continuous and distributed horizontal process in which viruses are constantly inserting themselves and recoding our DNA. In this world, the primary driver of continuous insertion, deletion, silencing, and activating of new genes is not our core gene code, not our epigenome, not our microbiome, but viruses. The implication is that an incredibly minute thing—a virus, a bag of DNA or RNA that cannot live and reproduce on its own, something between a simple collection of chemicals and a living cell—is the roadrunner of evolution. It also implies that only after a virus-driven, massive horizontal gene transfer occurred did evolution begin to operate primarily through vertical gene transfer mechanisms (heredity), as described and codified by traditional Darwinian geneticists.

We live in a world filled with viruses; they are everywhere that host species exist. Viruses infect and live among and within all species from bacteria to blue whales and redwoods. While there are an estimated 10 billion bacteria in a liter of seawater, there are also 100 billion viruses joyfully playing with them.[5] (Think about that the next time you swallow a mouthful.) Even more may live inside the soil and dirt that cover your hands. Viruses continuously shuttle in and out of your body, and there are at least ten times more of them inside you, every day, than there are bacteria.

The human virome includes trillions of viruses that live in and on our cells, plus even more that inhabit the bacteria in our microbiome.[6] The virome is poorly understood and could be considered the "dark matter of nature" and humanity; we know it is there but have a very hard time describing it or knowing what it is doing. The human virome is essentially our fourth genome; it interacts directly and indirectly with our other three genomes. Moreover, like your genome, epigenome, and microbiome, your virome is absolutely unique. Viruses live in our intestines, mouths, lungs, skin, and even in our blood, the latter being only discovered recently. But fret not; given that people are generally healthy day-to-day, the virome overall must be benign, and given the millennia of mutual coexistence, our viromes must provide benefits that we don't yet appreciate.

Viruses are champions of DNA mutation. A 2013 study of the human gut virome tracked the identities, abundance, and mutations of native viruses in one person over two and a half years.[7] There were 478 relatively abundant viruses, most of which had not been previously identified. A majority of the viruses were bacteriophage, the type that infects bacteria. Eighty percent of the viruses persisted for the entire 2.5 years, but they all mutated, some slowly, some very quickly. In some cases they mutated so fast that the virus would be deemed a new species within the 2.5 years. What came out of the body after symbiosis was very different from what went in.

So viruses, our ubiquitous interlopers, are an important part of our rapid evolution; they carry, exchange, and modify the DNA between cells or from one species to another. They drive evolution at all scales, in bacteria, plants, animals, and humans. The best example of this is the spread of antibacterial-resistance genes from one bacterial organism to its fellow species, and then to other bacteria of all types, in all geographies. Once a beneficial mutation arises in a microbe, viruses help spread it quickly throughout the microbiome and beyond.

A single sneeze propels 40,000 droplets, each containing up to 200 million individual viruses, across the room at speeds exceeding 200 miles per hour.[8] (Amazing how viruses created a way to make us sneeze so they could infect new people.) After you breathe in this viral code, it enters your cells, reproduces, releases trillions of copies of itself, and proceeds to take over other cells. Sometimes viral DNA simply embeds itself in your own human DNA, where it can lie dormant or sometimes come back to life when you least want it, as occurs with recurring cold sores, shingles from a long-past chicken pox, and even some cancers—particularly when our immune systems become weak. This is exactly what occurs with Kaposi's sarcoma in immunodeficient HIV patients. On some occasions, viral code can end up in the DNA in your sperm and eggs, which then gets passed on to future generations.

Over long periods three things can happen as viruses invade organisms. One, they kill the organisms (for example: Ebola, or the 1914 flu). Viral epidemics kill millions of humans, animals, insects, trees, and bacteria, but in doing so the virus is left with fewer and fewer hosts to infect and eventually tends to disappear. A second option is the body kills the virus and the virus stops propagating. A third option is the virus and the host species (for example, humans) coevolve; the virus is ever less nasty and the body stops bothering to attack it. It is this third option, the symbiosis between humans and viruses, that we know the least about. This is largely because investigators tend to focus their energies on the evil viruses that are contagious killers. (And it is hard to get grants to study the nice viruses.)

But it is this third way that solves part of the puzzle of rapid human evolution. Brand-new genes, or gene variants, arise all the time from viruses invading our genome. At least 8 percent of the entire human genome is viral in origin; more than 100,000 segments of retroviral DNA (i.e., originally derived from RNA viruses) have embedded themselves within our core genome, where they have propagated over millennia.[9] Most of these viral DNA elements are inactive retroviruses that sit largely within regions we still do not understand and call the "junk regions" of our genome.[10] But when an environment changes often these remnants of gene code reappear.

Over the next few years, cheaper and more accurate gene sequencing will begin to lift the veil on the virome, and our "dark matter" will be scrutinized and dissected. So stay tuned for breaking news and examples of how the virome affects health and disease. We will also see ever more antiviral drugs, which we should deploy carefully. The virome matters; let's better understand what we alter and evolve.

A Perfectly Modern Pregnancy

·································· ⁎ ··································

I t's natural, it's normal, it's common. We suspected it was coming. And yet when Steve and Stenie heard, "Surprise, we're having a baby!" from their daughter and son-in-law, Emilie and Alexei, the world stopped and they broke out into joy, tears, screams of delight. First grandchild on the way . . .

The next two months were tranquil and blissful, all reveling in the glow of a coming baby. Then came Thanksgiving, when Steve's daughter, after hearing about what we were exploring in this book, curiously asked, "That's fascinating, but what can I do about it? I don't want to know only about problems that I can't fix, give us some guidance."

Steve had done a lot of research but really had no idea what to advise these most precious of people, let alone anyone else. So he spent quite a few anxious nights on a slightly neurotic quest to find answers within seas of ambiguity and contradiction that he'd discovered while considering his daughter's specific situation.

A little unnatural selection had already come into play; the grandchild was not wholly an act of nature, because Clomid, a fertility hormone, played a critical role in conception.[1] This was different from

Darwin's day; in fact, back then, if two animals could not conceive and produce offspring, it was often considered as a species demarcation line. Since Steve was not quite ready to declare his son-in-law a nonhuman species, he had to accept, as do millions of other grandparents-to-be, just a little unnatural selection.

But let's get back to the question of the developing grandbaby. What advice, both sage and safe, could be given to the young expectant parents? What were their choices? One obvious choice, not available in Darwin's time, was for them to learn the gender of the baby. An ultrasound provided a baby photo pre-birth, but the parents decided to remain in the dark, not so much to avoid a premature Amazon-driven onslaught of pink or blue goodies, but because waiting seemed fun, mysterious, retro, and irrelevant to the baby's health.

In other areas they could choose to go far deeper. Consider the baby's four genomes. Were they all normal and healthy? In the case of the DNA genome, the parents had already opted for a DNA test to detect rare, recessive genetic diseases that would only manifest if both parents carried them. Only Emilie underwent the voluntary $100 blood test, and the results were negative, so Alexei was not tested because the baby would not be at risk for those eighty-five genetic diseases requiring that both parents be carriers.[2]

But was this enough research on the baby's core genome? Perhaps the parents should consider sequencing the baby's entire genome for a few thousand dollars? If they did, the question would be whether they should do this pre- or post-birth, and if the latter, at what age? Given the current state of the art, Steve counseled the parents-to-be that pre-birth testing might lead to more anxiety and frustration than real answers. No matter who you are, who your parents are, even if you are a descendant of Cal Ripken, you have innate genome flaws, and we do not yet know enough about the vast majority of these potential flaws, and their true risks, to make really intelligent choices. For instance, one study of entire genomes found that hundreds of absolutely healthy adults each carried approximately 400 really bad DNA mutations—mutations that seemingly should cause disease but didn't in these cases.[3] So the conclusion on the DNA genome sequencing was "Skip it; *que sera, sera.*" It is still very early in the genome-sequencing/bioinformatics-interpretation field, and red herrings swim everywhere.

As genomics moves forward, standardizes, and cheapens, it is certain the baby will be sequenced at some point later in life. If one has significant disposable income, it soon may make sense, on a speculative basis, to get an early genome read after birth, a comparative genetic baseline as the ultimate first baby photo. Early gene sequencing will be the norm, especially as we develop further technologies not just to diagnose but also to engineer around perceived gene defects.

The ethical quandaries posed by genetically engineering a child are becoming ever more complex. Some debates are relatively straightforward; if you could prevent Huntington's, likely you would. It's invariably fatal. Same for Tay-Sachs disease. But as we learn how to edit life code, the same technologies could be applied to enhance beauty or sports aptitude. Fortunately, given no dastardly gene mutations detected, and no real and safe enhancement options yet, no gene therapy for junior.

What about the baby's microbial genome? Mid-pregnancy, Emilie was informed that she was a candidate for C-section because the placenta was not positioned ideally. She knew that nearly one-third of U.S. births are delivered by cesarean section.[4] The procedure can be a lifesaver for mother and child, but no other mammal deliberately operates on its females to deliver a baby. So one might consider the long-term evolutionary consequences of this common but unnatural procedure, one that has spread broadly and is often overused. (The U.S. incidence of C-sections varies astonishingly, from 7 percent in some settings to 70 percent in others—just one indication that the procedure is not dictated only by an emergency but by different customs, aesthetics, priorities, and, dare we say it, even reimbursement regimes.) So, what to advise Emilie to help the baby develop a normal microbiome? Fortunately, father-to-be Alexei, a physician, had already come up with one "obvious" solution: somewhat akin to a fecal transplant, a different type of microbial transplant, so the baby has a full and complete initial microbiome. Down the road, mother-to-child transfer of vaginal microbiomes may become fashionable, but in the meantime we will leave it to your local OB or nurse-practitioner (and the FDA) to fill in the details and suggest current options. In fact, the field of a newborn's microbiome is in rapid flux. A 2014 study found that during pregnancy the placenta contains microbes that resemble those in the mother's mouth, and that this microbial community is transmitted to the baby.[5] So there is no pristine, sterile world inside the womb; rather, nature has found clever ways of providing the baby its third genome even before birth.

Next let's consider the baby's epigenome. An unborn baby gets a constant "news feed" from the outside world. Scientists know some of the big triggers that cause the epigenome to flip its switches include diet, toxins, infections, drugs, and stress. The latter can be emotional or physical, ranging from work deadlines and family worries to abuse, war, and famine. A loving, supportive family and comfortable environment can be beneficial to both mother and baby.

Mothers should avoid smoking, noxious fumes, and most recreational and prescription drugs, not just because of effects on the unborn baby but because of epigenetic effects on future grandchildren. The baby in utero, whether a boy or girl, develops cells that will eventually become the sperm or eggs that beget the subsequent generation. Thus, epigenetic insults that befall a developing fetus can become embedded in the genetic blueprint of subsequent generations.

But of course it is not just the mama who needs to take extra care; the father carries most DNA mutations and also has epigenetic memories in his sperm, so he too should stay healthy and watch his exposure to harmful elements in the years before conception. (Steve did not realize this early enough, otherwise might he have been tempted, like any slightly protective father, to put a tracking device on Alexei when he started dating Emilie?)

One ongoing concern, pre- and post-partum, is overall chemical exposure. For instance, bisphenol A (BPA) is an endocrine disruptor with a structure similar to estrogen; in sheep, it acts on reproductive health, raising their tendency toward hyperandrogenism (being more virile).[6] Mice exposed to BPA prior to conception had second-, third-, and even fourth-generation pups that were unusually anxious and nervous.[7] Yet BPA is still broadly used and deemed safe by the FDA and regulatory agencies in many countries. While BPA has been gradually removed from baby bottles and related infant products, in part because of consumer unwillingness to purchase and concerns about safety margins for infants, it continues to leach out of everyday plastics, canned foods, and thermal receipts (the kind you get at the grocery store). And it resides inside 93 percent of Americans' bodies.[8] If transgenerational BPA inheritance also occurs in humans, something yet to be proven or disproven, then imagine what multigenerational exposure to these types of chemicals might do to our progeny across time. In considering the causes of some exploding trends, like asthma and allergies, it is worth thinking about the potential influence of yesteryear's pre-regulation pesticides, plasticizers, paints, organics, fuels, and groundwater. We were a lot less cautious only a few decades ago and may be suffering some consequences today. It is very difficult to determine which substances, in what combinations and doses, are safe for a pregnant mother and fetus. What exactly are you supposed to do when the sign on an airplane door reads WARNING: THIS AREA MAY CONTAIN CHEMICALS KNOWN TO THE STATE OF CALIFORNIA TO CAUSE CANCER, BIRTH DEFECTS, AND OTHER REPRODUCTIVE HARM? And then you arrive at the hotel and see a similar sign. And then again at the gas station, and at the restaurant. (Not to mention the entrance of the hospital ER after Emilie was involved in a minor fender bender at six months.) Even if we were to test the substances deemed of most concern, they wouldn't necessarily be relevant, as they are often replaced by even more modern chemicals.

Evolution has been very pragmatic when it comes to toxins—trial and error; if you ate something poisonous, you died. Survival of the fittest meant those with the right smell, taste, and nausea sensors lived. (Morning sickness, for example, provides a natural reminder to avoid toxic foods early in pregnancy, when the baby is especially vulnerable.) The problem is that as we developed unnatural chemical arsenals, or concentrated substances that used to be really rare, like radiation and heavy metals, we did not have time to evolve the corresponding sensors/reactions. We can be gradually poisoned and our bodies won't know it, won't react. So pregnant mothers and children need to avoid industrial chemicals, radon, cleaning products, hair dye, and basically many things that smell, taste, look, or feel unnatural.[9] We need to rapidly develop new technologies such as "organs on a chip," which would test chemicals quickly and inexpensively on human tissues.[10]

During her pregnancy, Emilie ran into another hurdle that affects 10 to 20 percent of pregnant mothers today. In the sixth month of her pregnancy, her physician discovered that despite being normal weight she had gestational diabetes, a form of the disease that occurs only in pregnancy. Her blood sugar, and the baby's blood sugar, spiked high after every meal. Poor Emilie became one more statistic within a large emerging trend; a Canadian study showed that the rate of gestational diabetes doubled from 1996 to 2010.[11] She began monitoring blood glucose four times a day and adjusting carbohydrate intake, having to pass up celebratory cake at the baby shower. This was important to keep on top of because gestational diabetes, when poorly controlled, leads to large babies at birth. (Prior to unnatural C-sections, many extra-large babies and their mothers would not have survived birth, so their genomes, carrying diabetes genes, were lost.)

Nutrition modifies your epigenome both favorably and unfavorably. Poor nutrition early in life can affect a child's epigenome and lead to health-related conditions throughout life, independent of genetic background or parental weight. Infants or animals subjected in utero to low calorie intake, particularly low protein, as a consequence of famine, eating disorders, or emotional stress, are typically born underweight and often exhibit a "rebound" weight gain in early childhood as they get adequate calories.[12] Pre-pregnancy undernourishment of the mother transmits epigenetic signatures to the baby.[13] In mice, an epigenetic message of a mother's malnourishment can be passed to sons who, via their sperm, conceive offspring prone to obesity and Type 2 diabetes.[14]

Many of the vitamins and nutrients involved in the biochemistry of epigenetics, including B12 and B6, the amino acid methionine, choline, betaine, sulforaphane, and resveratrol, are naturally available in a balanced diet.[15] Supplements are usually unnecessary, and in excess can be harmful.[16] However, all mothers are encouraged to take prenatal vitamins, which typically contain DHA, thought to promote brain development (and may be responsible for enhancing intelligence, as seen in a study in Eskimo babies).[17] Steve checked Emilie's vitamins and found DHA on the label.

The exception to the usefulness of supplements may be folic acid, which plays a key role in epigenetic DNA methylation (a key mechanism in regulating our epigenetic switches) and helps prevent birth defects in the brain and spinal cord. Animal studies indicate that folic acid can also help overcome some of the effects of pollutants on fetal development.[18] Because many mothers were low in folic acid, in 1996 Congress mandated that breakfast cereals be fortified with it.[19] But overall, fresh vegetables and fruits, not heavily sugared cereals, remain among the most favorable epigenetic triggers.[20]

We can be exposed to different types and degrees of stress (or Stress, or STRESS). When a mother's "fight or flight" response is activated, hormones shoot up, particularly cortisol, the "stress hormone." This hormone crosses the placenta and acts directly on the developing fetus, as well as flipping switches on the baby's epigenome (the message comes across as "Watch out! It's horrid out here!"). The effects of any particular stressor depend on its timing (whether in early or late pregnancy), duration, and severity. Severe stress can reduce a baby's growth, disrupt brain morphology, alter sexual differentiation, and speed aging.[21] Acute stress tends to enhance the immune system, whereas chronic stress seems to suppress its activity. Interestingly, nature has given babies some defense against excess stress; an enzyme in the placenta limits the amount of cortisol the baby can experience from a stressed-out mom, but the activity of this enzyme varies among different babies (and the effects are turned off by eating black licorice!).[22]

In Emilie's case, living in peaceful and leafy Palo Alto, with enough resources, kept STRESS away. But a newly pregnant modern woman, maintaining her own high-powered career, with a high-octane MD helicopter-flying husband, away from family, can also face significant Stress. Many hormone systems are activated or suppressed by stress, and the fetus feels the effects. Fortunately, two practices developed thousands of years ago, yoga and meditation, work. A daily ten- to twenty-minute relaxation routine in healthy, non-pregnant human adults leads to beneficial changes in gene activity, including enhancing energy metabolism and chromosome health, and suppressing inflammation.[23] Working off your worries with moderate exercise during pregnancy also helps and leads to better language skills in kids.[24] (And Grandpa, if you want to help cut stress, send lots of humorous YouTube videos of cats—laughter helps.)

An interesting, and unanswered, question is whether too *little* stress could also be detrimental. A mother's immune and hormonal systems evolved over millennia when life tended to be, as Thomas Hobbes described it, nasty, brutish, and short.[25] And now, while we all face multiple challenges, risks, and pressures, for most people they fall well short of what has been the average level of flux, fear, and violence within human societies. As you remove obstacles and pressures on the survival of any species, you would expect to see big changes. For one thing, many who might not have otherwise entered the gene pool are surviving and having many descendants. Survival of the fittest is therefore an anachronism, and far more genomes are getting passed right along. Perhaps this is one reason our progeny are already evolving into something different from ourselves, right under our noses?

As the day of birth approached, Steve began thinking about the baby's microbiome. An infant's gut microbiome varies throughout early childhood, at a time when the immune system is in flux and learning. It adjusts to dietary changes during the first year of life and begins to stabilize in its makeup at about nine months of age, though it can change over the subsequent months as well.[26] One Canadian study showed that the stool microbes of infants born through cesareans and fed on formula vary substantially from those of children born and fed in "traditional" ways, even when delivered in the same antiseptic hospital environment.[27]

Breast milk also helps; breastfed babies get sick less often and tend to be less susceptible to allergies, asthma, diarrhea, ear infections, and pneumonia.[28] Mom herself receives some benefits from breastfeeding too; mothers who breastfeed have lower rates of ovarian and breast cancer.[29] A study in monkeys showed just how specialized and nutritious mother's milk can be: After birth, maternal milk adjusts to the needs of a boy or a girl; milk for a boy contains more fat, while girls get more calcium. (And, just in case you were wondering, the same holds true for cows.)[30] Humans? Don't know yet. But it would make sense that, evolutionarily, a human's milk also adjusts the same way. A final note on breast milk, because our newfound urban/domesticated/indoor lifestyle has negative side effects: Breastfed babies often require a daily vitamin D supplement because nursing mothers live indoors and use sunblock.[31]

During a baby's first eighteen months, the two dominant microbial communities inside the baby's gut battle each other for supremacy.[32] The idea, as you know, is not to keep all microbes out but to have the right combinations and balance. Love dark chocolate? Your taste buds are on to something. *Bifidobacteria* in your gut love dark chocolate as well, and they convert it to anti-inflammatory compounds that benefit your heart.[33] According to one fortunate study, mothers who consume chocolate daily during pregnancy have six-month-old infants with a more positive temperament.[34] What you eat does matter (although sometimes your microbes may be the ones to thank or blame).

External microbial exposure also matters. As mentioned earlier, children who grow up with farm animals have a very low incidence of allergies, a phenomenon called the "farm effect."[35] A baby's immune system becomes educated about what's benign—food, pets, pollen—so that it doesn't freak out and attack as if such agents were pathogens that need to be destroyed. Infants raised in pristine, antiseptic households, meanwhile, have higher incidences of immune reactions. The timing and amount of exposure are critical. Current thought suggests it is reasonable to avoid a farm when you are pregnant, but have pets around the baby after birth.[36]

An infant's microbiome makeup may also modulate brain development and function.[37] Microbe-free mice exhibit increased motor activity and reduced anxiety, both associated with specific gene activity in the brain.[38] Some probiotics reduce anxiety in mice and produce lower stress hormone levels in the blood.[39] And what about colic, that digestive nightmare that leads babies to engage in endless crying and breaks a parent's heart (and eventually patience)? Could it be a mismatch of gut bacteria and diet? Early and incompletely substantiated research found that newborns in Italy fed the probiotic *Lactobacillus reuteri* DSM 17938 for ninety days cried less and had fewer regurgitations and "evacuations" per day compared to babies who were placebo-fed.[40] Given the apparent importance of one's microbiome and epigenome, should you consider having them analyzed in your baby? Not yet. The methods for analysis aren't robust enough, so the data remains difficult to interpret, leading to more confusion and worry than anything, particularly given that most children are born healthy.

(But if you want baby to be able to look back and reminisce as an adult, "Wow, my baby epigenome shows my parents certainly liked to party"—maybe deep-freeze some DNA from the umbilical cord and a diaper.)[41]

As for Emilie and Alexei? After many nerdy nights of discussion and research, they welcomed a precious, healthy girl named Esme by a normal delivery. Ten fingers, ten toes, seemingly normal genome, epigenome, and microbiome, the latter expressed with an aroma only a new mother can love. Esme loves her cherry-flavored vitamin D supplement and Emilie eats plenty of nuts and naturals to educate Esme's immune system. Esme also enjoys meeting friendly pets who invariably insist on sharing a few of their own microbes, and she breathes in plenty of fresh-air allergens wherever she travels.

Should she develop allergies, ingesting allergens could help, according to Esme's great-grandfather, Kot. Raised in rural Poland in the early twentieth century with lots of farm animals, Kot cured himself of pollen allergies by drinking unpasteurized milk from a goat that lived locally and ate a variety of wild plants. Though Steve knew Kot was brilliant, he never managed to keep a goat near the woods or use raw milk to cure his own hay fever. Maybe he needs to reconsider, but meanwhile, please, please don't try this home remedy on yourself or your baby just yet. The effects of consuming unpasteurized milk can be far worse than those of allergies.[42]

Now team grandma and grandpa are focused on maintaining a highly stimulating environment, one that pays dividends in long-term social skills, happiness, and intelligence. Caring parents beget kids with fewer physical and psychological problems, something eventually transmitted, through epigenetics, to grandkids.

As Emilie and her dad reminisce about the past few months, Esme sits on the floor putting everything within reach into her mouth, feeding and altering her microbiome and virome. After a lot of research, Steve and his daughter were surprised by just how far we have come in altering when and how we give birth, and by how surprisingly little they really knew about a baby's development. But they did know that, globally, this is a very good time to be born; over the past few decades childhood mortality rates have fallen dramatically; between 1990 and 2010 global mortality rates, for kids under five, dropped from 88 per 1,000 to 57 per 1,000.[43] Average IQ scores are steadily rising, as is overall longevity.[44] Esme enters a world with many challenges, but she is also part of a generation that will likely be far healthier, and live longer, than any previous generation. Her epigenome reads, "Looks pretty darn good out here."

Bringing It All Together—DESTINY Is Propelling Evolution

·· :✳: ··

S o how is human-directed evolution playing out? You are born with
at least four parallel evolving genomes—core DNA, epigenome,
microbiome, and virome. Every human, plant, and animal possesses
these four genomes, which can be considered a "hologenome."[1] They
interact with one another, evolve at very different rates, and define
your basic biology and attributes throughout life. Eventually they
come together and encode the heritable traits and behaviors that you
pass on to your descendants and future generations. We are now
actively modifying all four genomes in many, many species, including
ourselves (as well as evolving "naturally," in parallel).

Humankind's core DNA genome has been essentially stable for
tens of thousands of years. Each generation historically experiences
tiny, random mutations; 50 to 100 of the 6.4 billion letters that make
up your DNA are different, at birth, from those of your parents.[2] This
represents a minuscule 0.0000016 percent of the human genome that's
altered each time we have a kid, and the rate is steady enough to date
how long ago specific species diverged from one another, including us
from Neanderthals, apes, and our weird neighbors.[3] (This glacially
slow rate of change, particularly when considering most mutations are

benign, meaningless changes, explains why Darwin, and even many of today's scientists, resist the idea of rapid and radical human evolution.)

In most cases, a one-letter change does not disrupt gene function, so a hundred mutations in a genome aren't devastating. It works much like our minds reading right thru misspellled words. But sometimes a single-letter change can and does make a big difference: Duck, Muck, Luck, Tuck, Buck, Puck . . . Even the exact same letters, with a different spacing or emphasis, can lead to big misunderstandings: Consider "The IRS" vs. "Theirs," or "KidsExchange" vs. "KidSexChange." (It is these types of small changes that can lead to two siblings who grow up in a very similar environment to develop very different cancers.)

Because it's relatively rare for beneficial mutations to occur in a person, and because we each have only a few kids, and those kids need a further twenty or so years to reproduce, traditionally it takes a long time for beneficial traits to ripple through a population. (Can you think of any people you know who have an extraordinarily beneficial trait that dramatically differs from their parents?)

The major reason for such slow evolution in the core DNA genome is that it encodes many of the fundamental functions of a species, which, if altered significantly, duplicated, or had pieces missing, would lead to bad outcomes, including crippling, serious diseases, infertility, or just plain death. And the very conservative evolutionary mechanism of DNA also ensures that deleterious mutations do not adversely affect vast numbers of individuals, thus preserving our species. Outside a few specific cultural niches, recessive mutations, those that manifest only if inherited from both parents, spread ever more slowly, if at all. This is particularly true post-caveman times, as societies enacted laws prohibiting you from marrying your sister and requiring an MD's permission to marry a first cousin.[4] Ultimately, significant changes in the core code can lead to new species.

So for millennia, most of our core genome mutated according to the rules described by Darwinian theory: slow, random evolution.[5] Aside from Europeans and Asians co-opting a smattering of genetic elements from our hominin Neanderthal and Denisovan cousins, as well as the emergence of a number of regionally important genetic mutations to ward off pathogens, survive food shortages, and adapt to less light, humanity largely shares one core DNA genome. This implies that from a survival, reproductive, and natural-selection perspective, humans have rendered most environmental influences relatively low-pressure and low-impact; over the past few centuries, natural selection has not been the only active process grossly altering humanity. Not surprisingly, Darwin and most evolutionary scholars following him ignored the possibility of humans speciating, much less speciating quickly. Until very recently, neither the mechanism of rapid change nor the environmental pressures to accompany it were there in overwhelming force.

Our core genome overlaps 99 percent with that of chimpanzees, and of the ~19,000 human genes only 60 are completely new, having arisen over the past 5 million years since we last shared a common ancestor with chimps.[6] But there is a whole lot of adaptation going on, just below the core genome. When you compare differences at the epigenomic level in the brain (that is, the "on/off" switches, technically termed CpG codes), then there are at least 474 genes that are controlled completely differently between humans and chimpanzees.[7] Human diversity and evolution is highly concentrated in our epigenetic switches; how genes are turned on or off, expressed with greater potency or silenced.[8]

And what of our third genome, our microbial genome? Within the microbiome, evolution occurs quickly; some bacteria can go through 2,600 generations in just over a month. As humans declare broad warfare on microbes, as they radically alter ecosystems, adding toxins, antiseptic soaps, mouthwashes, pharmaceuticals, nutraceuticals, global travel, urban lifestyles, changing diets, and leave rural existence behind, they guide/influence rapid microbial evolution. And in turn they alter their own bodies. So while you inherited your intial microbiome from Mom, they aren't your great-grandma's microbes anymore.

Finally, our fourth genome, the virome, mutates and evolves at a blazingly rapid rate. We have only begun to catalog the actual specific actors with the virome, so it is still early days with regard to understanding exactly how it affects us, from day to day or from generation to generation. One thing is certain: With our domesticated lifestyles, global imprint, and unnatural activities, the "typical" virome today must be very different from the one Darwin indirectly observed. And we are now also beginning to tame and deploy viruses, learning to rapidly edit them for our own purposes.

So as we alter our environments, how and where we live, we alter our bodies and our evolution. As we alter what lives within us, what enters and recodes our four genomes, we alter our own evolution. One way to summarize and recapitulate how the outside world, both natural and unnatural, changes our genomes and drives evolution is to think in terms of "DESTINY," an acronym for the six environmental stimuli (D-E-S-T-I-N) that you (Y) and your progeny adapt to every day, with both short-term and long-term consequences:

Diet (calories, protein, fats, micronutrients, vitamins) alters our bodies as well as what lives within us.

An *Enriched environment* full of information, music, toys, puzzles, schooling, and media enters and rearranges a child's developing brain.

The types of *Stress* we live with every day are very different from those of our ancestors.[9] We are far healthier, eat better, and suffer far less violence. But we also try to do so much more, faster, for so many.

Toxins are far more prevalent and common, though their effects can be invisible for years or generations.

Infections, and the lack thereof, continue to alter the recoding of our genes both directly and indirectly.

How we *Nurture*—how we love, cuddle, and raise our offspring within modern families—is very different from traditional rural/tribal settings.

And finally, *You* and your progeny are the recipients of this onslaught of the six environmental stimuli, but at the same time you are also becoming the decider, the primary driver of evolution, as humans develop ever more ways to recode their own DNA, epi-genomes, microbiomes, and viromes.[10]

Evolution no longer just "happens to you." In unnatural ways, humans are taking charge of and driving rapid evolution. And in so doing they are ever more responsible for their DESTINY. As we alter our outside world, the outside world, in turn, can alter our bodies, our four genomes, and that of future generations through at least three core evolutionarily conserved biological systems: The hormones and growth factors released by our *endocrine system* alter the functions and genomes throughout our bodies during our lives. The *nervous system* wires and directs neurotransmitters. And our *immune system* deploys its various weapons. The seven DESTINY stimuli, acting via or in response to these three systems, transmit events and conditions around us to our genomes, leaving memories that can linger across generations.

Your gonads are tuned in to these conversations and challenges and changes as well, reprogramming to give the next generations information about a changing environment. With pregnancy, the fetus joins the conversation and assimilates what Mom is communicating, affecting early development, and carrying memories into future generations—sometimes through core mutations, sometimes through softer drivers of evolution embedded within epigenomes, microbiomes, and viromes. And boy, is there a lot to transmit to future generations! The world we inhabit stimulates and challenges our bodies in ways that cave folk never experienced—no exercise, less violence, an abundance of calories, pesticides, lead paint, urban life, sixteen hours of light but little of it sunlight . . .

DESTINY provides a high-level, albeit rough map of what we have to consider as we charge ahead altering life forms. As our knowledge of DNA, epigenomes, microbiomes, and viromes rapidly accumulates, the map will become far more accurate and we will better understand where we are taking ourselves. And our overall design of the future of life should, hopefully, become ever more intelligent.

A WORLD OF NONRANDOM MUTATION

• • • • • • • • • • • • • • • •

Playing with the Building Blocks of Life

.... :✳: ....

It's one thing to modify our surrounding environment, deliberately selecting and breeding to promote unnatural selection, as has occurred over the past few centuries; it's an entirely different order of magnitude to engineer life code and rewrite it to meet our specific wishes. The direct and deliberate introduction, for a specific purpose, of new instructions into a living creature's gene code, in ways that will then be passed on to descendants, transforms evolution from a random, slow slog into a rapid, human-directed process. This particular branch of technology may just be our species' single greatest technical achievement so far.

Sure, viruses have been messing with our gene code for millennia, inserting their own bits and programs to help us, and them, digest and reproduce, or occasionally really hurting us. But finding a particular gene code, knowing what it does, and inserting it into a bacterium, plant, animal, or human, directly and deliberately, is a nonrandom mutation: an act that alters life and future generations through intelligent design.[1]

The scale and breadth of nonrandom mutation increases every year. At first it was exotic, rare, and scary. So scary that the first recombinant

DNA experiments, in 1972 and 1973, triggered a full-on voluntary moratorium on further experimentation and then a collective gathering of key scientists to hash out hazards and controls.[2] All kinds of guidelines were established: who could experiment, on what, and under what safeguards. Being just post-Watergate, in a very low-trust environment, everything had to be discussed and hashed out; one key set of guidelines came out of a gathering at Asilomar, a Big Sur, California, retreat still famous for its nude hot tubs. Various labs, using various biohazard containment methods, began to tiptoe into genetic engineering, but not everyone was convinced or cooperative. The mayor of Cambridge, Massachusetts, a most colorful Mr. Vellucci, almost succeeded in completely banning genetic experimentation in what is now the world's ground zero for gene research.[3] But eventually genetic engineering would commonly be performed on bacteria, fruit flies, and mice.

By 1982, humans were injecting themselves with insulin created through recombinant DNA techniques. Initially traditional pharmaceutical companies, whose medicines were based on chemistry and small molecules, scoffed at the new small, poor biotech start-ups. But companies making biologics—medicines derived from genetically engineered products produced by living things—expanded and exploded. By 2013, seven of the ten top-selling drugs were products of genetic engineering.[4]

Meanwhile, back in Cambridge, things have changed just a wee bit. The mayor's office began boasting that "over the past decade, biotechnology has emerged as a most important focus for our business community."[5] Almost every major pharma and biotech company had moved a substantial part of its R&D operations to the Boston area, and young biotechies ran loose in the streets igniting a "New Cambrian Explosion."

Every year hundreds of college and high school kids converge for a weekend at MIT for the International Genetically Engineered Machine competition (iGEM). The competition's final rounds look like any massive sporting event—hundreds of teams walking around wearing colorful T-shirts and weird hats, mascots, friends and family in tow.

But amid the cheers, huddles, and team high-fives you gradually begin to realize this is a contest unlike any other. These kids are designing competing *life forms*.

In a weird way, iGEM is a descendant of the 4-H Club; these clubs are a way of demonstrating and introducing new agricultural innovations into the heartland of America. Their slogan, "Learn by doing," has gotten thousands of kids to try hands-on farming techniques, eventually leading to massive increases in the productivity per acre of corn, the quality of pork, and the ever-popular "great pumpkins." The iGEM kids certainly adopted and adapted the 4-H philosophy, "To make the best better," but perhaps in a way that might disturb a few of their grandparents.

iGEM kids have been steadily building up a collection of standard DNA parts that can be inserted into living cells. Think of these various biological mini programs as bio-bricks or living LEGOs. When these pieces are assembled in the right order, in the right quantities, they enable living cells to do something new, something nifty. For instance, normal yeasts may be helpful in baking bread or making beer, but they certainly do not spontaneously self-assemble into a living color television screen, something an iGEM team from Valencia, Spain, showed off in 2009. Nor do earthly bacteria quickly evolve themselves so they might be able to survive on Mars, excrete oxygen, and eventually create an atmosphere for humans (Tokyo University iGEM project in 2009). And then there is the Baylor U team, which in 2008 attempted to create yeast that could make a kind of beer that contains resveratrol, the substance that makes red wine good for you. (Now, why might *that* particular project have occurred to a college kid . . . ?)

When iGEM was launched in 2004, only five colleges showed up, and the total number of DNA bio-parts available with which to design something new was fifty. But as the contest grew, as teens refined their bio-parts, learned how bio-instructions worked, how the outcome of one experiment could be applied to very different projects and designs, the variety of projects and outcomes exploded.

During the 2012 competition, teams showed up with cells designed to sense toxins, fight obesity, protect one's intestines, filter estrogen from drinking water, and become nanobots to turn the Great Pacific Garbage Patch (where ocean currents take plastic trash) into an island. In 2013, the 245 competing teams added a further 1,708 new bio-parts for the next generations to build upon.

What iGEM kids are proving time and again is that genetic engineering is not unlike computer programming. In their own words: "Simple biological systems can be built from standard, interchangeable parts and operated in living cells."[6] Yes, gene code is more complex, nonlinear, and often unpredictable, but it programs life forms; it changes and directs life. And unlike the digital world, in the bio world software makes its own hardware. So once you program a life form, it reproduces itself. (No matter how you program a computer, you will not have a thousand self-assembled computer clones by morning. But if you program a living cell . . . it self-proliferates and you have altered all subsequent generations.)

Life scales fast. Think of this in terms of an old traditional industry: publishing. When you ask an audience what is the most published book in history, in all languages and all editions, the answer in the West is usually the Bible. In other parts of the world surely it might be the Koran or the Tipitaka. (In still other, perhaps less reflective communities, perhaps the answer might be *Fifty Shades of Grey*.) All of these are wrong answers. Because the most published author of all time is a wonderful bearded, vegetarian gentle giant who suffers from narcolepsy, George Church.

A computer prodigy, George entered Duke's Ph.D. program after two years of college. He lived and breathed lab experiments but rarely bothered with class, which led one professor to give him an F. Which in turn triggered a letter from an associate dean that stated, "You are no longer a candidate for a Doctor of Philosophy degree . . . We hope that whatever problems or circumstances may have contributed to your lack of success in your chosen field at Duke will not keep you from successful pursuit of a productive career."[7] Perhaps the kind folks at Duke need not have anguished so much, because he did just fine. Now one of the top profs at Harvard Med, and a mentor to generations of iGEM kids, George helped create the field of DNA sequencing, was a leader in the Human Genome Project, helped launch dozens of companies, and was elected to both the National Academy of Sciences and National Academy of Engineering. By bringing together computer coding and gene coding, he automated significant parts of the life-editing process.

All of which brings us back to the topic of publishing a lot of books. In 2012, when he and Ed Regis published *Regenesis: How Synthetic Biology Will Reinvent Nature and Ourselves*, George thought it would be cool to encode the digital 1s and 0s of his Word file into the four letters of a DNA molecule. Then he could synthetically produce copies of his entire book written in gene code from four simple chemicals—G, A, T, and C. It worked. And soon George had a few billion copies of his book (which made it easy for him to provide comedian Stephen Colbert with 20 million copies as a small dot of DNA dried onto a piece of paper). Life scales.

But as the iGEM kids well know, copying something is nowhere near as big a deal as making something new. That is why they keep building upon the basic DNA parts, the basic tool kit with which they and others can rewrite life code. And as this tool kit gets bigger and bigger, so do human ambitions and dreams. After all, if you control and write the code, you control evolution.

Humans Hijacking Viruses

.... :✳: ....

Human genetic engineering is not new; it has been going on for a
long, long time—naturally. Ancient viruses are really good at
inserting themselves and modifying human gene code. Over millennia, constant infections would come to mean that 8 percent of the
entire human genome is made up of inserted virus code.[1] All this
gene recoding of our bodies occurred under Darwin's rules, natural
selection and random mutation. But nonrandom, deliberate human
genetic engineering is new, and it is a big deal.

As of 1990, increasingly genetically modified humans walk among
us. More and more gene therapies carry new instructions into our
bodies and place them in the right spots; in so doing, they modify our
most fundamental selves, our core, heretofore slow-evolving DNA.
We are still in the very early stages of effectively hijacking viruses for
human-driven purposes; just a few years ago it took a long time to
identify and isolate a single faulty gene and figure out what was wrong,
never mind finding a way to replace it with a properly functioning
alternative. Early gene therapy focused on obscure, deadly orphan
diseases like ADA-SCID (the immune disease that "Bubble Boy" had),

adrenoleukodystrophy (say *that* five times fast), Wiskott-Aldrich syndrome, various leukemias, and hemophilia.

In theory the technique is relatively simple: Take a neutered virus, one that is engineered to not harm you but that readily infects human cells to ferry in new DNA instructions, write a new set of genetic instructions into the virus, and let it loose to infect a patient's cells. And ta-da! You have a genetically modified human. (Think of this as deliberately sneezing on someone but instead of giving them a cold, you give them a benign infection that enters their body, recodes their cells, and fixes a faulty gene.)

In practice, it has taken decades for gene therapy to reach patients, particularly after one safety trial went terribly wrong and killed Jesse Gelsinger in 1999.[2] (Ironically, it was not anything the foreign gene did directly that was the culprit; it was a massive overreaction by Gelsinger's own healthy immune system that overwhelmed his body.) Gene-therapy trials were temporarily halted and regulatory approvals slowed way down. Safety increased, but costs ballooned.

So even today if you are sick and in need of immediate gene therapy, the overall approval process still feels ultra-slow and bureaucratic. But compared to a traditional Darwinian evolution time frame, things are moving at lightning speed. Every day we know more about how to alter the human gene code, and we are beginning to insert multiple genes into plants and animals to cure ever more complex diseases. Multi-gene disease cures in humans are likely a decade or more away, as we still have more to learn about safety and unintended consequences one step at a time.

Humans can now tame and redeploy infectious disease scourges as safe Trojan horses to carry and insert lifesaving genes.[3] By the end of 2013, more than 1,996 human gene-therapy clinical trials were completed or under way; 64 percent focused on cancers, 9 percent on diseases involving single rare mutations, and 8 percent on infectious disease.[4] In 2014 you could go to a pharmacy in the EU and buy the

first clinically approved gene-therapy product, Glybera, for treatment of lipoprotein lipase deficiency, which causes pancreatitis.[5]

Today more than 80 percent of gene-therapy trials use a tamed
virus, and half of these rely on two standard workhorses, adenoviruses
and retroviruses, to deliver and deploy new gene code into humans.[6]
But there are many, many ways to insert new genetic instructions into
everything from bacteria to Tea Party pundits, including "gene guns"
(an NRA-approved device?), electroporation (electrically shocking cells
to open pores and let DNA in), naked DNA, sleeping-beauty trans-
posons (promiscuous DNA that inserts itself into genomes), stem-cell
transplants, and even using neutered viruses as vectors, including her-
pes, lentivirus, measles, polio, and vaccinia.

Soon gene therapy techniques will go way beyond small patient
populations. There are hints of what is coming. Two separate trials
showed you can alter almost every blood-related stem cell in your
body.[7] And by altering your blood stem cells, which differentiate into
many cell types in your body, you can effect some pretty fundamental changes in nearly any organ in the body.

As human gene therapies get safer, they will begin to migrate
from the "must fix" diseases toward the "nice to fix" diseases. Moorfields Eye Hospital in London is ground zero for this transition; in
2009, they treated a healthy twenty-three-year-old suffering from
non-life-threatening inherited blindness. By 2014 a further nine
patients, suffering choroideremia, had been genetically modified and
their sight improved markedly.[8] And while the Europeans were working to restore human vision, the white coats at Washington University
"cured" color blindness in monkeys, and proved that the monkeys'
brains adapt, even as adults, to the new visual-color stimuli.[9]

Bring these two vision-restoring experiments together in your
mind, and you can imagine that all kinds of strange things will become
possible. Superhuman eyesight could someday transition from comic
books to everyday life. One might insert specific genes into normal
humans to allow them to see in other colors, say ultraviolet, as do
insects, fish, reptiles, and reindeer. We know this is humanly possible

because some folks, perhaps even Monet during his water-lily period, can see UV light due to surgery to remove the lens to treat cataracts.[10] And there are women who already carry an extra mutated red-light photoreceptor, allowing them to see in four colors when most of us see in three.[11]

Because viral gene therapies have been primarily focused on obscure human diseases, few people understand just what a broad influence these techniques could eventually have on our species. We will see more and more genetically modified humans; it will just seem common and normal, just as IVF babies are today. (Just think: Baby Louise, the first screaming-tabloid test-tube baby, is now thirty-seven years old, and her IVF sister Natalie just gave birth to her normally conceived daughter.)[12] But before we start to take these technologies for granted, we may want to reflect on our newfound powers. Soon gene therapies will likely be used for cosmetics, athletics, and longevity. We will begin to shape our own evolution by introducing "desirable traits" and editing out "negative traits" in ourselves and our kids. An easy decision when you are a parent facing a gene that will kill your child; much more interesting, complex, and nuanced when applied to how a human looks, grows, or thinks. And we best set some ground rules as we let a new technology, CRISPR, loose on the world. . . .

Editing Life on a Grand Scale

.... :✳: ....

As is true of many Christmas presents, once you have the basic
parts, then "some assembly may be required. . . ." Which is why a
new technology, "clustered regularly interspaced short palindromic
repeats," is so very powerful.[1] Fortunately, this complex name is commonly referred to by its simple acronym, CRISPR. It is a newfangled
technology uncovered while attempting to produce better yogurt.

Natural yogurt helps your digestive system by introducing live
beneficial bacteria; in 2006 the nice folks at Danisco were struggling
to control the viruses that attack, change, or destroy the good bacteria that make yogurt, wine, cheese, bread, and many other products—
a common problem in the food industry. As they set about finding
ways to stop this viral mayhem, they took the DNA out of bacteria
and read its sequence. To their shock they began to realize that a bacterium's DNA contains an identification mechanism, a series of "mug
shots," of viruses that attacked previous generations of bacteria.

When a bacterium carrying this "mug shot" of viral code in its
DNA faces a new attack, it can recognize the virus and deploy
CRISPR to defend itself. CRISPR identifies the malicious virus that
wormed its way into a genome, cuts it out, and sometimes replaces it

with some harmless code. Soon scientists realized that they had uncovered a biological version of a Norton or McAfee antivirus program that can identify and remove viral interlopers and replace them with the proper DNA code.

CRISPR can be repurposed to cut, paste, and edit any DNA sequence into or out of any genome quickly and easily, not just bacteria. And it goes way beyond just targeting viruses: CRISPR can effectively edit out any harmful DNA sequence (for example, a disease causing gene mutation) and replace it with a beneficial DNA code (a normal non-mutated gene). Then, when a repaired bacterium or other cell type reproduces by the millions, the genomes of the progeny carry the modified/repaired DNA—instant nonrandom mutation.

Unlike earlier genetic-engineering approaches, such as gene therapy, which introduce one new gene into a genome with many complex and tedious steps, CRISPR is rapid, large-scale gene-editing technology; it allows the quick modification of large sections of an existing genome, easily snipping out unwanted genes and replacing them with whole sets of new genes.[2] Think of this as transitioning from a mechanical typewriter, having to use Wite-Out, and retyping a word or phrase in one's term paper (gene therapy) to having a primitive word-processing program that allows one to swap whole paragraphs and pages in and out (CRISPR). And you don't retype it all; the printer does that for you.

Of course, while younger readers haven't lived through the pre-word-processor stage and may simply assume it was always easy to manipulate and edit large documents, they too will live through a similar techno shock when their grandchildren look at them incredulously and ask them to describe what were leading-edge cancer treatments back in their day: "Gramps, is it really true that when you were young and someone got cancer, doctors poisoned a patient's body with deadly chemotherapies, then bombarded them with DNA-destroying radiation, and even cut off their body parts? Didn't you understand that cancer is a genetic mutation and that by simply switching some genes on and off, or replacing them . . . ? How could you have been so primitive and ignorant? Are you sure you weren't also using leeches?"

In evolutionary terms, once you have a tool kit with the basic building blocks of life, and a rapid, large-scale editing system, you could re-create, edit, or collapse the 4-billion-year saga of life on Earth into a short documentary.[3] And then you can edit at will, or rewrite the story going forward. Almost any genetic mutation, benign or otherwise, man-made or natural, recent or ancient, from any species, can be re-created and introduced into a living cell, at will, sometimes in an afternoon. Because CRISPR can cut, remove, and replace hundreds, perhaps thousands of genes in the genome of a living cell per day, it is Lamarckian evolution turbocharged and on steroids.[4] (Remember Lamarck? He's the guy who a couple of centuries ago argued that beneficial traits can be deliberately acquired in one generation and then passed directly on to descendants and was laughed out of the room.) And while giraffes still don't (yet) grow much longer necks in a single generation, with CRISPR it's possible to engineer bacteria, and some plants and animals, with new traits within weeks.

The first experiments editing the genetic code of human cells has shown efficacy in repairing mutations and creating resistance to viruses.[5] In-vivo human trials are certainly not too far off. CRISPR is not squirrelled away inside top-end secret labs; it's in the hands of high school and college kids; it has permeated scientific research so quickly that Jim Collins, an eminent Boston researcher, facetiously apologized at a conference when he mentioned he would not discuss CRISPR.[6] The applications are endless; someone could study gene code left behind by past viral invaders in bacteria and be able to identify the oldest life forms on Earth (think of this like a carbon-14 test, but instead of measuring the radiation decay in some long-dead bones, scientists are tracing the genetic footprints left by the earliest infections). Others could study defense mechanisms in one of the most pervasive species (*E. coli*) and ways to take advantage of CRISPR to fight pathogens. But by far the most important impact of CRISPR will be on the modification and evolution of humans.

As broad-scale genetic engineering accelerates and decentralizes, animals and plants are already getting a little weirder. Scientists can build life forms to suit local tastes and needs; featherless chickens are running around the Israeli deserts, better able to resist the heat.[7] (However, not everyone thinks these bright pink creatures are good-looking.)[8] For good luck, the Vietnamese produce sea horses with gold dust in their cells. Cows, goats, sheep, and camels are made to produce medicines in their milk.[9] You can buy glow-in-the-dark cats, bunnies, fish, sheep, and silkworms. Or you can order a glow-in-the-dark plant for your living room—just the first step in a Kickstarter project that eventually aims to replace street lighting with glowing trees, powered by the sun.[10]

Not all of this is just fun and games. Some of this genetic engineering could upend economic systems. A key change, now apparent in algae and agriculture, but not yet in many other areas, is the ability to "stack" traits—that is, to insert multiple genes at once, replacing or repurposing an entire biochemical pathway or even multiple pathways. This used to be an expensive, tedious multistep process: Identify the desirable gene trait. Synthesize one gene by pasting together small bits of DNA. Test whether the new gene works in bacteria. Transfer the gene to the appropriate algae or plant cell. Add more genes to embed a multiple-gene pathway. Rarely did this work the first, second, or third time. After tens of millions, or sometimes hundreds of millions of dollars and years of work, you could sometimes produce commercial quantities of transgenic offspring, expressing the biology you designed into a plant.

Now, with CRISPR or through synthetic biology, multiple traits can be introduced into algae all at once—a new biological pathway confirmed to be in the cells in days, and commercial quantities grown in weeks to months. You no longer need to breed in one trait after the other in a linear fashion; just change it all at once. So near the salty brines of Calipatria, California, a very friendly and intense Jim Flatt, president of Genovia Bio, is trying to massively increase algae productivity and produce mountains of green gunk. Because the genes in these cells have been reprogrammed, they are beginning to produce fuels in one pond, vaccines in another, animal foodstuffs in a third. Eventually this could lead to big changes in the agricultural ecosystem and even global land use; engineered algae can theoretically produce a hundredfold more of a desired product, such as protein or oil, than traditional crops, using far less space. This may be the only way to supply feed to our expanding farmed fish industry, and provide animal protein for up to 9 billion of us.[11] It could also allow us to return a whole lot of cultivated surface back to nature.

As we get ever better at engineering nonrandom mutations, evolution compresses and really accelerates. Our newfound ability to engineer organisms on a systematic and massive basis is a game changer. Soon, if we find one particularly compelling and positive genetic mutation in one person, we may be able to spread it to many, without waiting a few tens of thousands of years.[12]

Rapid evolution is real. And with CRISPR, it's coming lightning fast. Mosquitoes are a case in point. Think back to the most basic Mendelian genetics charts, which show a bunch of girl and boy offspring represented by circles and squares, and in turn an even bigger bunch of grandkids. When you color in to mark a harmful gene trait in these boxes you find that these types of traits do not usually impact all of the kids and all the grandkids. If they did, and if they seriously limited reproductive fitness, they usually killed off that family line.

However, a technology called a gene drive can push a genetic trait so that each and every individual offspring and all subsequent generations acquire the trait. This means evolution could happen very, very fast across a species. In 2015, a team used CRISPR-Cas9 to insert about 17,000 letters worth of life code into *Anopheles stephensi* mosquitoes. Thereafter, 99 percent of these bugs could still bite you, but they could not transmit malaria.[13] In a sense this strategy is similar to that of achieving the widespread vaccination that eliminated smallpox. But there is a difference; with a gene drive, humans aren't deliberately going after just one microbe but modifying an entire population of insect species.

And why stop there? In the lab, an Imperial College team inserted three genes into sub-Saharan mosquitoes so as to deactivate egg production, making it possible to eliminate this species of mosquito altogether.[14] After all, as Bill Gates points out, mosquitoes kill at least 750,000 people per year, making them the deadliest animals in the world.[15] While many justifiably argue that this could alter entire ecosystems, the push towards rapid gene drive deployment took on a far greater urgency with the emergence of a terrifying new threat: Zika.

Mosquitoes, and most life forms, do not respect borders. Decisions made in one lab in one country can affect global ecosystems. Some advocate extreme precautions for doing any type of gene drive experiments with flying insects. These include: A single researcher being responsible and carrying out all experiments. Triple doors. Experiment only outside a bug's habitable range. Low temperature rooms. Air-blast fans. Separate the genetic components of a gene drive. An "off" switch after a few generations. . . .[16] All current major researchers in the field have agreed to follow these protocols, very sensible and smart, but as gene drive technologies cheapen and spread, not all may obey such stringent rules. We also need to prepare counterstrategies for a not so distant future where there is an accidental release or a deliberate attack.

Despite clear and real risks, gene drive wish lists expand rapidly; there are simply too many potential benefits to unnatural selection. Some have already engaged in gene warfare; in March 2016, Brazil, a hotspot of the Zika epidemic, began releasing hundreds of thousands of mosquitoes carrying self-destruct genes, in an attempt to stop microcephalic births and to protect soon-to-arrive Olympians and spectators.[17] At the same time, as the epidemic moved relentlessly northward, the usually staid FDA moved toward a Florida field trial for Oxitec mosquitoes.[18] The projects academics and companies are currently advocating include eliminating Lyme disease, eliminating ticks, stopping specific crop pests without using pesticides, eliminating invasive species like rats and toads on some islands. . . . As we accelerate our ability to pick and choose what lives, we should heed the advice of Professor Kevin Esvelt, a key gene drive researcher, who argues: "Just as we have informed consent in healthcare, people have a right to know about the development of gene drives and other technologies that could directly impact their lives."[19]

Whereas we used to be able to rationalize by saying, "God did this to us," or "Nature did this to us," or "Our enemies did this to us," now we are increasingly in charge. We are responsible for the outcomes of our choices on how life evolves, how our species evolves. Alongside the science, there has to be a broad ethical debate and education: This is what we can do, this is why we think it is a good idea to develop and deploy these new instruments, and these are the open ethical questions we need to address.

One place to begin to address ethical questions is when and how we choose to deploy gene therapies, domesticated viruses, and CRISPR to engineer the gene code of all of our descendants. The tools to edit life forms are here and widely distributed. Despite much international handwringing and real concerns, in 2015 the Key Laboratory for Reproductive Medicine of the Guandong Province, China, along with some colleagues, edited nonviable human embryos. The experiment, an attempt to correct a blood disorder mutation (beta-thalassemia), was a failure in a majority of embryos.[20] But there was a proof of concept in a few. As technology is perfected, the eventual ability, and the will, to intervene in the next generations is out there. Most reacted quite negatively to this experiment. Many developed countries prohibited or put a moratorium on editing the core genes of future generations.[21] But there are groups that believe that, given so many incurable hereditary diseases, it is unethical to stop research.[22] Others want to go further. If the editing techniques are deemed safe, they want to apply similar regulatory frameworks as those used in pre-embryo-implantation-diagnostics.[23]

It is one thing to choose to alter ourselves during just our lifetimes, but it is a decision of a different order of magnitude to alter the species going forward. Recently Chinese scientists demonstrated that

CRISPR can alter the genetic code of nascent human embryos, but widespread legal, ethical, and safety concerns, expressed by many, prevent implanting these embryos into humans.[24] Reasonably soon we will find a safe way to engineer long-term changes into our descendants. When we choose to do so, we will begin to shape the species according to our own set of instructions and desires. This is not just unnatural selection altering and shaping what already lives, this is nonrandom mutation rapidly creating and passing on something new. So let's now look at the completely uncontroversial topic of altering future babies. . . .

Unnatural Acts, Designer Babies, and Sex 2.0

.... ·✳· ....

Modifying even the most basic birth organs is now fair game. In the spring of 2013, nine women in Gothenburg, Sweden, got a very odd gift from their mother, sister, or aunt . . . a used uterus. Disease or deformation had made their own infertile, but transplant surgery had become safe enough, and in vitro technologies advanced enough, that it made it possible for these women to carry their own children. Seven of the operations worked well, and in September 2014 a thirty-six-year-old woman who was born without a uterus and implanted with a sixty-one-year-old's donated uterus gave birth to a healthy baby boy.[1]

While the uterus-transplant operation is large and flashy, and gets big headlines, even a transplanted uterus is not long-term "redesign" of the human body, of the core human code. But a microscopic and obscure mitochondrial transplant does alter future humans. In 2015, the UK may become the first country to allow transgenerational genetic engineering—changing a baby's gene code in a way that will then, in turn, be passed on to all future descendants. This type of procedure, known as germ-line engineering (sperm or egg), has so far been banned in most developed countries. But as technology and knowledge advance, the UK's chief medical officer, Dame

Sally Davis, now advocates lifting this ban, for a very specific use: to deal with mitochondrial diseases.[2]

Mitochondrial DNA (mtDNA) is strange stuff; it operates as a mini genome, separate from your core twenty-three chromosomes, and resides within the energy-producing units of your cells, the mitochondria. The 37 genes in mtDNA carry at least 400 disease-causing mutations, which are passed on from generation to generation. It is remarkably stable because it comes only from your mother; it does not recombine with your father's mtDNA, nor do sperm mitochondria get passed along to your kids. So mtDNA can unambiguously track your line of matrilineal descent. It's the reason we now know that every human alive today descended from one common ancestral mother, our mitochondrial Eve, who lived about 180,000 years ago.[3] While many other women lived at the time, only one was lucky enough to have kids who survived to have kids, through 7,200 generations, all the way to the present.

Our newfound ability to alter this, the most stable part of the evolutionary tree, could have long-lasting implications. We may want to proceed with mtDNA therapies so as to prevent an alphabet soup of really rare but generationally persistent diseases caused by mutations in mtDNA.[4] Patients, and their children, could then avoid persistent cyclic vomiting, some cancers, heart-tissue death, blindness, deafness, asphyxiation, and a host of other symptoms.[5]

The protocol for the treatment seems relatively simple and not entirely unlike IVF; take an unfertilized egg from the mother with defective mitochondrial DNA, remove only the nuclear chromosomal DNA from the egg, transfer it into the donor egg of a woman who does not have a mitochondrial disease and whose nuclear DNA material has been removed, fertilize in vitro with Dad's sperm, transfer into Mom's uterus, have a healthy baby, and therefore healthy descendants for that baby.[6] According to one of the scientists involved with the research, "What we've done is like changing the battery on a laptop. The energy supply now works properly, but none of the information on the hard drive has been changed."[7]

These kinds of mtDNA treatments are initially only applicable in a very small number of pregnancies; in the whole of the UK, maybe they save ten lives per year.[8] So why is it such a big deal? When one alters the core DNA of one's egg and sperm, not only does one alter the evolutionary development of the embryo, one potentially alters the future of a part of humanity. These technologies allow children to be born with genes from multiple parents, with an ever-increasing number of potential gene donors going into a single body. And there is a growing temptation, as we develop new techniques, to redesign not just an individual baby's genome but that of all its descendants.

One glimpse of where we might be heading is U.S. Patent #8543339: "Gamete donor selection based on genetic calculations." Owned by 23andMe, a genetics diagnostic firm funded by a founder of Google, and temporarily limited in its diagnostic services by the FDA, the patent purports to help those seeking a child via IVF to make informed decisions. The patent's crude drawings supposedly allow one statistically to opt for various body traits: Blue, green, or brown eyes? High or low risk of various cancers? (Gee, let me think about that choice . . .) Congenital heart defects? Long life span? The intellectual-property claims jumble things that are reasonably predictable, such as alcohol flush reactions, with much more complex diseases whose likelihoods are buried under complicated mathematical gobbledygook. While we remain a ways away from routinely implementing these diagnostic practices, a few aspects of such a system already operate today, due to our ever-increasing ability to test, pre- and post-birth, for deadly or seriously debilitating conditions.

Pre-conception, the Jewish Genetic Disease Consortium's advice is unambiguous: "All couples with ANY Jewish ancestry, including interfaith couples, should have pre-conception carrier screening for all Jewish genetic diseases."[9] Many others follow this advice as well, pre-diagnosing a broad set of potential conditions. Then, during the first trimester, many fetuses are also tested for yet another set of genetic conditions. Finally, at birth, the state of Massachusetts mandates immediate screening for fifty-two genetic disorders.[10] There are more and more questions asked at every stage, and as questions and diagnoses mount, there is ever more pressure to modify.

Every year the genetic casino gets tipped a little more toward eliminating more disease carriers and inserting positive traits; perhaps inserting the CETP gene, associated with a 69 percent reduction in Alzheimer's.[11] Or the DEC2 gene so you only need six hours of sleep each night.[12] A rare APOC3 gene mutation may become a popular addition, as it's been found to lower fat in blood by 65 percent in the test population of Old Order Amish and to greatly lower Alzheimer's risk in Ashkenazi Jews.[13] Japanese Americans who carry FOXO3A have significantly lower rates of cancer and heart disease than the average American.[14] Each of these discoveries increases the potential menu of desirable alterations to one's own genome, or that of one's baby.

Some fertility treatments are the Wild West of medical trials.[15] The most radical new technologies appeal to desperate couples, after natural means do not work and traditional IVF fails. They may be on a very tight "biological clock" time line and unlikely to voluntarily participate in traditional double-blind studies. Many hucksters peddle unproven, costly, and ineffective solutions. And often it is hard to track who is doing what to whom because most of these procedures are paid for out of pocket; insurance companies don't pay until efficacy is proven in a traditional human trial.

Great ethical challenges arise as assisted-reproduction gene-altering technologies become cheap, effective, widespread, and ever more powerful.[16] After all, many of these techniques can be carried out in a sterile office with inexpensive equipment and some increasingly simple recipes. We already witness lab, farm, and home engineering of bacteria, plants, and animals. Knocking out or inserting specific genetic instructions into many an organism is commonplace. Then the specific organism makes copies of itself time and again as a new and improved version. As the speed, knowledge, and power behind life-code manipulation increases, and as gene transfers and edits can be safely and reliably introduced into nascent embryos, we should expect a tidal wave of genetic upgrades in humans. We're well on our way toward nonrandom mutation and the custom design of our bodies and our babies. And now we are even starting to play with our brains. . . .

Boyden Brains

.... :✳: ....

Of all the organs and body parts we can alter, none will make as much of a difference, long-term, to the history and destiny of our species as the brain. It is what makes you "you" and an individual; it is what makes us humans as a species. As we map, study, modify, and perhaps upgrade our brains, as we experiment with how to stimulate neurons to do or not do something, we potentially alter consciousness; we change our fundamental selves. No one is closer to the bleeding edge of this research than Ed Boyden.

MIT is full of very, very smart people. Interacting with a genius is a daily occurrence. Sitting next to a Nobel laureate is not unusual. But even within this hallowed community of brains, misfits, creators, egos, and builders, Ed Boyden stands out. Many MIT folks think he is one of the most creative, constructive, imaginative people on campus. (Apparently President Obama agrees; when he launched his brain-research initiative in 2014, there were a lot of old, white-haired, top-of-the-science-food-chain-type folks running around at the ceremony . . . plus this one kid called Boyden, who looked more like a student than a professor.)

Boyden started out as a physicist, but soon found most projects too massive, expensive, and slow. Then he studied quantum computing, an area rife with uncertainty, philosophical questions, and doubt; what was fascinating to him is that quantum uncertainty could still generate some real, concrete mathematical and cryptographic solutions. But eventually the questions that really caught his imagination had to do with understanding, mapping, and modeling the brain: How do chemical and electrical stimuli interact to make you taste, love, think, and vote in particular ways? Can you build a map of specific human traits, actions, and beliefs? How does the mind make sense of the world? How does it repair circuits after strokes? What, if any, "humanity" is lost following a serious brain injury?

Drawing on his experience in physics, a field that requires building new machines to rapidly and accurately measure the minutest, shortest-lived, and most elusive of particles, Boyden entered brain science in the late 1990s, when there were very few ways to probe the brain and almost none to answer any of the questions he was interested in. Most of what was known about brain structure and function in living brain cells came from relatively macro-level observations of larger areas of the brain, measurements of just a few cells at a time, or observations of the effects of specific injuries on personality, memory, and behavior. In labs, blurred images from fMRIs provided the multiple colorful images many of us associate with brain research, but this type of study relied on ponderous machines and did not directly measure brain cell activity, and certainly not in real time. Microscopes were difficult to use on true living cells, particularly across large-scale networks. Directly stimulating specific brain centers, through electrodes, sometimes stimulated actions or desires, but it was pretty hit or miss.

Today's brain treatments remain singularly crude. A few central-nervous-system drugs sometimes work, but many trigger major side effects. It takes a brave pharmaceutical company to launch yet another brain drug trial, given the graveyard of failed and very expensive trials; 92 percent of central-nervous-system drugs fail, and they tend to fail late in the clinical-trial process, when it's really expensive. (On average a $2.6 billion "Oops, never mind!" every time . . .)[1]

One way to reset your brain circuits, even today, is through electroshocks. Or you can saturate your entire brain in a chemical soup of antidepressants, stimulants, and seizure moderators. Boyden took one look at this flailing around in the medical community and, instead of waiting for the right instruments to be built, he began to do what any physicist would do: He quit speculating and built new technologies to really understand what is occurring inside the brain, so he could measure and act accordingly as real data began to flow.

Domesticated viruses turn out to be a terrific way to get data. Most of the nasties that infect your body do not cross the blood-brain barrier, a very good thing. (A century ago, in the time before ethics panels, researchers filled some poor schlub's body full of blue dye. Soon all of the person's body tissues looked like a Smurf except the brain and spinal cord, thus proving there is a barrier that prevents many drugs and molecules in the bloodstream from getting to the brain cells.) Boyden, working with others, got around this issue by injecting engineered retroviruses directly into brains, where they soon colonize all the nooks and crannies. The retroviruses carry the gene code required to make an *opsin*, a molecule that allows microscopic sea algae to convert light into electricity.

Soon researchers were placing microscopically thin strands of optic fibers near living brain cells and watching minute flashes of light as individual brain cells turned on and off. This new method, known as optogenetics, means one can watch and map, in real time, what is happening inside a brain—exactly which neurons are active when an animal moves, eats, smells, or learns. And one can begin to control these neurons. After retroviral opsin molecules enter brain cells, they can be used as on/off switches.[2]

Because Boyden and his team send their protocols and reagents to any researcher who asks, many a graduate student is now busily filling brains with lentivirus "to produce high-titer, cell-specific neural labels." The result is a torrent of data illuminating what is occurring inside the brain of many different species.[3] All this provides an increasingly accurate set of blueprints as to what takes place inside each brain cell, networks of cells, and brain regions. And, just as occurs in physics, new machines and new technologies often provide answers to some of the most basic of questions: What kinds of neurons are in each region of the brain? How do they connect? Which behaviors are they linked with? New atlases now compare how particular brain regions interact and how entire classes of brain cells work.

One surprising and particularly exciting development is that opsin-transmitting light-encoded information does not just map the brain; the system can carry instructions into the brain. Using lasers and fiber optics, Boyden's lab can get mice to move in a particular way, feel particular things, or forget traumas.[4] What began as a mapping activity, a technique designed to query the brain, ended up birthing a way to instruct and control parts of the brain.

It goes way beyond mice. By 2012, using multiple colors of light and patterns, two groups of researchers demonstrated that one could use minimally invasive methods to alter primate behaviors.[5] Light could stimulate their neurons in such a way that the animals reacted faster. These discoveries could potentially alter how we treat various mental illnesses in the future; after mapping the active neurons in the brain of a mouse with post-traumatic stress, for example, Boyden's team could make an educated guess as to which specific regions of brain cells remembered and reflected the trauma. They then used fiber optics to stimulate or block specific neurons to see how the mouse responded.[6] Eventually they figured out how to block anxiety and stress with light pulses. The next step was to try out the same technique on humans.

One of the tragedies of the use of IEDs and other explosive weapons in war is that many soldiers suffer extreme physical trauma.[7] Prior to our unnatural and extreme medical interventions, very few would have survived the kinds of wounds one commonly sees today in VA hospitals. Piecing bodies back together sometimes requires implanting electrodes in wounded warriors' brains to instruct their bodies in carrying out basic functions and commands. Because the retroviruses that deliver fluorescent proteins to brain cells are safe and effective, surgeons could someday implant fibers in the brains of soldiers suffering from post-traumatic stress disorder (PTSD). Then, by lighting up these specific cells and activating them, these disabled veterans could block PTSD symptoms.[8]

New imaging methods, biomarkers, and a focus on precision measurement and engineering are altering our understanding of how brains work and how far we can alter them. Autopsies suggest that some distinct brain structures may make some people prone to suicide; individuals who committed suicide showed epigenetic changes that switch off the SKA2 gene, which controls the brain's response to cortisol and stress.[9] Not surprisingly, given a possible heritable component, suicidal tendencies can run in families. One strange, gruesome statistic: Biological relatives are six times as likely to commit suicide as are adoptees within the same family cluster.[10] By mapping brains, perhaps we may be able to tell who is at risk and institute preventive measures.

The ability to deliver very specific instructions to a brain keeps improving. Boyden's merry tribe realized that if they could make each brain cell sensitive to two colors, then they would have far more control over outcomes. That is, if a blue light pulses, do X; if green, do Y. One implication of controlling in two or more colors is that scientists could now use the binary code of computers (long strings of 1s and 0s, light and no light, or current and no current) to upload or download information into brain circuits.[11]

Light can be used to bypass traditional biological bridges between the brain and various organs. In a tour de force, Boyden's team took a blind mouse and labeled the bipolar cells in the retina, rendering the cells sensitive to light. Soon an animal that had struggled to find its way out of a water maze could easily identify the lighted path.[12] Companies like Bionic Sight and Gensight now want to take these discoveries into human trials to cure blindness, bridging damaged eyes and the brain. Humans are already undergoing clinical trials, but this is just the first step; brain light switches could have many uses.

Eventually one could conceive of a series of light-mediated connections between the brain and various organs—say, a cable linking the thought of moving one's left arm to the arm itself, thereby bypassing the severed spinal cord of a paraplegic. (In a sense this is like a heart bypass; you reconnect the parts that work. But you are using light to send impulses from the brain to the arm's nerves instead of to the spinal cord.) Neurosurgeons now envision ways of modulating specific brain regions during epileptic seizures. Or using light to bridge an injured region after spinal paralysis. Or perhaps even how ADHD might be light-moderated.[13]

While bridging the brain and various body parts, or enhancing our senses, is a big deal technically, these changes will have far less of an overall impact on humanity than using light and other instruments to control, modify, alter thoughts. Within the next few years, Boyden should be able to map, neuron by neuron, one cubic centimeter of the approximate 1,000 cubic centimeters that constitute the human brain. As more and more of the brain gets mapped, we will know what the effect is of modifying single cells, altering connections, or stimulating particular neurons. There are plenty of willing subjects to map and test; more than 250,000 people in the United States already have implanted electrical stimulation electrodes to treat epilepsy and other conditions. So it would be relatively straightforward to carry out various parallel, fully consented experiments.

What scientists are already discovering is that if brain chips/ electrodes are placed in the right areas, and stimulated just so, then they can modulate specific brain activities. One of the stranger, and potentially more than a little disturbing, implications of using electricity or light to control the brain is that firing off a few cells can sometimes create or erase a memory, including things that never took place.[14] Specific light pulses can trigger fear of an impending electric shock in rats; a different pattern of light pulses can make the rats forget these fears.[15] Or one might be able to insert and trigger traumas or pleasures. These emerging capabilities might just have a few legal, ethical, and moral implications.

Eventually 3-D optical neural micro-implants could become

standard and help us hyper-sense the world around us and interface with ever greater databases. We would need to develop very strict protocols as to what is controlled by whom. Even small, localized interventions can have massive effects. For instance, in humans, electrode recordings showed that groups of neurons sometimes fired moments before a "conscious" decision was made by the patient.[16] This implies that our brains often make decisions before they are "thought through" and rationalized. If one were to activate particular neurons, one might be able to induce or suppress actions before a person is ever conscious of making a specific decision. One might conceive of many potential benefits from this kind of intervention: reduce violence, limit strife, stop suicides. But what is to be controlled, when, and by whom are nontrivial considerations. They go to the heart of who we are as individuals and call into question what is free will and independent decision making. As we alter gene code and brains, our very consciousness may be at stake.

EVOLVING OURSELVES . . .

. . . .

Better Living Through Chemistry?

.... ☀ ....

Mary Lou Jepsen, a brilliant, bubbly, no-nonsense Ph.D., has had an extraordinary chemical journey. She and technology guru Nicholas Negroponte used their base at the MIT Media Lab to launch what initially looked like a completely quixotic effort to design, build, and distribute a $100 laptop, bringing computing and the Web into the poorest classrooms in the world. Along the journey, Mary Lou learned Chinese, earned the trust of Taiwanese hardware manufacturers, and became the first Westerner allowed into their fabrication facilities. After years of real-time co-engineering and literally sleeping on factory floors, she shipped the betas of the new computer in 2006. Her quest to provide the cheapest computing, full of tales of skullduggery, derring-do, and heroism throughout South Asia, eventually led her to rethink and simplify overall computing while taking on some of the world's largest and most powerful oligopolies. But even this achievement pales in comparison to Mary Lou's simultaneous struggles with her health.

Whenever Mary Lou jets around the globe, she is never more than a few hours away from dying, literally. During graduate school, she learned that her long history of extreme health problems was caused by a hidden brain tumor that, among other things, caused her pituitary gland to secrete numerous hormones in the wrong doses at the wrong times; this finally explained her chronic oversleeping, weight gain, vomiting, sores, and extreme mood swings.

Brutal radiation and chemotherapy led to the removal of many of the key hormonal pathways in her body. Though cured of her tumor, her body chemistry was so disabled and unstable, she couldn't live without constant chemical/hormonal interventions; forgetting to take a pill or injection could be fatal.

Her uncertain recovery and continual treatments didn't stop Mary Lou from finishing her doctorate, getting a top MIT job, launching the One Laptop per Child program, and getting married. Neither did it stop her from traveling all over the world, even though missing her meds because of customs, a bad connection, or a closed pharmacy could mean death. She established a triple-redundancy system whereby medicines would be pre-positioned, with her and with someone else, wherever she came from, wherever she was going; before every trip, she mapped out pharmacies that could quickly provide what she needed in an emergency, noting the local names for each drug and a doctor who could prescribe it, day or night.

At first her regimen was standard and brutal. Doctors prescribed; she took. Often the side effects were horrific. As doctors shifted strategies, Mary Lou began to realize that different formulations, dosages, and hours would lead to extreme mood swings and physical reactions. Being first and foremost a scientist, and having built a career in statistical analysis and coding, she began to observe, record, and run real-time experiments on her own body and brain. Her vast spreadsheets detailed what would happen as she altered the specific dosage and timing of every drug she was taking. She found that "average doses" were a recipe for disaster, and a personalized, self-administered prescription was far better.

Even when she found a workable regimen and her life could go back to almost normal, Mary Lou did not stop experimenting. Her drugs substituted for the master gland that regulates one's most fundamental feelings and desires, likes and dislikes—the pituitary; she began real-time tests on what different hormones might do to her emotions. By upping her dose of substances like testosterone she became more and more masculine. As she describes it, suddenly, within a few hours, she became the hormonal equivalent of a teenage boy. She couldn't stop thinking, continuously and obsessively, about sex, sex, sex. She became far more confrontational, arrogant, and aggressive. She began to see women and men in a very different light, understanding each far better.

Although it was fun to experience another gender, she decided within a few days to return to feeling like a woman and then began to experiment with various types of hormonally driven moods and desires. By the end of the experiment, she had lived several emotional lifetimes in one body, found a balance that she and her husband liked, and got on with changing the world's IT and computing systems. She took control of her brain and body. Had she been just a little less determined, just a little less brilliant, it is likely, she believes, she would have ended up, in her own words, crippled, drooling, imbecilic, in her mother's basement awaiting the end of a short life.[1]

The extent, breadth, and impact of hormonal adjustment carried out by Mary Lou is extraordinary, but we are already modifying millions of individuals' hormones on a massive scale. As we learn more about the interaction between our genes and the glands in charge of implementing their instructions, we are ever more consciously modifying our daily attitudes, desires, fears, and actions. And as we learn more about how our bodies feel and act under the influence of various pharmaceuticals and natural substances, more and more people are choosing to alter their complex hormonal balances. We enter a future where we don't just map our brains and thoughts but also consciously alter our feelings and emotions.

The desire and drive to alter, to feel more, to enhance is certainly present. About 7 percent of U.S. twelfth-graders have tried MDMA, aka Ecstasy or E.[2] Why? It makes them feel happier, more sensitive, connected, empathetic, and emotionally open. It accentuates music, colors,

and touch. For some, it significantly improves sex.[3] Ecstasy is but the tip
of the iceberg; the most-used drugs in the United States today, legal and
illegal, have one purpose: to alter your mood and feelings, to modulate,
attenuate, depress, and pacify your various demons so that you believe
you feel better. Drugs nonrandomly alter one's body chemistry: serotonin, norepinephrine, dopamine, epinephrine, octopamine, and a multitude of other biochemical systems.[4]

We still have much to learn about exactly how various drugs alter
your brain and how they affect different people. As demonstrated by
Mary Lou, "average" dosage means very little. Multidrug interactions
are rarely tested. And it's not just a drug itself, or the gene variant of the
specific patient; food consumption may affect the brain and drug interactions. More than 85 prescription drugs may interact with grapefruit
juice, at least 43 of these with serious consequences.[5] Food side effects
may last for weeks; adverse reactions to MAO inhibitors can be triggered by eating aged cheeses a full two weeks after taking the pills.[6]

We are still quite ignorant about how to target many drugs to do
only what they are supposed to do. That is why page after page of
small type follows every bright, cheerful drug ad with the happy
bouncing patient running through meadows or hugging a loved one;
it can all go horribly wrong in the wrong dose or the wrong patient.
Yet none of this stops us from basting our brains in chemical soups
to modify our feelings. In Rochester, Minnesota, and the leafy burbs
of Olmstead County, right near the Mayo Clinic, the population is
moderately Republican, mostly white, somewhat prosperous, and
apparently not completely happy; almost 1 out of 5 women aged
thirty to forty-nine is on antidepressants (about two times the rate of
males). And use of these "mother's little helpers" increases with age.[7]
Opioid painkillers are also quite popular, as about 16 percent of these
thirty- to forty-nine-year-old Minnesota women use them.

In 2014, use of benzodiazepine was correlated with a significant
increase in Alzheimer's. Turns out this is relevant to millions because it
implies the most popular antianxiety medications (including Ativan,
Klonopin, Valium, Xanax) may have very long-term effects; the longer

the patient took the drug, and the higher the dose, the greater the chances of Alzheimer's.[8] Keep in mind this is correlation, not causation. It may be that anxiety, not the drugs used to treat anxiety, is behind Alzheimer's. But there may be something to watch carefully here; it may be that some of the medicines we use today can fundamentally alter the brain decades later. Large portions of our modern societies are in the midst of massive experiments, far less radical and brutal than Mary Lou Jepsen's, but transitioning from tribes and rural enclaves to packed urban environments can make individuals feel as if they've ended up in a lonely sea of people. Many then turn to chemicals to alter or intermediate how they feel, act, live, mate. Better living through chemistry?

One of the ironic effects of chemists getting really good at building new medical compounds, ones that target very specific brain receptors, is the rapid displacement of many "all-natural" drugs of yore. Cocaine, opium, and marijuana grown in faraway jungles and then transported over long and uncertain routes are often displaced as drugs of choice by lab-made, nonnatural, legal chemical products.[9] Legal opioid sales throughout the United States have quadrupled in just over a decade.[10] By 2010, overall annual opioid production had reached the point where you could "medicate every adult in the United States, with the equivalent of 5 mg of hydrocodone every 4 hours for 1 month."[11] If you are middle-aged, the chances of your dying from drug use now exceed those of dying in a car crash.[12]

Other than in terms of purity and quality control, legal and illegal substances seem to be converging; according to the National Institute on Drug Abuse, "stimulants such as Ritalin achieve their effects by acting on the same neurotransmitter systems as cocaine. Opioid pain relievers such as OxyContin attach to the same cell receptors targeted by illegal opioids like heroin. Prescription depressants produce sedating or calming effects in the same manner as the club drugs GHB and Rohypnol. And when taken in very high doses, the cough suppressant dextromethorphan acts on the same cell receptors as PCP or ketamine, producing similar out-of-body experiences."[13]

Mood-altering chemistry is now part of day-to-day life for large swaths of our species. Our power to self-regulate mood could lead to far greater empathy, understanding, and well-being. We can modulate, control, or domesticate some behaviors and conditions by "taking the edge off." But this ability to constantly introduce unnatural chemistry into our bodies and brains also brings up significant political, security, ethical, and moral questions: How do we regulate and control our most basic feelings, and perhaps actions? Increasingly powerful and targeted mood drugs mean our kids and grandkids will be able to choose to experiment with what it feels like to be the opposite sex, gay, depressed, angry, happy, extroverted, introverted, and myriad other conditions that remain hard to empathize with without having experienced them oneself. But in experimenting we may also alter our fundamental selves, redesigning the moods and actions of large communities. Bioactive chemicals can affect how our bodies develop, how our sperm and eggs are epigenetically encoded, which microbes and viruses cohabit our bodies, and even with whom we choose to share genes. Unnatural and deliberate chemical interventions often are DESTINY's toxins, leaving long-lasting genomic memories that increasingly determine who we are, how our retirement years might play out, how our future kids develop, and what we will become.

Forever Young, Beautiful, and Fearless?

.... :✳: ....

Are we all doomed to age and die? The good news: We humans are so smart we have engineered life spans equivalent to five hundred human years. The bad news? So far . . . only in worms. If you genetically modify worms to suppress the insulin/IGF-1 signaling system, you can increase the creature's overall life span by 30 percent. Alternatively, if you suppress the nutrient-sensing TOR pathway, life span doubles. And if you modify both? That results in a fivefold life extension.[1] Now the quest is on to apply these kinds of discoveries to humans.

Seemingly forever, quacks and researchers have pursued the fountain of youth, sometimes through quite exotic treatments. In the 1920s, surgeons seeking to rejuvenate older men began transplanting the testicles from young sheep.[2] (Baa-a-ad idea?) All kinds of transplant experiments later, the odd quest continues. In 2013, Harvard researchers joined the circulatory system of an old mouse with that of a young mouse; within four weeks the older mouse's heart disease was gone and the organ rejuvenated.[3] (Fortunately, their next step wasn't starting to stitch young people to old people; it was to isolate the growth factor that restored heart health, GDF11, and begin moving toward human clinical trials.)

Over the past two decades, we have been promised that an unending alphabet soup of genes (IGF, GDNF, BDNF, EGF, IGF, VEGF, SDF, HGH, GnRH, TGF, EPO, FGF, NGF, GM-CSF, PDGF, PGF, HGF . . .) can help us grow, repair, or otherwise alter various body parts. Many of these experiments show promise, and each moves death a little further out. MDs are now starting to think of old age as a chronic disease. But the real question is whether, by bringing some of these treatments together, we will ever see a combined effect similar to the one we saw in worms—a fivefold life extension.

There are various hints. Our brains make GnRH, a hormone that controls our sex organs, regulating puberty and fertility. When older mice get infusions of this substance, age-related diseases decrease and life spans extend by 25 percent.[4] We don't yet know whether GnRH would do the same for humans, but we do know that Italian women who are still fertile in their late forties are more likely to live to one hundred.[5] Similarly, Korean eunuchs, deficient in testosterone, on average lived seventeen years longer than their peers.[6] Fountain of youth? Take your pick—a little GnRH or castration? Or perhaps a cocktail of other substances that might slow aging: NAD+, TFAM, or resveratrol?[7]

Geneticist Eric Topol's study of "wellderly" populations may provide a road map for what to add and subtract from our hormonal mixes. Close to 48 percent of those aged seventy-five and older in the United States never suffer significant hospital stays, and nearly another 30 percent only go in once for a major intervention (say, a knee replacement). These wellderly give us hope that aging may be a somewhat treatable and postponable condition. There are about 70,387 identified age-associated epigenetic switches that change as we age; about 56 percent of these switches tend to be flipped on and 44 percent are flipped off.[8] As we begin to modulate the right switches, we may be able to seriously alter part of the aging process (and perhaps even understand why men age 4 percent faster than women).[9]

Even without gene therapies, simply improving exercise, nutrition, vaccines, antibiotics, conquering a series of diseases, and reducing risk factors, we've made some dramatic gains in longevity. In Denmark the chances of an individual living to age ninety or above increased 30 percent per decade for those born in 1895, 1905, and 1915.[10] Throughout the world, social security systems are financially succumbing to what turned out to be way-too-pessimistic actuarial tables. With gene therapies added to many other interventions, longevity records and "the oldest person to do X" awards will be broken time and again. So don't feel so awful the next time an octogenarian passes you in a half marathon.

If we begin altering hormones and modifying genes to influence aging, we are also likely to deploy these technologies and techniques for other purposes, like modifying sexuality; GnRH regimens can block the onset of puberty in a transgender youth, or complement cross-sex hormone regimens.[11] Perhaps one day genetic surgeons will modify humanity's looks not with a knife, toxin, or filler but by silencing or promoting specific aspects of one's gene code. Maybe insert a variant of the ABCC11 gene so your underarms won't smell? Or how about an antiwrinkle pill to reduce crow's feet?[12]

DNA sequencing already allows one to statistically infer hair and eye color, racial ancestry, and various other physical features. But the maps get increasingly more accurate and granular. One study measured 7,000 features on faces of many shapes and varieties, and also measured DNA differences in many of the genes known to control anatomy. From this, they were able to deduce a crude image of what a person looks like, using only their DNA.[13] Anthropologists now use this knowledge to try to predict the looks of ancestors in cases where they have found a few bones with DNA but no skulls. Pennsylvania police used the same emerging technology to try to find a serial rapist.[14] But someday, if we get accurate enough, parents may want to adjust particular facial features based on genetic and epigenetic treatments in utero.[15]

Gene surgery could someday modify even deep fears and phobias. About 5 percent of Spaniards are prone to sudden panic attacks; imagine driving somewhere to pick up your kids and suddenly, out of nowhere, your brain chemistry dumps you into a state of extreme fear, chills, sweat, nausea, heart palpitation, pain. One of the causes of this recurring condition, in addition to the possibility of your in-laws coming to visit, is a faulty NTRK3 gene. This variant drives a miscommunication between the hippocampus, which directs the storage and context of memories, and the amygdala, which takes stored context and drives the appropriate response. The result is a series of inappropriate and panicked responses. However, a drug, Tiagabine, can sometimes help reset memories that lead to a panic response, which creates an intriguing set of questions.[16] Could we systematically modulate human fears? One could easily see why this might be something potentially useful, not just for those suffering panic attacks but also for an army or security apparatus. Eventually would there be circumstances under which you might want to genetically modify panic genes, like NTRK3, within an individual or a family?

Although specific and targeted brain modulation sounds like science fiction, as we have seen, there are more and more examples of brain control that go far beyond fear. Perhaps gene surgery will become as prevalent as plastic surgery is today. But it won't just modify looks and life span; potentially it will also change our most basic feelings and actions. One more set of instruments to unnaturally tune up one's self, and one's species. . . .

Unnatural Attraction

.... :✳: ....

W hom you have a child with is one of the most consequential decisions you will ever make. But as you pick your mate, beware of
unnatural selection; what you see is not necessarily what you get. As
humans develop ever more sophisticated camouflage, an increasing
amount of today's sex, mating, and reproduction is driven by deceit,
cover-ups, and outright lies.

Given how important looks are in mate selection, it's not surprising that personal appearance is one of the first places where humans
began to practice unnatural selection. From the earliest of tombs and
the oldest of mummies, be they in South America, Egypt, or China,
there's one constant: We try, desperately, to enhance, highlight, or
hide our appearance. Humans have been using tattoos for at least five
thousand years, for example, and Neanderthals may have used cosmetics for over fifty thousand years.[1]

One of the most common ways to incentivize someone we might
not normally mate with involves cosmetics and plastic surgery. The
human eye and brain (and nether parts) are exquisitely attuned to the
minutest signals of fecundity. For millennia we have finely honed
our senses to judge very quickly if that cutie walking by would be a

good mate. Do the skin and hair look healthy? How about women's hips, thighs, and breasts? Or a man's abs and gluteus maximus? Age? Facial symmetry? Choosing a mate is an especially important decision in monogamous societies. If you're to procreate with just one person . . . well, it truly should be Mr. or Mrs. Right if you wish to improve your descendants' traits, particularly given your own flaws.

Americans spend close to $250 billion on cosmetics in an effort to edit, enhance, or obscure our natural looks and signals.[2] And there is a dirty little secret behind many of the 14.6 million plastic surgeries that take place yearly: Not all of them were medically necessary. All this primping used to be illegal. As early as the 700s the British passed a law that read, "All women . . . that shall from and after this act impose upon, seduce or betray into matrimony any of his Majesty's subjects by the use of scents, paints, cosmetics, washes . . . shall incur the penalty of the law now in force against witch craft and like misdemeanors and that the marriage upon convictions shall stand null and void."[3] (Translation: "Argh, lass, go back to being a smelly, toothless hag . . .")

While these kinds of laws are laughable today, some feel cosmetic surgery has gone too far. Ethicist Kristi Scott believes "evolution continually strives to keep the best genes around to proliferate the species. Emerging cosmetic surgeries inadvertently attempt to cheat this by altering external flaws and ignoring the intact internal code where the flaws remain . . . with more and more people flocking to cosmetic procedures at younger ages, doctors and consumers need to understand and discuss the importance of this dramatic misrepresentation to the opposite sex."[4] The traditional clues and signals one received from hair, skin, or fat in the right places are being obscured or outright falsified. It's not just a Hollywood phenomenon; Iran has the highest rate of nose surgery in the world, seven times that of the United States.[5]

Ethical or not, cosmetics and surgeries often work. Those who are "put together" tend to be perceived as more beautiful and competent.[6] Even certain artificial looks, for example those clearly signaling exaggerated and grossly inaccurate mammary capacity, are no longer hidden but celebrated. The title of one of the most popular Colombian soap operas? *Without Breasts There Is No Paradise.* Many endure significant pain and expense to alter eyelids, smooth foreheads, shrink noses, tuck tummies, or paralyze skin in attempts to attract those who might have ignored their unenhanced version.

There's ample proof that as a species we don't just want to cover, we want to alter.[7] Now we are increasingly able to do so; enhancement can be rocket-boosted by emerging technologies that nonrandomly and unnaturally alter the way our bodies look and function. The fastest-growing cosmetic segment is "cosmeceuticals"—cosmetics with biologically active compounds and druglike effects. These range from prescription eye drops to grow longer eyelashes, perfumes you swallow, and antiscar creams. But sometimes the changes induced by these products can last a lifetime; human growth hormone is being repurposed from medical use to cosmetic use because if you are tall, you tend to have a higher income, a better-looking partner, and more power. According to one study, being 10 centimeters taller increased both men's and women's average income somewhere between $1,874 and $2,306.[8]

Altering chemistry is one more way to induce unnatural selection. Be it on a South Seas island or in L.A., Mexico City, or Rome, sometimes love just explodes. Instant overwhelming attraction—it does not dissipate even after seeing the other person without makeup, meeting the nasty potential in-laws, hanging out with the nerdy friends. There's sometimes just something there. We are only just learning how the chemistry of love really works, with a lot of the research particularly focusing on smells. And given that humans can distinguish more than 1 trillion different smells, there is a lot to work with.[9]

Smells can elicit all kinds of emotions.[10] In one of the more creative, and slightly perverse, experiments of 2013, a bunch of those white-coated folks engineered male fruit flies to release female pheromones. Then they let them loose in a box with normal fly guys, ensuring a huge amount of pent-up desire, but no way to mate. (Male fruit flies are driven to states of extreme desire in the presence of

concentrated female pheromones.) Soon the scientists had enough statistically significant observations to conclude: Smell can trigger extreme
lust. And at least in male flies, repeated *arousal interruptus* can have
long-term health effects: rapid fat loss, more stress, shorter life.[11]

In many mammals, sex and smell are also tightly coupled; the
vomeronasal organ detects pheromones and provides a lot of data
that tells them who's ready to breed and a good match. Arousal and
action soon follows. Given how many mammals use pheromones as a
signal/attractant/stimulator, it would seem that finding "chemistry"
in humans would be a no-brainer. But a wee detail complicates the
story: Supposedly humans have no vomeronasal organ, and therefore
no way to detect pheromones.[12] Even the existence of these molecules and their receptors within a human context has never been
conclusively found or proven.[13] That's why you see articles with titles
like "Facts, Fallacies, Fears, and Frustrations with Human Pheromones."[14] And in 2014, a research review is simply titled "Human Pheromones: Do They Exist?"[15]

There are plenty of (dare we say it?) tantalizing hints that human
sexual pheromones exist: In one experiment men found that the
unwashed T-shirts of ovulating women were far more attractive.
Women living together somehow synchronize menstrual cycles.[16]
Women prefer the T-shirts of men with different immune systems,
and tend to mate with those who present a very different repertoire
of antibodies.[17] There is a biological logic to this choice. By mixing
your immune system and HLA antigens with a very different set of
immune markers, you maximize your offspring's immunity.[18] Yet
one more reason not to marry a sibling or first cousin. The flip side
of this partner-selection pattern is that when your immune markers
are different from those of your mate, he or she is rarely a match if
you need an organ donation. But we didn't evolve to donate compatible organs to each other, we evolved to have children with strong
immune systems. (Organ transplants are a very recent phenomenon
and another good example of how the most "unnatural" of practices
can save lives.)

Yes, for centuries we have masked, altered, enhanced the way we smell and taste. But it has so far been based on a lot of image, not a lot of substance. Thirty percent of the top global beauty brands have been around for more than a century, and 82 percent more than sixty years: a market structure typical of products based on a lot of advertising and branding and little technological disruption. Four-fifths of cosmetic companies' budgets are used for advertising, not for R&D. But this could soon change, altering the entire industry, and perhaps even disrupting whom we mate with.

Temporarily covering stuff up with cosmetics is qualitatively different from truly altering the body and the underlying smell-taste signals it provides in terms of attractiveness and reproductive fitness. Modern chemistry can seriously interfere with the signals we send out to potential mates. For instance, a woman's ability to discern a man's immune-system markers goes kaput if she is on a birth-control pill.[19] So instead of choosing a mate who diversifies her offspring's immune system, she will tend to pick someone genetically closer to herself.[20]

As we learn to alter signals to the brain, or how those signals are interpreted, we may also develop the chemical means to alter even the most fundamental of sexual attractants and choices. Consider monogamy. Only 5 percent of mammals pair up for life. Only 27 percent of primates are monogamous.[21] But some prairie voles do seem to be strictly monogamous, at least until some of their brain receptors are stimulated, silenced, or fooled to prevent the hormone oxytocin from working; soon thereafter faithful voles began to covet their neighbors.[22]

Although humans like to believe in complete free will and choices, basic chemistry often drives particular behaviors and traits. The levels of "bonding hormones," including oxytocin, that flow through one's body can affect everyday choices and behaviors. For instance, a subset of Swedish women who have a specific genetic subtype of the oxytocin receptor were revealed to have suffered a 50 percent higher incidence of marital crisis in the year prior to the study.[23] (Another screening criterion for Match.com?)

Humans also have very high levels of oxytocin receptors in regions of the brain involved in addiction, and it so happens that "monogamous species are loaded with oxytocin receptors in areas of the brain associated with addiction whereas the non-monogamous are not."[24] As the song says, we may well be "addicted to love." The excess or lack of a second bonding hormone, arginine vasopressin (AVP), is also a reasonable predictor of how well a man will do in a monogamous marriage.[25] And while you are in diagnostic mode, you may wish to test a potential son-in-law or husband for the AVP334 genetic variant; men who have one copy of this variant are less attached to their partner and are more likely to have considered divorce in the past year. And those who carry two copies are twice as likely to experience a marital crisis versus those who have no copies.[26]

Soon we may add chemical treatments to couples therapy. Start with the initial diagnosis; women express high levels of oxytocin during relationship distress, whereas men have high levels of AVP.[27] So if you are a man who finds your girlfriend or wife often screaming at you, and you are not being more of a jerk than usual, perhaps consider a gift of oxytocin nasal spray for her.[28] As we understand more about how to modify brain chemistry and body chemistry, we may begin to influence some pretty fundamental beliefs, actions, and desires. Humans are getting ever better at designing ever more powerful oxytocin variants for very specific purposes and tasks.[29] For better or worse, we may find ourselves able to live with and mate with people we might not have naturally tolerated or been attracted to.

But even though hormonal receptors are absolutely key determinants of sexual behaviors, it is an area where funded researchers often fear to tread. PubMed, the U.S. government's meta site for scholarly medical research, lists 22,580 articles involving oxytocin. A grand total of sixteen of these articles dare mention the words "oxytocin," "human," and "sexuality" together.[30] Sexuality is a really understudied, underfunded area of scientific research, so if you want some indication of where sexual/chemistry/hormonal research is headed, you have to go somewhat far afield, which brings us to sexologist Emmanuele Jannini, a man of rather unique research interests.[31]

When Dr. Jannini wrote an extensive review article on a completely uncontroversial subject, "Male Homosexuality: Nature or Culture?" he was struck by how common and prevalent homosexuality is, across time, cultures, and even species.[32] Almost 1,500 species practice it, including gut worms. In humans, there's a strong inherited and/or biological link to male homosexuality; the preference shows up regardless of culture or religion. (Though its overt expression may remain far more discreet where there's severe repression and societal rejection.) But doesn't homosexuality seem to contradict Darwin and evolution? Wouldn't families more prone to produce homosexual males tend to die out?

Quite the contrary. The same genes, hormones, womb environment, and/or brain structures that may predispose toward male homosexuality also appear to lead to higher fertility in female relatives.[33] Birth order also seems to affect homosexuality. The more male children a mother has, the higher the chances that the next child will be homosexual. Each additional male has a 38 percent greater chance of being gay than his previous brother.[34] This does not seem to be a culture-induced phenomenon; if the younger boys grow up in different households from their elder siblings, their chances of being gay are not reduced. So it is not growing up with an older brother or a mellower parenting environment that makes the difference; it is fraternal birth order. Adopted kids and stepbrothers have no increased chance of being homosexual, even if reared in a household with a lot of older boys. All this points to a genetic predisposition for male homosexuality. (Which means discriminating against those born gay makes as much sense as discriminating against someone who was born left-handed or redheaded.)

But as is true for intelligence, even though there's clearly a strong inheritance bias, specific genes that fully explain homosexuality have not been uncovered. Early clues, carefully worded in science gobbledygook to avoid rousing rabid fundamentalists, argued that the brain's "suprachiasmatic nucleus" was twice as large in homosexuals.[35] Another found the "third interstitial notch of the anterior hypothalamus" was two to three times smaller in homosexuals.[36] But recent research focuses less on core genetic or anatomical differences and more on biochemistry and epigenetics during fetal development. It is not our core genetic code but the other genomes in our bodies that may be at work.

Different chemical cascades at different times can lead to a broad range of sexual preferences and feelings. One explanation as to why some people elect to alter their sexual organs is that they simply feel trapped inside the body of the wrong sex. They may be right. Sexual organs are determined, and begin to grow, when a fetus is two months old, but the brain's male/female orientation appears to be determined later; a boy's surge in testosterone occurs during the second trimester of a pregnancy. So if the chemical waterfall is off, a body can develop in one direction and the brain in another, leading to transsexuality.[37]

Sex turns out to be really complicated. As more and more people feel safer to express what they feel, perhaps the most interesting and surprising aspect of recent research into sexual orientation is how many possible answers there are to sexual-identity questions; we are now way beyond the "simple days" when the only question was "Boy or girl?" Or even the more modern "Gay or straight?" For many, sexual orientation is a continuum that may tilt in different ways at different times, depending on circumstances. Facebook recognized this continuum of options on the day before Valentine's Day 2014; now anyone can self-define according to a wide variety of gender identities, including: Cisgender (birth identity matches sexual identity), FTM (female-to-male trans), MTF (male-to-female trans), genderfluid, neutrois (genderless), androgynous. . . .

So while the U.S. political far right and far left stand atop the ramparts fighting over gay marriage, there may be far more issues to address than who does what with whom behind closed doors. We know that changes in our environment, stress, water supply, pesticides, estrogen, chemicals, and our own body weight can alter our reproductive fitness. We like to think about sexuality as a choice, but there is also another big "C" present: chemistry. And we may, as we alter the chemistry of our bodies, also be altering our own sexual orientation and desires, as well as that of our kids and grandkids. In fact, we may already be running a massive worldwide experiment in chemical/environmental/reproductive fitness; our genomes are always alert and listening, transgenerationally.

Because sex research is an academic and grant minefield, we still know very little about the biology of sexuality, including the epigenome, microbiome, and virome. But our very ability to reproduce, as well as our desires, are under extreme selective evolutionary pressures. We see this in epidemics of early puberty, childlessness, and older parenthood. They are all symptoms of unnatural selection. As we learn more about how we alter our most intimate development and desires, we will better understand how we are shaping our descendants and our species.

Sports Quandaries and Beyond . . .

·········· :✳: ··········

The Olympic motto: *Citius, Altius, Fortius*. In constantly pushing the envelope on faster, higher, stronger, we push the boundaries of unnaturally evolving and altering the human body. And there is a lot of financial pressure to do so.

It's not just the Olympic "amateur" athletes who care about the outcomes because of potential endorsements. Those are relatively small personal stakes. The system of sports is big dollars; NBC pays $4.4 billion to transmit just the Olympics, but billions more are at stake in a variety of other so-called games.[1] Every year U.S. couch potatoes bet the equivalent of everything produced, during an entire year, by every citizen of Thailand, Colombia, Iran, or South Africa on a variety of sports.[2] There is more than a little economic incentive to improve, to be competitive, to cheat. And when sports, economics, and individual ambition come together and demand that the human body improve, fast, then evolution itself can become turbocharged.

As we learn more biology, as we find new ways to fix and upgrade humans, the bright and clear lines that determine who is eligible to compete in what categories quickly blurs. One rapidly approaching dilemma: Who can compete where during the Tokyo Olympics (2020)?

Among the mildly controversial questions on the table is how to define gender. What is "natural"? Who is handicapped and who is enhanced?

Only two types of Olympics exist at present—the Olympics and the Paralympics—each with only two gender categories, male or female. But as we sequence the genes of more and more people, and as laws liberalize and allow many of the seriously oppressed out of their closets, the gender picture takes on many hues. For instance, what if male/female differences aren't as clear-cut as we think they are? Traditionally, female chromosomes are XX and males are XY. But what should Olympic judges do when a competitor has ambiguous external genitalia? Or what if the external organs signal one gender but the internal organs another? Or if the external genitalia had been "corrected" at birth so as to conform to one category or another? Does it make a difference if the hormonal or chromosomal signatures vary significantly from what is typically physically considered male or female? What do you do with XXY-chromosome competitors who have female genitalia but "aren't really considered women"?[3]

Australia, one of the first countries to recognize gender ambiguity, allows its citizens to choose M, F, or X on their passports. As of 2013, Germany became the first EU country to allow "indeterminate" as an option on its identity documents. Should the Olympics do the same now that exquisitely calibrated and sensitive drug-testing practices discover women competitors afflicted with hyperandrogenism? In layman's terms, this means they have male bone structure and musculature, no breasts, labial fusion, and enlarged clitorises.[4] Already there are fierce debates over sex testing in the Olympics and, ultimately, who should be allowed to compete in which category.[5] The issue involves so many variants and permutations that one specialist blogger writes, "Even genetic testing cannot confirm male or female. In fact, it is so complex that to do proper sex determination testing, you have to take a multidisciplinary approach, and make use of internal medicine specialists, gynecologists, psychologists, geneticists and endocrinologists. I am afraid that dropping your pants is not proof at all."[6]

As more chemicals that mimic the action of hormones enter our ecosystems, we may have to amend both reproductive and gender classifications. Many substances affect your fertility, muscle strength, and perhaps even your Myers-Briggs "personality type."[7] Chemicals, whether deliberately taken or not, can alter reproductive systems; for example, a common herbicide, atrazine, feminizes male frogs.[8] Dioxins, created when carbon and oxygen burn with chlorine, can depress both sperm count and immune function.[9] The list of potential gender-change agents includes organophosphate pesticides, mercury, arsenic, lead, and glycol ethers that can shrink a rat's testes. A lot of the results of our extensive environmental engineering first show up in the increasingly sensitive and calibrated sexual eligibility and blood-doping tests for athletes.

As endocrinologists (the doctors who fiddle with and adjust our most basic hormones affecting instincts, desires, and actions) better understand hormonal balances and tipping points, they can provide more and more options to young parents and teens. If a baby has too little 21-hydroxylase, for example, one consequence can be "excess virility"; depending on the degree of chemical imbalance, symptoms can range from a female being born with a slightly enlarged clitoris to having almost full-blown male sex organs. The condition is somewhat correctable, sometimes through drugs and surgery, but employs a treatment unapproved by the FDA. Dosages can vary significantly per patient, or even within the same patient, and when the individual is under stress, doctors sometimes double or triple doses.[10] At what point, and taken when, does too much or too little tip the balance in classifying an athlete as male or female?

There are real incentives and pressures to be "just a little more male" in many sports. As scientists alter our bodies with chemicals, birth and fertility enhancers, and a host of other inputs, they can also alter some people's sexual identities and athletic abilities. Right now, we don't have the knowledge or subtlety to be able to act in a pinpoint manner, and the results are pretty blunt. But as genetics and sports medicine go forward and map out the exact differences and mechanisms that sometimes create a top athlete, the temptation to intervene early and upgrade increases. Sports is just the leading edge. As pharmacologists get better at dosing, modifying, and modulating, society, parents, and kids begin to have a lot of chemical choices: Want a particularly tomboyish girl? A gentler boy? In our increasingly unnatural world we can evolve, change, and modify the answer to what used to be considered an unambiguous binary boy-or-girl question.

And sex is far from the only complex topic facing Tokyo 2020. Given how much is at stake, one needs to consider who has a "natural" body and who does not. Consider follistatin, a natural human hormone that helps embryos develop. In mice and monkeys, extra doses of this supplement enhance strength and appear safe.[11] In humans, follistatin treats patients with degenerative muscle disorders . . . and some elite athletes claim it helps build muscle mass—something that might be relevant to more than just a few Olympians. There are more and more emerging hormones and growth factors to enhance humans and complicate future Olympics.[12]

Sports and medical needs converge and incentivize the discovery and development of compound therapies and gene therapies that could significantly alter body types, personalities, longevity, and mating patterns for generations to come. Female mice engineered to express the PEPCK-C gene in skeletal muscle, for example, become hyperaggressive and hyperactive, and postpone menopause for a long time.[13] There are athletic consequences as well, given that PEPCK-C mice run twenty times farther than normal mice, at high speeds. And the effects are longlasting; these mice reproduce and live twice as long as normal mice. How long will it be before adventurous humans, in search of competitiveness and longer lives, begin to experiment on themselves?

In sports-mad Australia, seizures of various types of performance enhancers at the border increased 106 percent from 2010 to 2011.[14] On the top-ten use list: hormone-releasing peptides that aid in the growth of muscle and bone, and maybe in the repair of soft tissues.[15] Then there are hormone variants like AOD-9604, in clinical trials for fat-burning and cartilage repair. Or selective androgen-receptor modulators that increase bone and muscle mass (but that, in women, lead to maleness and in men lead to baldness). Or how about a dose of mechano growth factor (MGF), which leads to rapid muscle repair following weight lifting. There are ever more ways of pursuing and ensuring *Citius, Altius, Fortius.*

As chemical synthesis improves and as the basic composition of new drugs closely or exactly matches natural human hormones, detecting drug use gets harder and harder. Traditionally, big pharma built medicines out of small molecules—simple chemical substances created in test tubes—so it was easy to detect these foreign substances. But then the wild world of biotech programmed living cells and manufactured biologics, far more complex medicines that match natural human proteins. Erythropoietin (EPO) is a favorite of Tour de France bikers because it increases the red blood cells that carry oxygen, can be fully synthesized from lab chemicals, and mirrors our natural EPO.[16] So, in Tokyo, the question is how to detect and regulate synthetic biologics that can look identical to what naturally exists in our bodies, only in excess.[17]

Even the unenhanced, the "all natural" but really unusual mutations, can produce serious Olympic conundrums. In seeking gene dopers, sometimes officials come across extremely rare, naturally occurring beneficial gene variants that lead to extraordinary performance. Are normal people who spend tens of thousands of hours training and striving really competing on equal terms when they enter the track to compete against these one-in-a-million folk? More than two hundred genes and gene variants correspond with greater or lesser athletic prowess. Sometimes the natural and unnatural reinforce each other. For example, when suspicious Olympic judges tested one of Finland's greatest athletes for blood doping, they indeed found some chemical enhancements.[18] But they also found that skier Eero Mäntyranta had a mutated EPO receptor that in turn allowed him to naturally produce extra red blood cells and boost oxygen carrying capacity by 25 to 50 percent.[19] About twenty-nine of his relatives also carried this rare gene mutation.

Genetics matters. A UK twin study found that heredity determines 66 percent of elite athletic ability.[20] Everyone who has summited Everest without an oxygen tank has had a favorable ACE gene variant.[21] Almost every male Olympic power athlete carries a 577R variant of the alpha-actinin-3 gene. (But if you carry two copies of the 577X variant, maybe your dreams of fame and fortune best be redirected toward computer coding rather than pro athletics.) So how should future Olympics treat "mutant" athletes with extraordinary biology? This Rubicon was partially crossed when some female athletes, found to have naturally high levels of testosterone in their bodies, were required to have surgery to remove endocrine tissue and reduce their testosterone levels.[22] Humiliating, embarrassing, and certainly unnatural: No medical problems warranted such intervention.

Here are three options for the 2020 Olympics: (1) Be a showcase for really hardworking mutants. Only those lucky enough to have had the right parents and trained really hard viably compete. Or (2) have gene handicaps so that everyone competes on an equal basis. This practice already occurs in golf, sailing, and other sports. If you have the "slow variant" of a particular gene, then you get one-tenth of a second's head start. Or perhaps, as gene therapy becomes safer, (3) make some athletes eligible for an "upgrade" so they can compete on equal terms with the genetically fortunate.[23]

We are getting more and more sophisticated at identifying, parsing, and isolating that which begets extraordinary performance. And we are designing drugs and genetic "inserts" to mirror what were once almost unique performances and attributes. Not to mention "exercise pills" that could mimic the beneficial effects of physical exercise, even for couch potatoes.[24] Thereby we are upgrading the human body and its capabilities. As long as the focus of "upgrades" is limited to sports, many scientists are comfortable and happy to continue the debate. But a room turns decidedly frosty, and the conversation gets tense, when one dares ask whether gene variants may have implications beyond athleticism. Are there positive or negative gene variants that correlate with traits other than athletics? Such uncomfortable questions have been buried, with good reason, for a long time. Historically, seemingly well-intentioned scientists attempting to answer these questions generated horrors. But we are getting to the point where the data may show that because of life code, genetics, epigenetics, or microbiomes, specific and identifiable groups of people are more or less talented at specific tasks. We need frameworks and forums to address the resulting questions and their repercussions. Especially now that we are able to voluntarily engineer specific traits into people.

Designer Organs and Cloned Humans

.... ✳ ....

The ink-jet printer sitting over in the corner looks normal: four ink cartridges moving back and forth over a flat surface. But if you look a little closer, you'll see there is a laser in front of the print head, measuring minute changes on the surface where the machine prints, which means it is printing in 3-D. And then if you examine what's being printed, you'll see something that looks an awful lot like skin.

Sitting next to the printer, with a big smile, is the Willy Wonka of human organs, Anthony Atala, director of the Wake Forest Institute for Regenerative Medicine. He and his team grew and implanted a human bladder in 1999, and that was perhaps one of the most prosaic of their efforts.[1] His lab is filled with all sorts of human organs, being cultivated with varying degrees of success.

Human skin is the largest, and one of the more complex, organs in our body. It keeps dirty stuff out, regulates temperature, facilitates touch. To multitask in these extreme ways, skin has multiple layers with various functions. Today, when someone gets seriously burned, doctors graft skin from another part of the body. In the future, perhaps due to Atala's research, a scanner will measure just how deep a burn is and an "organ printer" moving in 3-D pass by pass will deposit each of the correct types of human skin cells to rebuild the various layers of this complex organ. The right cells, in the right quantities, in the right places; the scanner will determine whether this particular portion of a body suffers from a first-, second-, or third-degree burn and will fill in the wound with the correct cells, depending on the depth and severity of the burn.

A growing cadre of regenerative medicine scientists is engineering even more extreme organs. Nina Tandon seems like a sweet, shy yoga practitioner; not many realize she spends her days bioengineering heart valves and figuring out how electrical currents help shape the design of a retina. At night she's busy building an open-source "plug-and-play" bioreactor to create "massively parallel human tissue cultures," which means many others can join the effort to rebuild bodies one part at a time. Because of Tandon, Atala, and a host of other researchers, before too long hospitals may be fitted with supermarket shelves full of substitute-organ scaffolds; these biodegradable molds can then be covered with cells taken from each patient's own tissues and programmed to regrow whatever organ is needed. Once the cells grow into the right shape, function, and size, the scaffold dissolves, and voilà: Here is your new regrown kidney, ready for transplant. Because the organ is grown with your own cells, your immune system will not think something new and nasty has entered your body. No need for immunosuppressant medicines that decimate your defenses.

As we continue to unnaturally evolve, we are learning how to reproduce each of our organs. Perhaps this sounds a bit strange or creepy until you consider teeth. We are all born with no teeth (for which Mom is most grateful). Gradually we grow a full set, only to lose them a few years later. Then we build a second full set. But if we lose any of this second set, they don't naturally regrow (unless you are either a shark or a lawyer). However, your body has already shown it can make a set of teeth—twice. So why can't it build a third set? After all, every instruction and body part required to make a full set of teeth lies within each of your cells inside the human genome. Researchers at Harvard's dental school have already rebuilt copies of many people's teeth in small glass dishes. And if you can rebuild teeth using the genetic instructions in every person's body, and you can decipher the instructions and create the right scaffolds, eventually you can rebuild any human organ.

Just as we refurbish our cars, boats, or airplanes by gradually replacing parts as they wear out, we will begin to do the same with our body parts. Eventually our bodies will contain a series of organs we weren't born with, but that we cloned and grew externally. With such capabilities, the whole notion of age begins to change. However, it might be wise to look to the story of the ancient Cumaean Sibyl: She cheated death for a thousand years after the god Apollo granted her wish that she live for as many years as the number of grains of sand she held in her hand. But she neglected to ask Apollo that she also stay youthful. Eventually, her body shriveled so badly that she was kept in a small jar, and when asked what she most wanted, she said, "I want to die."

To wish for immortality, the ancients tell us, may be a fool's dream, but what if you could remake your entire body from scratch and are youthfully reborn time and again? Eventually organ engineering will seem trivial compared to what is already in the works in a few labs. Go back to the very beginning of your conception story. Your father's sperm and mother's egg came together and formed a single-celled ovum, which then divided and divided and divided until it grew into you. For the first sixteen divisions, every fetal cell was pluripotent; that means every cell was not yet specialized and could still become any organ type. With the next cell division, cells begin to differentiate into specialized cell types to create the various organs and tissues.

One way to imagine this progression from being a pluripotent stem cell to being a differentiated tissue type is to picture skiers taking a chairlift up to the very top of a mountain (each skier equals one pluripotent stem cell). When they get there, they cannot go any higher but they can choose from a very broad range of potential branching paths on their way toward the bottom. As each skier picks a path and descends, there are still many choices available farther down the mountain, but it is ever harder to get back to the top, or even to the last branching point, without a chairlift. Eventually you run out of options and end up on the wrong side of the mountain after the lifts are closed. (That is your body in old age after it has spent a lifetime trying to rebuild some worn-out body parts but now cannot rebuild the most basic structures of the body itself.)

In 2006, stem-cell researcher Shinya Yamanaka inserted four genes into mouse skin cells and turned normal skin cells back into pluripotent stem cells. These reprogrammed cells, known as induced pluripotent stem (iPS) cells, were then implanted into a surrogate mouse mother.[2] Three weeks later, identical cloned baby mice began scurrying around a cage. These mice were not conceived, nor born from sperm and egg, but from reprogrammed skin cells. (Virgin birth, indeed . . .)

Yamanaka's discovery was recognized with the 2012 Nobel Prize (alongside John Gurdon, who learned how to clone tadpoles from adult frog cells in 1962). The discovery effectively built a cellular ski lift, which could take differentiated cells back to the top of the mountain and reset these cells back into a stem-cell state, thus allowing the entire development of a body, of a life form, to begin anew from even a single adult cell.[3]

How could this happen? Well, each cell in a mouse's body contains the animal's entire gene code; therefore, each cell knows how to build every basic part of an entire body.[4] If you can manipulate mouse cells this way, you can do the same with fruit flies. And you can take human skin cells and turn them into undifferentiated human stem cells. Now scientists are growing human liver cells and other organs from skin.

Perhaps someday the goal won't be replacing organs one by one, but cloning one's entire body. What now seems like science fiction is ever more feasible; after all, we have already cloned buffalo, carp, cats, cattle, deer, dogs, ferrets, frogs, fruit flies, gaurs, goats, horses, ibex, mice, mouflons, mules, pigs, rabbits, monkeys, sheep, and wolves.[5] By 2014, a single mouse had begotten twenty-five generations of identical clones. There are 580 copies of her, at different ages, wandering around nearby cages.[6] If things proceed apace, from a technical perspective, humans might be able to order up identical twins of themselves every couple of generations. (But beware: "Genetically identical" is not the same thing as truly identical. We know that the epigenome in a mature adult skin cell is not exactly the same as it was when that person first entered the world; it is modified, perhaps for good or perhaps not. And what about the microbiome and virome? Technically, these genomes should someday be cloneable too. Stay tuned. . . .)

Human cloning could ultimately upend all natural selection and random mutation. No more messy, random recombination with traditional sperm-and-egg procreation to see what comes out. Instead, humans could become WYSIWYG: What you see is what you get. Just a series of reruns that look very, very similar but whose epigenomes and microbiomes could lead to significant individual quirks. (So be careful what you wish for: Kardashian lookalikes could haunt Hollywood for centuries.) Is this really cheating death and the beginning of immortality? Or is it just creating a series of younger and younger identical twins?

Would your answer to this question change if you could further enhance your newly minted body by adding in other things—say, your memories?

Evolving Brains Revisited

.... ❉ ....

If you lose a kidney, an arm, or a leg, no one would declare you "less human." The same is true of organ transplants. You are no less "you" if you carry someone else's heart. But that was not always the assumption; when the first heart transplants were done, some asked if the recipient would fall in love with the donor's wife. Today we know the heart is an extraordinary and exquisite muscle, which carries none of the emotion we have historically attributed to it: "She broke my heart," "I gave her my heart," "With all my heart"—they're all just metaphorical.

The brain is the last possible receptacle of all emotion, consciousness, and humanity. Legally we treat people differently when they are brain-dead; we are even allowed to kill their bodies, by unplugging life support, because they are no longer considered quite human.

The flip side is that changing the brain changes the human; and changing the brains of humans changes the species. If we were able to download and upload significant amounts of information into a brain, or if we could modify the structure and wiring of the brain, that would be a fundamental game changer in terms of who we are, what we decide, what we think.

Some of the same questions that came up in the first heart transplants, regarding memories or emotions such as love and attachment, will likely arise again when we see the first human brain transplants. Would memories and emotions transfer along with the brain? Would the recipient be attracted to the donor's wife? Or will it turn out that the brain, too, is simply a type of electrochemical organ and not the custodian of all emotion and consciousness?

Likely many moral, ethical, and medical considerations will, and should, block wholesale transplants of significant human brain materials for a while. But when they eventually take place, we will get answers to some essential questions about the nature of our humanity.

Meanwhile, we can get some early hints and indications of what it means to mirror and deconstruct the programming of a brain. *Caenorhabditis elegans* is a very small, transparent nematode, otherwise known to civilians as a worm. These creatures are relatively simple to observe and model, so the heroically patient biologist John Sulston spent years steadily, intently, and completely recording the fate of each cell of *C. elegans*, from conception to full maturity.[1] As he observed each and every cell division through his microscope, he recorded exactly when and how each of the worm's 959 cells grew, and in what order, and how each and every bit of nerve and muscle developed. This included every detail of the 302 neurons in the worm's brain. (He won a 2002 Nobel Prize for this work.)

On the shoulders of giants . . . enter computer scientist David Dalrymple; not exactly a slacker, he was MIT's youngest ever graduate student, age fourteen. He amused himself by working on a Ph.D. at Harvard while mapping the activity of all those 302 worm neurons Sulston had found. This "merely" required his inventing a new subfield of mathematics that would allow him to visualize the function, behavior, and electrochemical action of every neuron.[2] Eventually, having the biology plus the specific mechanism of neural actions might lead Dalrymple and the scientists who follow him to a model of an entire worm brain—a virtual one that could live outside a body, in a silicon chip.

Building a virtual brain, even that of a small worm, will likely teach us quite a bit about how information moves back and forth, how it is stored, and how it is altered. It may be a way to eventually store memories. Then one would have to develop an interface that would allow the data/emotions/actions/memories to be re-uploaded into a person's brain. Scientists, meanwhile, are focused on at least five ways to attempt to achieve a fundamental shift in cognition, but all of them boil down to one principle: Store memory externally, upload later.[3]

The most brute-force option is to do what we have done with many other organs: Simply transplant the brain itself. In the repertoire of extreme and gruesome surgeries, perhaps the poster child is the whole-head transplant. It ain't pretty. One way of doing this is to sever the entire head of a mouse, dog, or monkey and graft on another body. A second alternative is to graft on a *second* head/brain. Both seem cruel and "yucky" until you consider this research may provide critical answers to questions like "Could you ever reattach a human spinal cord?"

As surgeons get closer and closer to being able to attach or bypass severed spinal cords, some are beginning to consider the protocols and techniques required to attach a human head to a different body.[4] Given the physical characteristics of the brain, and its incredibly complex interconnections, in many ways it may be easier to transplant an entire head and brain than to attempt to insert and graft on pieces of just the brain. A whole-head transplant might save those whose other critical organs have failed, or who have had their bodies destroyed in an accident.

A human head transplant would have to reattach bone and muscle. This has already been partially done. In 2002, Marcos Parra was hit by a drunk driver so hard his entire head almost came off. The only thing really holding it in place was the spinal cord and some veins and arteries. Fortunately Dr. Curtis Dickman had been preparing for such an eventuality. After inserting screws to reattach the vertebrae to the base of the skull, Dickman used a part of the pelvis to bring neck and skull together.[5] (Parra fully recovered and was playing basketball within six months.)[6]

Constant improvements in heart-transplant procedures and finger, limb, and face reattachments means that arterial and venal grafting is ever more sophisticated and is likely getting to the point where a full-head reattachment to the venous and arterial systems is feasible. That leaves the final and most difficult part of the process: reattaching the spinal cord and reconnecting nerves to the brain. But even here, experiments involving peeing rats provide hope.

One of the many horrific consequences of a severed spinal cord is loss of bladder control. In 2013 a Cleveland team reattached severed mouse spinal cords through nerve micrografts. The procedure was so successful that not only did the animals recover some respiratory functions but even the ability to control areas much farther down their bodies, including their bladders.[7] Such an operation makes it conceivable that a full mouse-head transplant might someday be successful. And if it was successful, we might begin to be able to test various hypotheses: If a mouse had learned in detail how to ask for food or navigate a maze, would the mouse's head take that knowledge to a new body?

No one has attempted a whole human brain transplant, nor should they. Nascent technologies and knowledge make the procedure far too risky and speculative, and the chances of success are minute, not to mention the ethical challenges of identifying and qualifying a donor. But as science progresses, if one became able to transplant a human brain or portions of a brain, then one could begin to answer some fundamental questions about the nature of consciousness, memory, and personality. (Never mind whose passport should the person carry, the one with the fingerprints or the one with the photo?)

While we wait for full-brain transplants, there is still a lot of data flow; even mini transplants can make a huge difference. Because ethical constraints limit the kinds of experiments we can perform on humans, scientists get ever more creative at blurring the lines and distinctions between animals and humans. The most basic of human cells, stem cells, which program all functions in our bodies, are being inserted into species far and wide. As we blur species lines, as we "humanize" parts of animals, we begin to see blind mice that grow human corneas.[8] And because some of the organs and biological structures in pigs are so close to those of humans, there are more and more efforts to modify these animals' immune systems, humanize some of their organs, and transplant them directly into humans.[9]

In an attempt to find cures for various neurological diseases, more and more human brain cells are entering animal bodies, which often results in significant and noticeable upgrades. Alzheimer's researchers found that transplanted human stem cells led to mice with improved spatial learning and memory.[10] When one inserts human glial stem cells into the brains of newborn mice, the new cells grow and eventually overwhelm many of the original mouse brain cells. Soon you get mice that can learn much faster, retain memories longer, and whose brains transfer certain information three times as fast as normal mice.[11] (Of note is that this latter procedure is transplant of glial cells, the cells that preserve, feed, and protect neurons. It is not yet a neuronal transplant, so while it is unlikely this kind of transplant would transfer memories, it does seem to significantly enhance cognition.)

If we can transplant human cells into animals' brains and significantly improve their cognition, it is also reasonable to think that one could transplant and develop enhancements to the average human brain; recent stem-cell transplants into Parkinson's patients' brains show some promise, albeit inconsistently.[12] Whose brain cells we get, at what stage, through what procedures, may end up making quite a difference. (They may also give rise to a slew of ethical and access issues; would you want a transplant from an average brain or a genius brain?) As we continue to seek cures for various neurological diseases, we are likely to find more and more examples of interventions that significantly alter and enhance various brain functions. And this will give us more choices in how to enhance, evolve, and build up the most human of our organs.

Meanwhile, we are continually attempting to "upgrade" our brains through electronic inputs, both internal and external. Sophisticated electromagnets placed on the skulls of monkeys can direct them to pick out any one of 5,000 random objects 10 to 20 percent more accurately than nonenhanced monkeys.[13] Early tests on seven human epilepsy patients, through already-implanted electrodes, showed improved navigation through virtual mazes.[14] Soon the "handicapped" became better at this task than "neurotypicals." Human deep-brain stimulation will likely enter early clinical trials in 2015, to try to boost memory in Alzheimer's sufferers. But if techniques like this really work—a big "if"—then they could be broadly deployed to enhance the memory of the species. (While the Beatles used to sing "All you need is love," Ed Boyden might counter that maybe all you really need is a "very simple, off-the-shelf laser and viral injector system for *in vivo* neuromodulation.")[15]

Drugs provide yet another path to enhance/modify human cognition. While we regularly approach a Starbucks vaguely hoping a triple shot of caffeine will upgrade our mental capacity, the effects of tea, energy drinks, and other caffeinated boosts are temporary. Modafinil may be different. A pill originally designed to help you sleep better, this drug may have the side effect of memory upgrades that last for significant periods.[16] As we understand the biochemistry of the brain better, we will likely find more and more ways to boost, refine, and improve cognition, once again unnaturally altering the species.

And then there is the external cognition option. Back at the MIT Boyden lab, they are busily building tiny computer chips, embedded with thousands of needles 1/1000th of an inch wide, which allow measuring, and perhaps altering, activity inside individual neurons. The volume of information coming out of these experiments is such that the team had to completely redesign computer interfaces; they can now download brain data directly into computer memory (which might be useful for small start-ups like Google and Twitter, which covet this kind of massive direct-download capability for more pedestrian purposes like efficiently siphoning fire hoses of data off the Net and into their servers).

If brain "co-processors" can connect single brain cells, or large groups of cells, having an efficient internal-external interface for memories means we may be able to record and control memories outside the brain. Although this sounds like science fiction, memories of performing a particular task have already been transferred from one rat to another.[17]

Remember just how clunky early computing was? Massive digital machines interacted with punch cards, then teletype, keyboards, touch screens, and now voice commands and EEGs. Then humans miniaturized from mainframes to desktops, laptops, tablets, smart phones, Google Glass, and electronic contact lenses, the interface becoming ever more useful and ubiquitous. And now we just take it all for granted, the cell phone as an extension of our lives, our memories; the average person checks his or her PDA 150 times per day.[18] We may live through a similar process with brain interfaces. Already the giant PET scan and MRI machines of yore are being displaced by miniaturized electronic, chemical, genetic, and optical interfaces to store, instruct, and/or control certain brain functions.

As we better understand our brain we delve ever deeper into the most essential part of what it means to be human. No organ contains and embodies our human identity to a greater extent. As we re-create, store, and export brain circuits, as we grow or alter tiny three-dimensional bits of brain tissue, we push evolution forward in completely nonrandom ways. These technologies can provide inklings of answers to truly fundamental questions: What is consciousness? Do we alter free will and our humanity when we chemically, electrically, or optically alter our brains?[19] How far can we alter the human brain before we begin to make these changes permanent, a part of normal living for everyone, and begin to deem ourselves a different species?

A core human aspiration is to be smarter, to know more; it is, after all, what got us to the top of the species pyramid. Technical, safety, regulatory, and of course ethical hurdles will likely delay the widespread use of brain chips or other implants. But brain upgrades will eventually occur; the upside and incentive is simply too compelling. Our brain architecture and function will be ever less naturally selected and randomly mutated. Our thoughts will be increasingly intermediated and dictated by extreme bionics, synthetic neurobiology, neuralmechatronics, and neural biomaterials, not to mention gene therapies. We will begin to store thoughts through chemistry, electricity, light, biomolecules, neuro chips, and synthetically created living cells.

Given the increasing likelihood of cloning our bodies, if one were to be able to transplant emotions, memories, or consciousness by transplanting a brain or externally storing and downloading, then one could see a path toward a far longer-lived species. As our bodies wear out we could copy them, and as brains and memories faded we could renew or transplant them. In generational terms we are a ways from this reality. On an evolutionary time line we are close. In fact, we may just be a blink away from thinking ourselves into a successor species, one that can gather, synthesize, and use the power of billions of memories and emotions. And as we begin to think about really long-range travel we may even choose to outsource a part of our brains or cognition to non-carbon-based life forms, like robots. . . .

The Robot-Computer-Human Interface

.... :✳: ....

The first "Arc Fusion" dinner in San Francisco was quite a do.[1] Its producers, journalists David Ewing Duncan and Stephen Petranek, took over a theater, laid out an enormous table in the shape of an *X*, theatrically lit the whole place, and invited the weirdest and most interesting people they knew to talk about human evolution. Attendees were given one of five colored name tags, depending on how they answered a multiple-choice question: "How far would you go to enhance yourself if these technologies were possible?" (1) Minimal enhancement as my knees and organs age, otherwise I'm fine the way I am; (2) Take a pill daily that increases your concentration by 25 percent and doubles your strength by 50 percent; (3) Implant stem cells in your brain that have been genetically modified to make you quicker and smarter; (4) Replace your legs and arms now with prosthetics that are better than flesh and blood—more durable and stronger; (5) Transfer your mind into a durable robot or into a virtual reality world that will last for centuries—or longer.

So . . . what would be *your* answer?

As dinner progressed from foraged nettle soup through organic Aztec chocolates, the debate got pretty heated. Robots and mechanical enhancements can be a Rorschach test reflecting fundamental divergence of opinion on how far to enhance, how long to live, and whether to create paths toward broad human speciation. Can a human be a human with consciousness enhanced and without much of a physical body? We may someday be able to, and want to, store a part of our consciousness outside the brain, because robotic design is improving fast; humans are increasingly comfortable with various implants; and soon we may wish to travel very, very long distances.

There is still much deserved skepticism. Only one-fifth of the Arc dinner's participants chose the most radical option on the questionnaire: robot fusion, partial immortality. Endless waiting for useful everyday robots has made many people wary of the promises of robotics; not surprising since *The Jetsons* so long ago showcased Rosie, the perfect maid. *Lost in Space* told us robots could warn us of any impeding dangers. *Star Wars* made us wish we had C-3PO as a buddy. And then *The Terminator* showed up. Suddenly robots didn't seem so cuddly anymore. And meanwhile, back in the present, the only semi-robotic thing that seems to work somewhat well is a Roomba. Most of us don't yet think of robots as particularly relevant or real.

But as occurred with iGEM, many high school and college kids don't really care about what we think we know, so they are busily building a present far different from what we think exists. Throughout the 2013–14 school year, more than 350,000 kids figured out how to build robots that could pick up and shoot large beach balls. These are not pre-made, pre-designed, off-the-shelf machines. They were built from scratch, without instructions, for a very specific challenge that changes yearly.[2] (These kids also had to recruit 64,000 adult mentors and 3,500 corporate sponsors.)[3]

In typical engineer-nerd fashion, the founder of this contest, Dean Kamen, came up with a tremendously catchy name for the competition: For Inspiration and Recognition of Science and Technology, which he then fortunately shortened into FIRST Robotics. The contest became a movement and an ethos, a mechanical-educational cult, reflecting

Kamen's "don't tell me why you can't do it" attitude.[4] (After all, Kamen himself conceived 440 patents, including the Segway, a wheelchair that climbs stairs, and diverse medical devices, including most of the insulin pumps currently on the market.)[5]

Robotics is one of the few major college competitions where almost all participants can, and do, go pro. FIRST kids are ten times more likely than other college students to land an internship at a company. And after almost a quarter century of FIRST contests, literally millions of adults now feel very comfortable designing and coexisting with robots. Many have Ph.D.s and are already beginning to change our relationship/interface not only with robots but also in some cases with the human body itself.

So when the background conversation during the 2014 FIRST finals turned to the question "Which will be the first event in professional sports where robots will beat athletes?" it seemed wise to pay attention. It's one thing for robots/computers to win at chess, but imagine a world in which a pro team takes on a robotic tennis, golf, soccer, hockey, or basketball team and loses—something that will likely occur within the next decade or two. The level of design, strategy, and adaptability required for such feats goes way, way beyond a Roomba.[6] (And, ironically, in a few decades many of the players on the human team will already be mechanically enhanced with titanium hips and elbows and perhaps eye and brain implants.)

Already, in field after field, robots are beginning to demonstrate extraordinary skills. No human pilot shows greater talent than one who consistently lands on a carrier at night; the runway is moving, forward, sideways, and up and down, in irregular swells, in pitch-darkness. The task is so hard that pilots' heart rates can exceed those of the same pilots during air combat. Yet on July 10, 2013, the last air-robot/drone deniers were stunned into silence as they witnessed large unmanned drones execute landing after landing at sea.[7] To see automated machines preforming the task better than the best of the best humans represents a critical Rubicon.

Robots are here, even if we don't realize it yet. Typically by the time robots are tested, approved, and deployed in critical functions, they perform any given task at least an order of magnitude better than "average humans." Twice as many aircraft accidents are now due to pilot error as opposed to mechanical or autopilot failure.[8] Automation and human-robot coexistence have driven the chances of dying on a commercial flight down to 1 in 45 million.[9] Pilots will touch the controls less and less, until eventually a flight becomes, like the trains one takes from terminal to terminal in an airport, driverless.

Robotic cars will soon follow planes because overoptimistic, unaware, and incompetent make a bad combination behind the wheel of any automobile; yet 93 percent of U.S. drivers judge themselves "better than average."[10] (This disparity is accentuated by the Dunning-Kruger effect: The most incompetent among us are also the ones who most tend to overestimate their abilities.)[11] To help some of these delusional folks, car designers are incorporating more and more computers and robotics into new models, including collision avoidance, self-parking, lane maintenance, and adaptive cruise control. (This will somewhat reduce the chances that the loudmouth cutting you off while talking on the phone and eating a hamburger will make you yet another roadkill statistic.)[12]

As we begin to realize just how good driverless cars are, the anachronism will be a human driver. Some early adopters, likely those with teenage kids or elderly parents, will breathe a sigh of relief, as will many soccer moms and *Wall Street Journal*–reading commuters. But some will feel their liberty compromised, that driving constitutes an essential and permanent freedom. Perhaps one way to think about this inevitable transition is to mentally go back five or six generations. Some of our ancestors would be outraged and furious that their lazy descendants quit teaching their kids to travel long distances on horses. Within decades "driving a car" may be a fanciful novelty, recreation for those who can afford the extra insurance required for a human driver.

One very unfortunate consequence of autonomous vehicles, besides having everything moving at a constant speed, predictably, with seamless merging and far fewer traffic jams, will be the job losses for many ambulance-chasing lawyers. We will see fewer billboards on highways with advertisements like INJURED?!!! CALL BOBBY JOE RIGHT NOW TO GET YOUR PAYMENT IN COURT! Collision shops would also suffer, insurance rates would go down, and drunk drivers would fade away into the backseat. Does this mean no more accidents, ever? No. But as occurred in aviation, accidents will be ever diminishing. No distracting texting or phone calls. Fewer fender benders. Many fewer careless and distracted driver mistakes. Robots mean almost every driver will be average, and the average will be very high and improving all the time.

As occurs in planes, there will still be some mechanical failures and software misfires in cars, but overall road traffic safety will begin to converge with air traffic safety—a good thing, given that flying has a fatality rate of 0.003 per 100 million miles while the automobile death rate is 20,333 percent higher (0.61 per 100 million miles).[13]

Why is this relevant to overall human evolution? Because we are gradually getting very, very used to living side by side with robots, needing them, improving them, holding them in our hands, and even implanting them. Three million Americans are alive today because of an internalized, body-hybridized robotic device we call a "pacemaker." We not only tolerate robots, we increasingly demand them. And the graduates of FIRST, and of many other programs, are designing more and more robots that are not just useful but essential.

There is increasing need and pressure to interface with, use, and integrate robots into our daily lives, internally and externally.[14] Short-term, what we do daily, for how long, and with whom, is likely to change in a similar way to how mechanized agriculture changed the habits, landscape, and bodyscape of millions. Abrupt changes in physical labor patterns, physical abilities that evolved over hundreds of thousands of years that are suddenly useless, can have real consequences. Already it's true that eight, ten, or more hours at a desk significantly increase mortality from all causes, particularly heart disease, diabetes, and cancer.[15] Not moving even has reproductive consequences; men who watch more than twenty hours of TV per week have half the sperm count of those who do not watch TV.[16] And if you exercise at least fifteen hours per week, your sperm count is 73 percent higher than that of couch potatoes. (Meanwhile, some gyms in Florida have installed short escalators leading up to their front doors.)

Gradually we may demand that robotics be a part of our bodies. When MIT professor of engineering Hugh Herr enters a room, he bounces. He has the strength, grace, and agility of an extreme rock climber, so unless he is wearing shorts, most people don't realize he froze off both of his legs during a winter climb. What started out as an effort to "help the handicapped" has led to ever more competitive physical alternatives; Herr still climbs, using the feet, ankles, and calves he designed in his lab, which allow him to climb routes few others can touch. And a year after the Boston Marathon bombing, Herr created a completely new and radically redesigned leg for a professional ballroom dancer and had her dancing onstage at TED.[17]

The large and clumsy cones our great-grandparents used to improve their hearing gradually morphed into the large boxes our grandparents wore above their ears (which emitted high-pitched whines during dinners and phone conversations). Now our parents wear almost invisible and quite accurate hearing aids. And even the deaf now have cochlear implants with twenty-two channels that allow almost full hearing in English (this is not sufficient for tonal languages like Mandarin, and certainly not for music, which involves hundreds of channels).[18] As implants continue to improve, the deaf will likely end up with more sensitive, focused, tunable hearing than normal-hearing people. Beyond just music, some implants could enable one to hear sounds in tones currently audible only to dolphins or bats. Eventually one might choose to go beyond "bridging" for the injured and into outright sensory enhancement. Perhaps symphony orchestras of the future will hire only the hearing enhanced, and audiences may be more discerning.

Sound too far-fetched? Some "handicapped" runners and jumpers face accusations that they are too good, overly enhanced, and should be barred from competing against "normal" humans. For years no one paid much attention, except to express sympathy and support, to Markus Rehm's efforts to compete against the able-bodied in long jump. Until, after years of training and effort, and thousands of jumps using his one "normal" and one prosthetic leg, he won the German Nationals (2014). Within twenty-four hours the sports authorities' reaction turned to fear, and they immediately dropped him from the squad going into the European Championships.[19]

At various stages in life, for different professions, human-robotic interfaces will become increasingly common. The same type of appendages, actuators, enhancers, and interfaces that Herr designs for the handicapped may be used by the "able-bodied"; humans can get used to enhancements and begin to take them for granted very quickly. In 2013, Herr placed an external exoskeleton over soldiers' calves and boots that made it easier for them to walk carrying heavy packs for a couple of hours. Thereafter, the soldiers found it very hard to go back to walking without the enhancers; most reported feeling sluggish and clumsy without the apparatus; assisted walking is an especially easy enhancement to accept because, by nature, humans do not run on four legs in a stable manner. Instead they are continuously adjusting their balance as two legs take them through continuous, controlled forward crashes while leaving their arms free for other tasks. (And while on the subject, you may want to review Daniel Lieberman's research on why pregnant women don't tip over and other fascinating paradoxes of the human body.)[20]

Eventually we will likely see a gradual mental and physical acceptance of hybridized human-mechanical interfaces. When one military veteran who had lost both arms showed up in Dean Kamen's workshop, the team built the injured soldier two new arms that were so accurately controlled he could pick up grapes without breaking them and then feed himself a soft-boiled egg. (His wife was crying nearby and said it was the first time her husband had been able to feed himself in years.) But the most interesting part of this enhancement is that the arms were controlled by the veteran's thoughts, not by his muscles. The human-mechanical interface was wired through the brain, not through muscle actions. Hugh Herr now wants to take this integration one step further by growing muscles and nerves around the artificial legs he creates so there is no real conscious difference in how one walks and runs, whether the leg is natural or not.

Strange coalitions are rapidly advancing the machine-human interface. Groups like Humanity+ coalesce a like-minded community of top academics, Wild West do-it-yourself biohackers, Singularity University entrepreneurs, transhumanist ethicists, antiaging activists, and extreme engineers. Their overall objective aims to change, redesign, push, prod, and pull the human body in various ways so as to improve it.[21] Short-term, they aim to fight aging and ever deteriorating bodies. Long-term, they are far more ambitious and seek to integrate human brains and nervous systems with various mechanical enhancements. They argue that a far better alternative to cloning, or increasing human longevity, is to integrate our water-and-carbon-based life form with much longer-lasting materials. An ultimate step in unnatural selection?

Perhaps an Ethical Question or Two?

.... ✳: ....

E ugenics, a "breeding science" that purported to select and preserve "desirable human traits" was adopted by Hitler, but it was not German-bred and -born. It was birthed by Darwin's cousin, Francis Galton, in his book *Hereditary Genius*.[1] Then it was lovingly nurtured at Stanford University, Cold Spring Harbor, and the Carnegie Institution.[2] In Indiana, in 1907, way before the Nazis, eugenicist Harry Laughlin wrote up model laws for sterilization of "the inferior" and those "misfit to reproduce." Soon, thirty-three U.S. states were making sure that thousands of people they considered "socially inadequate," plus those who were maintained "wholly or partly by public expense," had no kids.[3]

State governments could and did take control of the bodies and reproductive organs of anyone they deemed "feebleminded, insane, criminalistic, epileptic, inebriate, diseased, blind, deaf; deformed; and dependent." And for good measure they also threw in "orphans, ne'er-do-wells, tramps, the homeless and paupers."[4] The Nazis thanked Laughlin with an honorary Ph.D. for the "science of social cleansing" in 1936, and then celebrated his accomplishments by sterilizing 350,000 people.

Despite all evidence of harm, many within the United States didn't learn from this history, and they continued sterilizing the mentally handicapped through the 1970s. Until 1977, in South Carolina, social workers were officially deputized to decide whether a person should ever be permitted to have children. The last forced sterilization in the United States? Oregon, 1981. Because billions of people have been judged, oppressed, or categorized as superior or inferior based on skin color, income, gender, national origin, and so on, the default position taken by most geneticists is: We are all equal, period.

Understandably, there is enormous aversion, anger, and fear about reopening a discussion on any topic having to do with genetic differences between peoples. Touching this debate while aspiring to tenure is academic suicide. And after spending billions of dollars studying the ethical, legal, and social implications of genomics research, the National Human Genome Research Project rarely addresses eugenics or the issues involved. (Of the twenty total times the topic comes up on genome.gov, it's often addressed with bromides about how we should learn from the past and avoid making the same mistakes.) Only the Cold Spring Harbor Laboratory Web site takes the issue head-on and unflinchingly, showing images and rationales used by their past academic superstars for obscene outcomes.[5]

But fear of controversy means we avoid the underlying issue while databases catalog more and more differences; there are some very specific gene variations among people. Sometimes these variants have to do with sex, race, ethnicity, family background, and other controversial topics. So when they arise they are often explained away or ignored. Meanwhile, the evidence accumulates.

Start with a politically fraught concept: Men and women are not equal. Few academics would dare take a position like that in public, which is one of the reasons biologist David Page isn't just smart but also quite brave. As director of the Whitehead Institute, his aim is to build the best research genetics faculty in the world. Along the way, with humor, guts, and smarts, he skewers sacred cows, including researching why men and women get diseases at differential rates. There are obvious answers for things like testicular and ovarian cancer, but what intrigued Page is how many non-sex-organ-related diseases affect men and women very differently. Yes, men and women get lung cancer and heart disease in different proportions because of different lifestyle habits, but how about things like lupus, multiple sclerosis, heart disease, Parkinson's, autism, schizophrenia, and stroke? Sometimes particular diseases affect women five times more often than men. Could there be fundamental differences in how men's and women's bodies and immune systems deal with disease?

If we take Page's questions seriously, and leave dramatic gender-bias/how-dare-you-ask-these-questions denunciations aside, then we must modify research significantly. Should early clinical trials for a particular medicine distinguish between male and female cohorts in mice? We know women and men react differently to many drugs; low-dose aspirin and some insomnia drugs, for example, should be administered very differently.[6] And yet, even for diseases that disproportionately affect women, like depression, neuroscientists studied 5.5 male rats for every female. Pharmacologists used the ratio 5:1.[7]

What about the gender of basic human cells in petri dishes that are used in the earliest stages of drug development? We know that in the brain, male neurons pay more attention to "excitatory neurotransmitters" and female neurons are more sensitive to the type of stimulation that leads to programmed cell death.[8] Yet most studies ignore whether the original cell line is male or female.

Once one really considers sex differences, real consequences follow the potentially offensive themes raised by Page and a small number of others. For example, should we treat all peoples, races, blood types, and sexes equally?[9] Before we enter these thorny canyons let's begin with one fact: Our species is relatively new and, because we outcompeted our closest hominin relatives, we are a remarkably monocultured group. A group of West African chimps has more genetic diversity within it than do our billions of Facebook friends.[10] Even the genetic elements that allow diversity within pluripotent stem cells—those that allow a species to rapidly adopt and adapt—are more prevalent in chimps and bonobos than in humans.[11]

But sometimes small differences count and, if recognized, can improve a group's chances of a healthy and satisfying life. Yet the majority of the scientific research community wants nothing to do with these questions. Period. Some argue that all studies based on race or IQ should simply be banned.[12] A poll taken among the readers of *Nature* magazine, an educated, science-savvy audience, showed that were it up to them, only 5 percent would allow studies linking genes and violence. Only 6 percent would allow studies on sex and genes. Only 8 percent approve of carrying out any type of genetics study tied to intelligence. And only 9 percent would support a study based on race.[13] Might one say there is a strong, predisposed research bias?[14]

Yet uncomfortable differences keep popping up in unexpected places. Statins reduce blood LDL levels (bad cholesterol), but they are considered less effective in African Americans.[15] One blood-clotting agent happens to be far *less* effective for African Americans than for whites in part because African Americans express higher levels of phosphotidylcholine transfer protein (PCTP) in their blood.[16] And one of the reasons for higher incidence of cancer and cardiovascular diseases among African Americans may be low vitamin D.[17] The list goes on: Some Hispanics and Latin Americans may be more susceptible to certain diseases like diabetes. Even within these populations there's huge variation; the incidence of diabetes is twice as high in Mexico as in Chile and Brazil.[18] Just as we need to understand specific environments better, we also need to focus on differential gene expression.

African Americans ages forty-five to sixty-four are two and a half times as likely to die of heart attacks as their white counterparts. Researchers consider diet, access to health care, and exercise among potential culprits. But one company, NitroMed, looked at its initial clinical-trial data and identified another factor. They had tried out their new medicine on a broad, random population, and it rarely worked well; however, it did appear to help one subset of the population: African Americans. NitroMed ran a new trial using only black patients. While quite controversial, the results were compelling enough that the trial was stopped early and in June 2005 the FDA approved the first race-specific drug after an "all-black clinical trial found that patients taking BiDil experienced a 39 percent reduction in the rate of first hospitalization for heart failure and a 43 percent increase in survival rates."[19] Despite these positive, lifesaving results, the company soon suffered brutal attacks by physician, research, and civil-rights groups. The drug flopped.[20] The company tanked. The lesson learned? Even if you can segment and cure a large group of people . . . don't go there.[21] Ideology and fear trump results.

Graduate school constantly reminds aspiring Ph.D.s that correlation is not causation. The brutal knife of "true, true, but unrelated" has sunk many a research presentation. But when unexpected correlations do exist, even if they are imperfect, it would seem appropriate to take advantage of information that might benefit the well-being of specific individuals or groups. Perhaps a researcher could be supported in attempts to delve deeper into the scientific basis of the correlation between skin color and other uncomfortable topics/differences, in order to engineer a therapeutic response?

Clinical trials could be designed to evaluate various subgroups that suffer disproportionate side effects, but there is currently not much incentive to do this; focusing on differences can explode in controversy and reduce market share. Few academics are willing to tread into such minefields; there are even explicit warnings in top science magazines telling you how to act and think if you dare enter these debates.[22] It is far easier to get a grant, generate outrage, and put on a white hat if you point out the multiple, and real, disparities in *access* to health care by race, income, and region than to study whether there might be any biological differences among groups and dare attempt to differentiate treatments.

In our egalitarian times, those hunting for and identifying certain types of genetic differences are rarely celebrated or rewarded. Geneticist Bruce Lahn published a couple of very carefully worded studies in *Science*.[23] They showed two gene variants that affect brain size commonly appear in Europeans but not Africans. Their conclusion was ultimately refuted with further research.[24] Lahn's original study unleashed a firestorm of debate and criticism by daring to suggest that the brain can evolve and vary across human populations.[25] Separately, researchers Gregory Cochran and Henry Harpending argued that Ashkenazi Jews carry not only gene variants for rare diseases (e.g., Tay Sachs), but also gene variants tied to higher IQs. Needless to say, they too have been crucified by fellow academics and the media.[26]

But despite criticism, fury, and misinterpretation, researchers continue to discover gene variants that differ among various ethnicities.[27] They've found that African Americans are four times more likely to suffer from end-stage kidney disease, perhaps because of the high prevalence of a noxious APOL1 gene variant.[28] This same variant provides some protection against sleeping sickness, which makes sense in evolutionary terms: Carrying the gene mutation enhanced survival and reproduction in places where a Tse Tse fly–transmitted disease prevailed, and the evolutionary fact that it also caused kidney failure later in life was irrelevant to their ancestors, who had lived long enough to reproduce.

Lahn still (cautiously) focuses on recognizing human diversity and has developed three key arguments: Promoting biological sameness in humans is illogical, possibly even dangerous; ignoring the possibility of group diversity is poor science and poor medicine; and a robust moral position embraces diversity as one of humanity's great assets.[29]

Out of fear that we have already lost crucial human genomic data, in 1991 Dr. Luigi Luca Cavalli-Sforza suggested that researchers cast as wide a net as possible in collecting human DNA so that the Human Genome Project would reflect all of humanity. He explicitly included various native tribes and isolated groups; soon he was being pilloried for "exploitation" and "stealing" instead of being celebrated for preserving what was rapidly assimilating and disappearing.[30]

Today, as we sequence, release, and analyze tens of thousands of human-genome files, many scientists will compare gene patterns, variants, alleles, to diseases and outcomes. But there are few acceptable frameworks or forums in which to present and discuss the differential results. We leave such a vacuum of information that when masses of data do become available, there won't be a scientifically and ethically sound framework with which to discuss or address the implications, allowing demagogues freedom to interpret the new data and use statistics to suit their own agendas. And tragically, we may seal off directions of inquiry that could improve treatments and medicines, enabling them to get to the right patient, at the right time, in the right dose.

Technically Life, Technically Death

············· :✳: ·············

The survival rates of prematurely born infants today have more to do with human desires, discoveries, and technology than with natural selection. In Darwin's world, very few preemie babies survived. In the United States, between 1990 and 2006, the number of late preemies increased by 20 percent.[1] Survival has increased at a staggering rate, and not just in the Anglo-European world; Qatar's rate of neonatal mortality fell 87 percent between 1975 and 2011.[2] Its preemie mortality rate fell 91 percent. Granted, Qatar's is an extreme case fueled by oil and gas discoveries and a rapid wealth increase, but in 2010 about 11 percent of all worldwide births were preemies (less than thirty-seven weeks). That's 15 million people in one year.[3]

The extremely young and fragile sometimes survive, including 17 percent of babies who weigh just over two pounds and are in the womb for only twenty-three weeks.[4] Sometimes even twenty-two-week-olds, a full eighteen weeks before term, survive. Without extreme human intervention, all of these children would die. So, from an evolutionary perspective, preemie survival rates are a great example of our unnatural, domesticated world—a world where human-driven selection dominates.

Saving so many, so young, is a blessing, but it can also have serious long-term consequences. In the UK, only 1 in 5 early preemies end up with normal health. The rest suffer moderate to severe mental and cognitive impairment by age six.[5] Even babies born one to three weeks short of term suffer more respiratory distress, hospitalizations, medical costs, and first-year mortality. Preemies typically need expensive care. In Finland, health-care costs for very preemie babies can be 4.4 times more than the cost of caring for full-term births in the fifth year of life.[6] There are also consequences for the couple; one study in the Netherlands determined that caring for seriously premature babies doubled the chances of a marriage dissolving and significantly increased the chances of becoming poor (even within a society with a large and generous social safety net).[7]

We continue to alter and, most believe, improve upon the traditional brutal weeding out of humans. We continue to lessen the constraints on "survival of the fittest dictated by natural selection." Instead, we select and explicitly advocate for the survival of as many humans as possible, including some of the most varied and vulnerable. We built our laws, moral codes, teachings, and institutions to fulfill this ideal.

The core idea, the operating principle, is no longer single-minded optimization and perfection but variety and multiculturalism. This is a human-driven desire and trait, not one traditionally practiced by the sharp knife of natural evolution. While completely against the principles of nature as described by Darwin, what we are doing is profoundly *humane*; it is our choice, our desire, our idea of what our population should look like and reflect that guides who survives and reproduces. And we have become all the better for this series of choices, unnatural and costly as they may be.

The extreme effects of unnatural selection, both good and bad, are also ever more apparent at the end of life. Thanks to cholesterol-lowering medications, hip replacements, and new knees, a healthy gramps can be around to coach his grandkids far longer. Many people enjoy a full decade or two or three of travel, swimming, cycling, or running that would have been inconceivable for their own grandparents, who saw no point in saving for retirement.

But sometimes practicing unnatural selection means we keep loved ones alive long after the body would have given up on its own. Yes, it's great to have Grandpa and Grandma around for a decade more, but sometimes we treat our elders in ways we would never dare treat our dogs. Many pet owners have gone through the absolutely wrenching experience of putting a long-term member of the family, usually a beloved dog or cat, "to sleep." Although upsetting for the family, it's not a brutal experience for the pet, who quickly falls into a coma and dies peacefully almost immediately thereafter, suffering over. Only the seriously disturbed would dream of forcing a dog to crawl around for years in ever more pain, and then artificially extending his life even further so he could suffer some more, and then intervening in more and more extreme ways such that his life and suffering continue on though there is no hope of recovery. And yet that's often what we do with Grandma. . . .

At some point we almost all will likely say, "I want to go quick, in my sleep, without pain." But today at the end stage of many diseases, where there is constant deterioration and no real hope of recovery, death doesn't come peacefully in one's sleep. Rather, it's prolonged with brutal suffering punctuated by "breakout pain"—pain that cannot be stopped or controlled by most existing medications. Even then many families and doctors seem congenitally indisposed to recognize the natural end of life. Many hospitals, seemingly incentivized to increase their revenue by doing "everything possible," whether it really helps the patient or not, continue an undignified assault on the body, bringing an agonizing onslaught of useless procedures and stacking up bills for naught. We spend 17 percent of the $550 billion annual Medicare budget on the last year of life, and much of this spending does not lead to good or peaceful deaths.[8] Doctors know this (though they readily point out that no one is ever really certain when a patient's final twelve months actually begin).

While doctors often don't talk to their patients about end-of-life options, they do know what they want for themselves; 64 percent of U.S. doctors have written out explicit instructions for their own end-of-life care. The vast majority want pain meds but would refuse CPR, ventilation, dialysis, or chemo. But when it comes to patients, despite all the evidence, doctors and families usually overestimate the chances of making someone healthy again or being able to provide a comfortable and dignified end. So doctors are often doing things for their patients and their families that they would never do for themselves because they have seen firsthand what the effects are, what it costs both personally and financially, and what the outcome is.

As we deploy ever more extreme technologies that keep bodies alive unnaturally, a growing percentage of the population is asking for death with dignity. Whereas 36 percent supported a physician "ending a patient's life by some painless means" in 1950, in 2006 this concept had 69 percent support.[9] Ironically, the same types of culture wars fought over birth control and abortion are also playing out at the end of life, and even the words ring similar; the motto used by Compassion and Choices, a death-with-dignity organization, states, "My life. My death. My choice." Their argument is that everyone should be free to follow their own religious and moral beliefs, but one should not have the right to force their beliefs on other mentally competent adults who want death with dignity.

The "pro-choice" controversy, both at the beginning and at the end of life, tragically intermingled in November 2013, when Marlise Muñoz collapsed and her brain died. Being a paramedic, knowing such things can occur, she had previously prepared a living will. Her husband knew, and her lawyer knew, that she did not want to be kept alive artificially. Muñoz, of all people, knew what she was doing and asking for, having witnessed firsthand the consequences and costs of extreme resuscitation measures, as had her husband, also a paramedic and firefighter. Both were competent adults, with clear directives. But Muñoz's accident happened in Texas, and as it turns out, in Texas others can snatch control of your body, especially if you are pregnant—and Muñoz was.[10]

The hospital continued to keep Muñoz artificially alive, against her own and her family's explicit wishes. Her fetus, fourteen weeks old, had likely already suffered irreparable brain damage because the mother's pulmonary embolism meant neither Muñoz nor the fetus got any oxygen for close to twenty minutes. No matter—the operative words were "pregnant mother," not "viable child." Against Muñoz's will, her baby was forced to remain in a brain-dead incubator for twenty-six weeks while the hospital was sued for the "cruel and obscene mutilation of a corpse," until the court finally recognized that the baby had malformed and that the mother should be taken off life support.

Then the John Peter Smith Hospital thought it would be a good idea to send Mr. Muñoz a bill for over $300,000 to pay for unwanted and unsought procedures. Just to add insult to injury, some major Texas gubernatorial candidates began to bleat that they should tighten their laws so this could never happen again; with the lieutenant governor leading the charge, these men were seeking to use their power to ensure that no woman is ever taken off life support for any reason if she is pregnant. Too much medical technology meets too little common sense, with a huge dash of testosterone arrogance.

Moving away from nature, deciding for ourselves when it is time to reproduce, survive, or die can have significant consequences; sometimes there's a major clash between what we do and what we should do, or between what's natural and what's possible. Unnatural selection at both ends of the life spectrum can lead to choices none of us would want to face but that we will need to openly discuss, because others are often deciding who lives and for how long, even when we have made a different decision. Meanwhile, more and more bodies, of the very young, and very old, are warehoused in hospitals in completely unnatural states.

Trust Whom?

.... :✱: ....

The questions driven by and the powers granted by our increasing ability to decode, encode, and engineer life forms will challenge religions, corporations, and governments. Each will seek to earn our trust and help guide us as we face some thorny issues. In turn, each of these core institutions will also need to adapt and evolve existing dogma; each will have to re-earn its legitimacy to provide guidance on some truly complex options and dilemmas.

Globally, when faced with the most basic existential life questions and moral dilemmas, 4 out of every 5 people turn to priests for answers.[1] Many practitioners of religion argue that the answers are already there, that beliefs and scriptures came directly from the wisdom of a particular God, which implies an absolute and unchanging word and truth. But few successful religions are unchanging doctrines; the smart ones, those that survive a long time, evolve.

Abraham's word speciated into Christianity, modern Judaism, and Islam. Within each of these branches, there exists a great deal of further sub-speciation; Christianity begat Russian Orthodox, Greek Orthodox, Roman Catholic, and Protestant. Eventually each of these branches in turn sub-speciated; Roman Catholics can follow Jesuits, Dominicans,

Franciscans, and many other cassocks. A few centuries later a large branch of "heretic" Protestants also began to sub-speciate; now there are at least forty-two varieties of Baptists. The process continues with some Catholics opting to cluster in Opus Dei, the Legionaries, and, as of 2004, the Sisters Adorers of the Royal Heart of Jesus Christ Sovereign Priest. A similar process occurred with Judaism speciating into Orthodox, Reform, Conservative, Hasid, and Kabbalah. And again with Islam branching into Sunni, Shi'a, Ibadi, Sufi, Ahmadiyya . . .

The emergence of a new religion, or a major branch, typically begins with a "divine intervention": Abraham bringing down the Word; God sending his only son; Allah speaking to the Prophet Muhammad. Soon thereafter the interpretation and application of God's commandments becomes a long and constant process of adapting to changing political, ethical, and moral beliefs, as well as scientific discoveries. Good process: Otherwise, if we were to take the literal words of the Bible to heart, we would now be painfully and slowly killing all stubborn and rebellious sons, all who practice premarital sex, those who touch Mount Sinai, and anyone who curses.[2]

Morals and ethics have to adapt to changing circumstances. For example, why might Americans and Europeans react so differently to extramarital affairs? In Europe, for a significant portion of the twentieth century, they were almost blasé about them, while in the United States they often led to divorce. At least part of the answer may be the much higher numbers of young European men killed during the two world wars. In 1917 in England, the senior mistress of the Bournemouth High School for Girls told her sixth-form class that only 1 in 10 could ever hope to find a man to marry.[3] With few eligible men available after two great wars, the reduced probability that women would have an exclusive, age-appropriate partner changed society's understanding of what was acceptable. In early Islam, during periods of enormous violence, multiple wives may have been a way for men to rapidly spread the religion and ensure powerful allies.[4] Necessity may be the mother of invention, but it can also be the mother of acceptance.

Religions provide a common belief and rules system for large populations to organize behaviors and attitudes, and they attempt to improve their followers' fates. A people's health can suddenly improve, even in the midst of plagues and epidemics, because they adopt a new religion and new customs that require washing hands, face, and feet five times a day. Both Muslims and Jews, in an era of trichinosis and swine flu, decreed pork dirty and to be avoided. Christians often drank wine instead of contaminated water during times of little sanitation. Many of these customs and rules led to each population feeling a little more protected vis-à-vis its "unclean" heathen neighbors.

Increasingly, when religious beliefs conflict directly with new science, there has to be some evolution in belief or at least part of that faith may erode. As the cheeky astrophysicist and science popularizer Neil deGrasse Tyson likes to say: Science is true whether you believe in it or not.[5] When there is a clear geological, fossil, and molecular biology trail leading back billions of years, it gets ever harder to argue that the world is but 7,000 years old. Not that many don't try to argue this very thing, but it puts them in conflict with overwhelming and mounting evidence. Which is precisely the type of behavior that eventually leads to educated people abandoning irrelevant and outdated religions. After all, most gods, most religions, have gone extinct; much of what the world so reverently worships is but a few thousand years old.

As we face truly difficult choices on reproduction, altering humans, cloning, longevity, speciation, and myriad other challenges created by nonrandom mutations, some religions will adapt, learn, and remain relevant guides and actors. Others will ossify in fear and lack of knowledge and understanding. What is clear is that life sciences can and will fundamentally challenge, transform, and evolve many a religion.

So how will increasingly secular societies cope with the tremendous conundrums and opportunities presented by our newfound abilities to guide evolution? Many will still turn to their pastors for guidance. Many others, who absorbed and incorporated many of the core beliefs common to various religions and secularized them, will no longer turn to a "Higher Power" to decide, but they will debate and reexamine their moral codes with their friends and community.[6] Life science forces us to think about complex issues, including reincarnation, extreme longevity, and self-stored versus outside-stored memories. As we grow to realize how many beliefs and assumptions are about to be challenged by the reality of what we can do, it is not just religions that will be challenged.

Other big institutions are also under tremendous pressure to answer, lead, explain, and control powerful technologies. One group desperate to earn your trust, and your dollars, is composed of corporations. Many individuals instinctively react, "I would never, ever trust any corporation with my data." Their answer: You already do.

There is good reason to worry; our genome, health, lifestyle, and habitual data are among the most personal stuff we have. How many calories you consume, with whom, and when; whether you exercise or not; what medicines you take; how intimate you are with someone else . . . But in a big-data world a whole lot of folks know a whole lot about you: your doctor appointments, specialists, Internet searches, prescriptions. Beyond health care, in the world of big data we leave behind an extraordinary trail of digital exhaust through toll booths, pharmacies, supermarkets, airline loyalty cards, Facebook, Twitter, blogs, Yelp! and TripAdvisor reviews (not to mention credit card and loyalty programs). Those who are interested can then easily begin to infer your lifestyle and with this lifestyle profile make a whole lot of statistically accurate predictions about your health and habits. Company after company wants to obtain, interpret, and use your data and help you shape your future.

Determining which companies we trust with what information is going to cause an interesting debate. We can now know so much about individuals, down to the molecular level, and we can know so much about their families, that anonymity is getting close to impossible. The limits we place on the use of this data, even if supposedly "anonymized," will ebb and flow. So far, Europe seems far more concerned about protecting the individual's information and reputation than does the United States. Recent directives in the EU, such as the right to "erase" part of your past life from search engines, tip the scale toward individual choice and away from corporations.[7]

And then there are governments . . . So far, many governments' default advice, when faced with the question of how you should handle any personal health-care or genetics data: Trust almost no one, just us bureaucrats. Strict HIPAA compliance laws can cost anyone who releases personalized medical information to unauthorized parties up to $50,000 per incident, plus criminal penalties.[8] These restrictions make it very hard for patients themselves, never mind researchers, to access and share their own medical records. The default answer and standard is "no sharing." But such a broad policy can stigmatize many diseases, be very costly for society as a whole, and harm individual patients.

Taking the extreme opposite position on privacy, vis-à-vis the U.S. government, Jamie Heywood reminds many people of the Tasmanian Devil cartoon: a perpetual tornado of movement, great ingenuity, occasional destruction, and enormous creativity. He described himself as a serious engineer—a nerd, in fact—before his brother was diagnosed with ALS. As of that day, Jamie changed his life's focus and mission: He would tear down the whole of the medical establishment to help find a cure.

While learning all the facts about ALS, Jamie also took on medical secrecy. In his view, the only rational way to combat a deadly disease with few answers is to share everything, with everyone. Your bowels were really loose after you took medicine X? Don't just tell me, don't just tell your doctor, tell everyone. Share it with the entire world. Create very large databases of information from hundreds of thousands of patients containing every last detail of their lives so that every variable, each step in a disease process, can be teased out.

In a sense, Jamie and the team at PatientsLikeMe are emulating one of the single most successful ongoing clinical studies in the world, the Framingham Heart Study.[9] In 1948, doctors recruited 5,209 people and began recording every aspect of their lives they could think of, including diet, exercise, stress, blood work, and family history. They continued to do so every two years, from 1948 through the present. In 1971, they added a further 5,124 patients, primarily the kids and spouses of the original group. In 2003, they included grandchildren. Consistency and scale led to remarkable discoveries on disease causation and risk factors: tying cigarette smoking to heart disease (1960), high cholesterol and blood pressure to heart disease (1961), high blood pressure to stroke (1970), the benefits of HDL cholesterol, and more.[10]

But there's one big difference between Jamie's PatientsLikeMe database and the Framingham research. Jamie asked everyone to voluntarily provide the many intimate details of their life and health, not just to researchers but also openly to each other and even the world, often with their real names. To date, 250,000 people, principally those with a neurological disease, have agreed to have any researcher or other patient contact them, ask them questions, or share with them. In addition to information on disease progression and clinical trials, the data includes questions like, How do you feel? What is your quality of life? What really helps you feel better? How much sex do you have? All the questions a modern doctor rarely has the time to ask, much less collate and address, but that are essential to researchers and can help patients in real time.[11]

Not everyone wants to share all of their teenage or adult peccadillos with everyone else, due to embarrassment or fear that if they share, they may lose a job or their insurance. But keeping all data in the dark, with only the government to oversee, protect, and determine exactly who sees what and when, may not be the best answer. As we accumulate masses of data far and wide, patterns emerge; Harvard bioinformatician Isaac Kohane ran a retrospective patient study that correlated the use of a particular drug to a much higher incidence of heart attacks.[12] Soon the drug was off the market, not because of an FDA review but because the transparency and public accountability brought about by Kohane's study forced action; many lives were saved by transparency. Having huge companies know what the government knows but promising to "keep it all confidential" while simultaneously putting the huge databases of information to work for their own purposes may not be the best answer either.

In the end, the imperfect but perhaps least costly answer may be to accept less and less privacy; for Jamie, openness is more democratic. As databases grow, even if we wish to keep total privacy it will be harder and harder to do so as our relatives and neighbors release digital exhaust or genetic test results. Disease after disease that was rarely mentioned in polite company is already a topic of everyday discussion thanks to brave leadership by people like Betty Ford (substance abuse), Kitty Dukakis (depression), and Magic Johnson (AIDS). We are all better off because the unmentionable is now public and far more transparent. Shame and disease are not words that belong together.

Transparency, accountability, and evidence—a willingness to say yes I did this or decided that—in public may become a better standard for priests, patients, bureaucrats, entrepreneurs, and corporate types alike. We need more and broader discussion, not more privacy and fewer people making decisions. The shy and embarrassed should be afforded protection, but establishing robust, large, parallel, opt-in, open data sets will demystify and destigmatize. In the end we need to consider more ways to trust ourselves and our communities.

THE FUTURE OF LIFE

. . . .

Throughout this book, we have tried to take some really complex science and provide enough to tell the story, but not make it unintelligible.[1] Doubtless there will be folks who will say we should have qualified X more, explained Y better, noted Z was just one study or that a particular article we cite has been superseded. We could have added a lot more supporting evidence, and detailed a lot of the emerging science with further caveats and qualifiers. We are just at the first stages of discovery in how life code works, how it can be deployed. A lot will change. But here is what is *not* going to change soon . . .

Humans increasingly tip the balance of evolution away from what nature would dictate and toward what we want and decide. As long as humans are alive, unnatural selection and nonrandom mutation are here to stay. The quiver of instruments we have created to redesign and drive fast evolution is so powerful, effective, and dominant that we are not going to give them up, or even curb them much. Any country that does so on a large scale risks losing races in health care, longevity, agriculture, industrial production, education, information storage, and many other fields. It would be the equivalent of a nation suddenly giving up all computers and electronics and going back to pen and paper. (This does not mean that for specific moral, religious, or political reasons, particular technologies won't be banned or limited in some countries.)

We will continue evolving bacteria, plants, animals, and ourselves to our particular desires. So now is the time to ask: Having put ourselves in charge of our own evolution and that of other species, what will we choose to do with this extraordinary power?

I Don't Remember You ... De-Extinction

.... :✳: ....

Not that we are all on our way to becoming saints, but one of the first results of applying the laws of unnatural selection and non-random mutation is re-creating, time and again, one of the great Christian miracles: resurrection.

Darwin recognized the fundamental role of extinction. It's simply the way it is, the way it has been for all time: The natural world thrives on creative and cruel destruction. Nature has so very many ways to kill off whole species completely; it occurs over and over. And once it happens, there's no reasonable scenario under which natural selection and random mutation would ever lead to the revival of an extinct species. The possibilities of recombination, changes in the environment, food, shelter, predators, and changes to all four genomes would likely lead to the emergence of a different species, not a carbon copy of a predecessor.

Humans have accelerated the extinction of many a beast; we eliminated and modified whole species because they scared, bothered, amused, enthralled, fed, clothed, or decorated us, or in some other way served our purposes or annoyed us. Over just the past five hundred years, human activities led to the extinction of 869 major species, including auks and dodos.[1]

But now we may be about to reverse this trend; one of the strangest consequences of being able to read, write, and reprogram life code may be the ability to run evolution in reverse. Not only will we create new life forms, we will re-create old ones as well. Nonrandom mutations and intelligent design make wholesale resurrection a near certainty. Just ask iconoclast, rebel, mapmaker, and all-around rabble-rouser Stewart Brand; he was one of the original '60s rebels, one of the Merry Pranksters, along with Ken Kesey, who was in on the Electric Kool-Aid Acid Test described by Tom Wolfe.[2] He was an early explorer of better living through chemistry. When you meet Brand, he comes across as the most unlikely of radicals.[3] A distinctly mature, very smart, quiet man with a twinkle in his eye—he doesn't boast. Often he just listens. So you may have a nice chat, leave, and have no idea you just interacted with one of the most forward-thinking minds on the planet. He birthed the *Whole Earth Catalog*, one of the precursors and catalysts of the environmental and green movements. But not even his craziest trips could compare to what Brand attempts today: a large-scale reversal of diverse extinctions.

Now in his seventies, Brand is a strong advocate of biotech and with his wife, Ryan Phelan, recruited a coalition of the world's top life scientists to launch a "de-extinction" movement. In his new paradigm of Conservation 2.0, you not only preserve and protect that which you haven't managed to kill off, you also redress past tragedies. This "Revive and Restore" strategy births a new field known as resurrection biology, which is full of new rules and options that would surprise, and perhaps delight, Darwin.

Brand's new passion builds on and accelerates a movement that has already produced surprising results: Alberto Fernández Arias briefly brought back an extinct Spanish breed of ibex, bucardos, using cryo-preserved tissue. Australia's Michael Archer revived an early-stage embryo of the extinct gastric brooding frog. How could these feats be possible with no living specimens? Through the retrieval and copying or cloning of DNA from frozen specimens. Another way is knowing the current and past DNA makeup of the extinct and their descendants, systematically back-breeding and modifying, which is what the Netherlands' Henri Kerkdijk-Otten, an activist/farmer/curator, is attempting in his quest to bring back the European auroch, an animal last seen in 1627. Meanwhile, Harvard's George Church edits wholesale the genomes of band-tailed pigeons to reverse-engineer the extinct passenger pigeon, a creature once so common it used to blacken the skies for North American observers for hours, sometimes days. William Powell applies similar techniques to try to bring back the nearly extinct American chestnut tree.

Thanks to people like Oliver Ryder, who spent a lifetime creating one of the world's greatest frozen-DNA "zoos" in San Diego, there are samples from more than 1,000 threatened and near-extinct species to play with, revive, and restore. For older extinct creatures, such as woolly mammoths, there's the all-natural frozen zoo of Siberia. What the permafrost preserves is astonishing.[4] In early 2013, a Japanese science expedition discovered an almost perfectly preserved woolly mammoth. Not only was the hair preserved, so was the blood in its veins (Note that this discovery was only possible because we drove mammoths to extinction so very recently; they were alive when Egyptians were building pyramids.)[5] While the skeleton ended up in a Japanese science center where visiting children can "touch the hair," the blood went to South Korea to a private biolab that hopes to implant an elephant egg with the mammoth's DNA, though most scientists are highly skeptical.

As the world warms, the remains of creatures such as steppe lions, woolly rhinos, and giant deer surface from melting ice with growing frequency. When scientists and explorers can get to remains before wolves and carrion feeders, they can preserve samples of ancient DNA. One expedition recently extracted the DNA of a 700,000-year-old stallion known as Thistle Creek from the Yukon permafrost, where its contemporary ancestors roamed beginning about 4 million years ago. Thistle Creek's progeny evolved into modern-day horses, donkeys, and zebras, which disappeared from the American continent about 7,600 years ago, only to be reintroduced to the Americas once again with the arrival of Columbus.[6]

But what about those species so long extinct that there's little chance of finding them in the permafrost? Bringing back a *T. rex* remains the stuff of *Jurassic Park* melodrama, but as our ability to sequence, read, and reassemble DNA from fossil specimens grows, even reviving long-extinct species may become possible. DNA-assembly technologies are advancing so quickly that, with infinite patience, scientists are able to reassemble a whole genome even out of minuscule quantities of severely decomposed fragments of DNA. Think of this like trying to rebuild a fallen brick wall from a pile of bricks that look similar but are nonidentical. By stacking bricks that have matching overlapping features on top of each other, eventually you can scaffold the pile of bricks into the original long wall. The wall of bricks represents the reassembled end-to-end genome sequence of the long-lost specimen. But with DNA the assembly process is done with a computer.

UC Santa Cruz's Beth Shapiro and McMaster University's Hendrik Poinar have taken every fossil and DNA specimen they can get hold of and slowly built up a large database of genomes from long-extinct species. Once one has a reasonable computer-rendered facsimile of the original genetic code of a species long extinct, one can begin to build a road map of how today's descendants of that creature evolved. And then one can map the gradual genetic transitions that occurred across generations. Having a general sense of the basic underlying gene code that led from an ancient extinct species to today's descendants provides a blueprint of what once was.

Eventually this database of ancient DNA codes will provide clues as to what types of surrogates need to be bred to revive creatures that have not walked the earth for hundreds of thousands of years. One might devolve an existing species by understanding what the creature's DNA looks like today and what DNA code it evolved from; remove one gene, add another, slightly modify a third, silence a fourth . . . until one gets an approximation of the original creature's DNA. (Think of this as walking back from the edges of tree branches toward the trunk on an evolutionary tree.) For instance, if you know that a snake's ancestors once upon a time had limbs, and you insert or modify a few genes, you could re-create snakelike creatures that walk (imagine letting one of *those* loose around Halloween).

In a world where we begin to control evolution, the arrow of time can run backward, or in many different directions. We know that birds are descended from dinosaurs, and it's just conceivable that by deleting or modifying certain genes from today's chickens, one might be able to advance quite a long way toward re-creating the genetics of a dinosaur. Along this path we may produce hens with teeth, tails, and scales. (And to quote a wonderful *Wired* magazine cover on de-evolution . . . "What Could Possibly Go Wrong?")

All of which brings us back to human evolution. The same technologies being developed to revive extinct animals might be used to push our own evolution backward so as to revive and restore various extinct hominins. After all, the difference between humans and Neanderthals is a minuscule 0.2 percent of total human gene code.[7] We have the sequence in hand, so it should eventually be possible to synthesize a Neanderthal genome. And a recent analysis explored the Neanderthal and Denisovan epigenomes as well.[8] And why stop there? As we map out the gene code of various other ancestors, we might be able to bring them back as well and have substantial sections of the human evolutionary tree walking around at the same time.[9] There would have to be specific tweaks to a few genes, like FOXP2.[10] A single-letter/DNA base change in this particular gene in humans can lower IQ and lead to the loss of language. While mice and humans share the gene, there are only three amino-acid differences between these species, and one amino-acid difference between humans and chimps.[11] If one were to alter the FOXP2 in a monkey so as to "humanize" it, the change could modify a variety of brain-related traits that separate humans from apes, and one might conceivably begin to build up part of a missing link in the evolutionary chain between apes and hominins.[12] Or by extension, one could transfer traits from humans to other species and vice versa.

Diversity has helped our survival. While Tibetans are well adapted to living in a high, cold, oxygen-poor environment, few Han Chinese are. At higher elevations, infant mortality is far higher for the Han, among other things because their babies are not getting as much oxygen.[13] At these elevations Han babies die three times more often than Tibetan babies.[14] In 2014, detailed comparisons of genomes from both groups led to some surprising and extraordinary results. Tibetans are now thought to be unique among living humans because they carry a particular variant of the EPAS1 gene, a variant that likely came from interbreeding with a long-extinct hominin known as Denisovan.[15] This gene variant enhances the oxygen-carrying capacity of the blood.

Using unnatural means to revive and restore is simply running back the evolutionary movie to an earlier, more "natural" state. But if we revive lost cousins, we would face myriad complex and interesting questions, including what rights and legal responsibilities one would ascribe to and demand from various hominins: Does it matter how evolved they are? Where would they live and under what conditions? And if we beget more evolved hominins, how do we want them to treat us? Though the legal, ethical, and moral obstacles are daunting, we should remember that historically it is normal and natural to have multiple varieties of hominins walking around at the same time. Do we really want to bet the future of all hominins on the gene code of one particular species?

Humanity's Really Short Story

.... :✳: ....

Before we consider where we are going, let's stop and consider how we got here. About 14 billion years ago, a single minute point of unimaginably concentrated energy fluttered, and caused the greatest explosion we can conceive of, ever. As plasma spread and began to form enormous clouds of dust, a few clumps of dust reached critical mass and gradually gravity took over.[1] Immense quantities of dust began to compress until atoms began to fuse, which in turn ignited thermonuclear reactions, giving birth to stars.[2] Trillions of new stars formed galaxies in wondrous shapes and sizes.[3]

We live in the Milky Way, a relatively ancient galaxy; some of its stars are almost as old as the universe (more than 13 billion years old).[4] Within our galaxy lie 200 to 400 billion stars. (A mere uncertainty level of +/-200 billion stars, just within our own galaxy, gives you a small sense of how very ignorant we still are about the numbers.) Astronomers currently estimate there are 80 billion galaxies containing 30 to 70 sextillion stars.

Within the billions of planets and moons in our galaxy lies our sun, a teenager at a mere 4.57 billion years old. Leftover floating chunks of matter from the great explosion formed a few planets.

Including a tiny planet called Earth, 4.4 billion years ago. That's right: For more than two-thirds of the universe's history there was no Earth.

Relatively soon after Earth was created, it was crawling with life.[5] Then almost all life went through at least five major cycles of extinction. The Permian-Triassic cycle alone eliminated about 83 percent of all genera on Earth.[6] (Talk about extreme spring cleaning!) It was only after the last major extinction, the Cretaceous-Tertiary about 65 million years ago, that mammals, and then humanoids, gradually and tentatively began to spread.

Put this all in context: 99.96 percent of the entire history of the universe took place before the first hominins, never mind the first humans, showed up. Then, after the rise and fall of at least twenty-five proto-humanoids, we, *Homo sapiens*, somehow escaped almost certain extinction, survived, and thrived.

Within this overall picture, do you really believe that we, the self-named *Homo sapiens*, are the be-all and end-all of the entirety of evolution? In other words, do you think the entire reason and purpose of the past 14 billion years of the known universe . . . and the sole purpose of 4.5 billion years of Earth's history . . . and the sole purpose of 4 billion years of life's evolution . . . and at least five cycles of extinction . . . was to create the likes of us?

Our story, our purpose of being, continues to unravel and evolve. When Darwin was around, the only humanlike fossils available for study were a few lonely Neanderthals. Now we know we branched off from chimps and bonobos about 6 million years ago, and eventually evolved into *Ardipithecus ramidus* about 4.4 million years ago. By 2009 we had more than 110 specimens detailing a few aspects of *A. ramidus*'s life and habits. Then the sexier, less monkeylike but still unibrowed *Australopithecus* appeared around 3.7 million years ago.[7] He began using tools to butcher and build more than 2.6 million years ago (thereby genetically conditioning subsequent descendants with an innate desire to troll around Home Depot). But despite the burst of recent discoveries of ancestors, potential ancestors, and kissing cousins, we are just beginning to understand fossil genomics and

DNA-encoded genetic trees well enough to begin to really understand who did what to whom, when, and what that led to.

None of the dozens of other ancestral *Homo* species made it. *Homo sapiens* itself came within a few thousand individuals of complete extinction.[8] But unlike all other versions of humanoids, we survived. And over time, we began to take some control over what nature could do to us, how it could "naturally select" us. Bit by bit we began, for better and for worse, to master more and more of our environment and guide it toward our own purposes. And now we have changed the nature of our world to such an extent, and developed such profound capabilities for recrafting our bodies and environment, that we are birthing our successor species.

Evolving Hominins . . .

.... :✳: ....

One of the greatest evolutionary experts, Ernst Mayr, concludes his 1964 book, *What Evolution Is*, by arguing: "What is the probability that the human species will break up into several species? The answer is clear, none at all."[1] He based this conclusion on two criteria: We occupy every geographical niche on Earth, Arctic through tropics. And there are no truly isolated human populations.

Most people believe there's only one human species. And that there has always been only one human species. The first assertion is still likely true today; the second is demonstrably false.[2] At least three of the "other species" were alive when our ancestors emerged from Africa 50,000 to 80,000 years ago: Denisovans, Hobbits, and Neanderthals.[3] Many of today's Asians share Denisovan ancestry, as do modern Tibetans and Spaniards.[4] One big global family.[5] For better or worse, humans have evolved quite a bit.[6] We know of dozens of prototypes who represent the absolute pinnacles of human speciation, such as Elvis, Michael Jackson, and the Kardashians. (That's it? Is it over? Have we really stopped evolving?)

If having multiple hominins running around at the same time has been the historical norm, then there is, of course, a corollary: It's historically highly unusual, and quite rare, for there to be a single solitary hominin species on the planet. This is hard to picture, given how dominant we now are, so try a thought experiment. . . . Imagine a single species of bird, everywhere. Just woodpeckers everywhere and nothing else—no robins, hummingbirds, finches, cardinals, peacocks, eagles, crows, parrots, toucans, vultures, flamingos, hawks, pigeons, ducks, canaries, geese, owls, sparrows, swans, pelicans, bluebirds, warblers, not even pesky seagulls. That would be weird, no?

So why believe it's normal and natural for there to be only one single species of *Homo* alive today? This counters all of evolutionary history and the fossil record. Not having multiple closely related species tends to make an entire evolutionary branch far more vulnerable to extinction. (In fact, that's almost what happened to us less than 100,000 years ago when it is estimated that at one point only two thousand humans were alive in the world.)[7]

Which brings us back to this troubling question of separate human species. . . . The more we find out, the more similar many hominins seem. Neanderthals used fire, ate a lot of meat, used spears, buried their dead, cared for the sick, made great art.[8] Until just a few years ago, we thought we were really beginning to understand Darwin's evolutionary tree for humans. As far as we were concerned, Neanderthals were a separate but coexisting species with *H. sapiens* and were not a precursor to humans. Little did we imagine that Neanderthals were truly our kissing cousins.[9] Using a molecular clock, we can guesstimate when humans last had sex with Neanderthals—some 50,000 to 60,000 years ago.[10] (It was cold and dark, there was a cave, maybe some booze . . .) In 2013, the story got even more complex. A bone found in Spain shocked researchers, first because it put the existence of early sentient hominins back another 100,000 years, and second because its DNA showed it was part Neanderthal and part Denisovan (the latter of whom were thought to reside only in Russia and Asia).[11]

So you may think your weird cousin is a Neanderthal, and you are partially right.[12] If you have European or Asian ancestors, even if you do not live inside the Washington Beltway, 1 to 3 percent of your genomic DNA is inherited from a Neanderthal who lived 50,000 years ago. (Steve took a basic consumer gene test and found that his DNA is 2.9 percent derived from a Neanderthal.) But evolutionary geneticists found that our modern X and Y chromosomes are devoid of actual Neanderthal ancestry.[13] This provides a genetic fig-leaf rationale for why we are a different species.[14] Phew!

Darwin might be surprised by the revelation about interbreeding (never mind Bishop Wilberforce, Darwin's nemesis in the evolution debate), but he would nonetheless predict that the 1 to 3 percent of today's non-African DNA that came from Neanderthals must have conferred a survival or reproductive benefit. Indeed, Neanderthal DNA had an influence on skin pigmentation, neurons, immunity, vision, brown-fat metabolism, and olfaction.[15]

DNA mutations can identify historical patterns of human migration.[16] Based on a genetic analysis of his Y-chromosome DNA, Steve discovered that his paternal ancestors lived in China about 20,000 years ago and from there they migrated into Siberia, western Russia, and eventually Scandinavia, landing in Finland, where they ultimately begat 60 percent of today's population. (How would *you* like to have a coauthor disguised as a mild-mannered Finn with the dark heart of a Hun-Neanderthal?) Language experts confirm these genetic migrations, finding the origins of the Finnish language in eastern Russia. (And while we are on the topic of ancestry, someone should tell the nice ladies who are Daughters of the American Revolution, and descendants of those who came to America on the *Mayflower*, that, so far, the only relatively "pure" race of *Homo sapiens* is . . . African.)[17]

Humans can and do diversify, recombine, speciate. There are significant, identifiable differences among us; for example, we can't just randomly donate blood to accident victims. If you transfuse the wrong blood type, you can cause an acute hemolytic reaction, in which the immune system attacks the new red blood cells.[18] The result is particularly bad when A-type blood enters a type-O person. As new blood cells break apart, the recipient's urine turns red and before long the kidneys fail, often resulting in death.

We're not suggesting that type-O people are "a different species" from type A and so on. The point is to highlight again how fundamental biological changes can emerge and spread even within one species. A/B/O blood types vary from country to country and region to region. The most prevalent type in Iceland is O+ (46.7 percent), Norway is A- (42.5 percent), India is B+ (30.9).[19] Why such extreme differences? And why does anyone have a non-O type if it increases your risk of arterial and venous thromboembolisms? One reason for these relatively recent mutations (20,000 years ago) may be that they produce thicker blood that reduces the risk of hemorrhages and infections. On the other hand, people with type-O blood are also more susceptible to severe cholera, so it's far less common in India and Bangladesh. Some people with type-A blood may have some protection from ulcers and stomach cancer, likely because they are resilient to *H. pylori*. In other words we continuously evolve, our blood types adapt to genetic accidents, weather, culture, different diseases, and environments. (We still don't know what might have led to the B type.)

Blood diversity sometimes intrudes even in the most intimate of bonds; in Europe, far more so than in Asia and Africa, a mother's immune system attacks her baby. This often occurs when a mother has type Rh- blood and father has Rh+. (This affects about 13 percent of all European pregnancies.) The more pregnancies, the greater the likelihood of a mother's quasi-allergic reaction to the fetus.[20] It's why prospective fathers should not donate blood to their partners during childbearing years.[21]

Many kinds of changes begin to occur, and in some cases are physically visible and obvious, long before formal speciation. Consider that Chihuahuas and Great Danes are technically the same species. Speciation can occur as a gradual continuum, or sometimes as a sudden, punctuated break.[22] It all depends on whether geographic, environmental, physical, disease, religious, and/or cultural changes are the primary driver. (Until eventually you can't, won't, or shouldn't have kids with the Other.)

As we consider human speciation, it's important to also consider gene mutations that give a subgroup in a specific environmental niche a selective advantage, akin to the beaks of Darwin's finches in the Galápagos Islands. By comparing Asians to Africans to Europeans to South Americans to Inuit and so on, scientists are combing human genomes to find specific beneficial mutations, which in some cases are regionally focused. For instance, had you been wandering around Spain around 7,000 years ago looking for tapas, you might have run across a strange chap, one you no longer see in Europe, one with dark African skin and deep-blue eyes. If you had found a common language and been able to go out for a meal together, his diet and yours would vary substantially. Fettuccini Alfredo would have been a no-go, given that this ancestor could not digest starch or milk because he did not have the appropriately mutated enzymes.[23] Overcoming lactose intolerance is a recent beneficial mutation that swept across Europe with the advent of domesticated goats and cows. Similarly, a mutation of the HERC2 gene for blue eyes arose in the Baltic about 10,000 years ago.[24] Various genes account for fair skin, blond hair, and straight, thick hair as well as altered skin function and disease resistance in various Asian populations.[25]

Any speculation on future human speciation generates much sound and fury. But it's going on. Not because there's some Dr. No hidden away on a Caribbean island plotting the end of *H. sapiens* but because of two overwhelming forces: unnatural selection and the emergence of nonrandom mutations. We are massively altering our environment, which is altering our four genomes across generations. We are also taking control of our genetic code. In addition, we are creating enhancements for human bodies that some people will eventually decide they cannot live without and that could become integral to our survival or reproduction in the future. We are discovering so much so quickly in so many fields.[26] Let us stress . . . there's not one lab, one experiment, or one technique that leads directly and deliberately to speciation. It's not one technology, government, company, region, or discipline that's driving change in our species. Discovery is widespread and decentralized. An avalanche of modifications, changes both small and large, accumulate and accumulate until we are no longer the same.

We are already in a period of extreme disequilibrium. Over the past 10,000 years human evolution occurred ten to a hundred times faster than at any other time in our species history.[27] Part of this is a sheer numbers game. If the average child each carries 100 DNA mutations, how often some rare mutation occurs, whether beneficial or disastrous, depends partly on how many people there are. About 1.2 million years ago before humans there were only about 18,500 total hominins alive on the entire planet. They were more endangered than today's gorillas.[28] That implies that our ancestors had a couple of million total mutations to recombine. Today there are about 7 billion folks on the planet, which would represent potentially 700 billion mutations. Since the human genome is only 3.2 billion bases long, this means that statistically every G, A, T, and C in the genome has been mutated in someone, and if the mutation is not lethal, someone alive today probably has it. So, once we sequence enough people's genomes, we will have a full catalog of all individual mutations.

Perhaps one way to envision the new normal in genetics is through a talk that Del Harvey, Twitter's head of Trust and Safety, did for TED.[29] In January 2014, users tweeted more than 500 million times per day, which means that the one-in-a-million outlier happens five hundred times per day. When this is applied to life code, the chances of finding the outliers that might allow one to survive under very strange conditions become far more common. Very few researchers are deliberately trying to build new hominins, but as they search for cures to various diseases, they are finding a lot of variants in the genes that regulate our ability to eat, smell, and reproduce. Each of these discoveries is a genetic raindrop. Raindrops build rivers, lakes, and oceans. . . .

New genetic-engineering instruments turbocharge rapid evolution. Now with the right editing tools—say, a genome assembler plus CRISPR—one could individually silence any human gene to see which one gene or combination of genes is an evolutionary game changer. There is the very real possibility, if not near certainty, that as we accumulate and implement knowledge at a breakneck speed, we will evolve the current human species into multiple human species.

Why would multiple new *Homo* species be more likely to emerge over time as opposed to one new superhuman all of a sudden? Because people make choices. Some will embrace all change, some will adopt partial change, some will want to allow no change at all. The technologies in question then allow rapid divergence based on personal choice (recall the Arc Fusion dinner questionnaire or the T-shirt that reads: FINE . . . I EVOLVED. YOU DIDN'T). As the menu of ways to adapt and alter our bodies expands, as the technology gets cheaper and safer, our great-grandkids will have fun playing bio-LEGOs with our descendants.

But bringing back this small planet to its more traditional state, one in which various versions of hominins live side by side, raises all sorts of hairy questions. We've struggled mightily with the problem of so many humans finding it somehow acceptable and moral to treat those they perceive as being from different races in extraordinarily cruel ways. Those biases and brutalities have persisted even after various analyses showed that individuals of different races may well be much closer to us genetically than we are to some of those within our own supposed race.

As with all things in life and in evolution, there will be winners and losers. Economic and cultural divisions within and between countries, nations, tribes, and societies will likely lead to very different ways of adopting and adapting. Some people, or cultural groups, will opt out for religious or moral reasons. Others may not be able to afford various adaptations. And there will also be groups who prefer to be isolated and left alone. Geography, religion, education, income, and culture are all factors that may lead to diversity in our evolution, and ultimately to speciation.

Does this seem far-fetched? New environments change species. Everything from plants on Mount Etna changing post–volcanic activity, to African fish, Hawaiian cave-plant hoppers, snails on Crete, coregonine fishes, and southeast Australian invertebrates bears it out.[30] So to suggest that humans are not also changing, you would have to assert that there has been no change in the human environment over the past millennia, over the past centuries, over the past decades, and therefore no adaptation/evolution is necessary.

Or perhaps we are immune to change in our environment? University College London's Steve Jones argues, "Things have simply stopped getting better, or worse, for our species. If you want to know what utopia is like, just look around—this is it."[31] (Wow, now just pass the Prozac and dust off your existentialist texts!) This point of view, these acolytes of the conservative past, argue that even as we drastically alter our entire environment, our food, climate, predators, epigenomes, microbes, bodies, and brains . . . nothing changes. Interesting thesis, given that it goes against every shred of evidence in the fossil, DNA, microbial, and environmental record. But for those who observe and believe in rapid change, who understand the Earth is no longer just a Darwinian world, the expected outcome is rapid hominin speciation.

Synthetic Life

·· ·✳· ··

Perhaps there's no better example of how humans are driving rapid nonrandom mutations than synthetic biology. You generally start with something that exists, and then create a new product or creature, one that would not exist but for deliberate human invention. . . .

Like many good stories, a key branch of synthetic biology begins in a bar, almost a decade ago. Imagine a Nobel Prize winner, a rogue scientist, a hotshot lawyer, and a venture capitalist sitting in an Italian bar in Alexandria, Virginia. . . . After a few scotches, Hamilton "Ham" Smith, Craig Venter, Dave Kiernan, and Juan began reminiscing about the last decades and asking, "Could we ever program cells in the same way as we program computer chips?" Somehow, who knows why, as the single malts piled up, the project seemed more and more viable.

A computer chip doesn't read music or letters, or see pictures, moving or still. All that flows through a computer chip is binary code: current or no current, 1 or 0, light or no light. But these two digits are incredibly powerful. They can collapse every word written and spoken in every human alphabet, every bit of music in every tonal scale. Photographs and film are stored inside the same two-letter language, making it possible for you to carry all your messages, documents, e-mails, photos, music, and film in your pocket phone. This language gives a street vendor in Mumbai as much information as the president of the United States had just a couple of decades ago. Want a map, biography, history, article, image? Just tap, tap, tap.... The chip doesn't care what you're processing, sending, reading, doing—as long as it can be coded in 1s and 0s.

We take it for granted now, but it's hard to overstate how much this digital transition changed every aspect of our lives. It's ever harder to take an elevator, open a hotel room, turn on a car, get music, take a picture, communicate with friends and colleagues, research, or do countless other daily tasks without digital code. Most of the wealth and jobs created over the past few decades come from this transition into the digital world.[1] Over several decades, scientists and technologists have developed various ways to interface between our world and the digital world. Computer chips have been miniaturized and now contain billions of transistors to seamlessly take phone calls while playing video games, and update our calendars and photos. It was not one key discovery but a layering upon layering of ways to apply and manipulate digital code. This language now accounts for 99 percent of all the information and data transmitted on the planet. Dare you try maintaining your routine without using a single digital device or input? See how many hours, or minutes, you last.

Since the 1930s, in parallel to the development and deployment of digital code, we have also been learning how the code of all life forms is spelled out. Recall, all life forms are written out in the four letters of DNA. Once you can understand these instructions and create blocks of DNA to your specifications, then you can "program" DNA to execute life code. The cell then becomes the equivalent of a computer chip, but it uses life code—GATC—instead of digital code. And just like a computer, the cell can theoretically be programmed to make cells that produce a lot of things, including foods, chemicals, and fuels.

Nothing's ever as easy as it appears late at night in a bar. It took years, around forty million bucks, some clever insights, and a huge amount of frustration to boot up the first synthetic cell. An extraordinarily creative biologist, Dan Gibson, devised a different way to assemble very large molecules of DNA; this technology is now known as a "Gibson Assembly" and has become standard for the emerging synthetic biology industry. In making the synthetic DNA, biologists Ham Smith, Clyde Hutchison, and their team painstakingly checked and rechecked every one of more than a million letters (that is, base pairs) of DNA to figure out why an initial version of the synthetic DNA molecule did not work and "boot up" the synthetic cell. (Turned out that just one single letter of the DNA code was wrong in a critical spot, holding back the project for months.)

The first human-programmed cell, named Synthia 1.0, was born by replacing the native DNA in a bacterial cell with a synthetic DNA molecule designed, assembled, and inserted in a lab—kind of a species switcheroo. (The new DNA mimicked the gene code of another species, and also included the scientists' names, plus some poetry.) Through very specific human wishes and design, one species of bacteria became a different species, then divided and propagated. (One way to conceive of this transformation is to imagine an engine that could be inserted into that old VW in your barn and would then turn every part of the old car into a new and complete Ferrari.)

Those in the field of synthetic biology had been expecting Synthia's arrival for years. As various DNA manipulation techniques and discoveries piled up, it became almost inevitable that humans would design a computerlike chip, just as in the 1950s it became obvious that they would soon discover the structure of DNA. As the new cell was being built, the Venter Institute reached out to all of the world's major religions in hopes they could work together to address key spiritual questions and concerns, foster understanding of emerging technologies, and "bless" the use of life code in a programmable bacterium. On the other end of the spectrum, security agencies were briefed on synthetic cells and required certain safeguards in the making, programming, and deployment of these organisms.

There was quite a bit of excitement at the Newseum in Washington, DC, on May 20, 2010, the day Synthia's birth was announced. It was heralded "the Immaculate Creation."[2] Not many science stories have both the White House and the Vatican reacting and commenting favorably, almost immediately. Her arrival made the cover of every major U.S. newspaper, and was featured in about 4,800 other major global publications and news outlets. (Perhaps even rarer, the announcement got Nico Enriquez, Juan's teenage son, to wear both shoes and a tie.)

As the ability to program life expanded, so too did the deployment of the technology. Over the past few years the company that deployed Synthia's life-programming technology, Synthetic Genomics, Inc. (SGI), developed ever-expanding business partnerships.[3] British Petroleum bet that cells could be programmed to execute complex chemical reactions. Then a "little start-up" in Texas named ExxonMobil decided this forty-two-person company, SGI, would be a good partner to genetically program algae to make fuels. Food companies began speculating that engineered cells might substitute for millions of acres of agricultural crops as a way to provide proteins and oils. Drug company Novartis formed a joint venture with the goal of producing flu vaccines for the entire world in a week instead of in a year. United Therapeutics used SGI's technology to begin humanizing pig lungs—a project that could eventually help save the 200,000 people who die every year waiting for an organ that never comes.

Having programmable cells introduces a very powerful programming language into the library of human knowledge; the consequences, mostly good and occasionally bad, will change history in many ways. A synthetic cell is a game changer just as the computer chip was a game changer. Placing embedded instructions, using unnatural genetic codes, inside new cells can serve very specific human desires: biologic drugs that are more specific and stable; genetically engineered organisms that only survive in the presence of a non–naturally existing amino acid; industrial enzymes that withstand heat, high pressures, toxic environments, harsh chemicals, or perhaps the high radiation of outer space. Our grandchildren will someday take the ability to build and evolve life forms for granted just as our children cannot conceive of a world without the Internet, mobile phones, and texting.

Change is inevitable. In 2013, Yale's Farren Isaacs and his team redesigned an element of the three-letter "words" contained within the ATCG genetic code for an entirely new and unnatural purpose; this went beyond genetically modified organisms (GMOs) and began to produce genetically *recoded* organisms (GROs). In creating and inserting a new three-letter code, one that begat a functional twenty-first amino acid, one not naturally used by living cells at all, Isaacs altered a language that is hundreds of millions, or more likely billions, of years old. Until then the DNA instructions required to make functional proteins, like hormones, receptors, and enzymes, had all used one of twenty amino acids.

So for the first time in perhaps a billion years, the three-letter genetic code was modified to enable a new amino acid building block. A fundamental new building block, designed by humans, used to make entirely novel, unnatural proteins with new functions and properties. By analogy, imagine adding an entirely new letter to a written language; you could make new words; new domain names . . . or imagine adding a new phoneme (unit of sound) to the library of forty sounds that comprise spoken English; you could make new words, new rhymes. . . . Or add a ninth note to the musical scale; you could make new music, possibly making even heavy metal sound good. Having a way to change the repertoire of amino-acid building blocks available to make proteins has broad and far-reaching implications.

In the near term, programmable cells could make biofuels, chemicals, IT storage modules, nanomaterials, and novel antibiotics that overcome resistance. We will build a massive "biotechonomy" in the same way that programmable computer chips built our current tech economy, or techonomy. But the key implications of large-scale genome synthesis and editing technologies, as well as changing the core building blocks for life, will be direct and deliberate nonrandom species engineering and evolution. In a sense, these discoveries revive memories of Lamarck, making evolutionary convergence and divergence feasible within a very few generations.

Life could soon get far weirder. We will likely create whole classes of animals and plants that are resistant to viruses and that only grow and reproduce in very specific environments, eventually opening the doors to entirely different branches of life that would not have grown naturally on Earth—life forms that are human-designed and begin to branch off of new evolutionary trees in interesting ways. Such changes in the genetic code wouldn't readily arise from any natural process in a living cell today, or perhaps in a million years. Especially now that mankind is rewriting the operating manual. . . .

Humans and Hubris: Does Nature Win in the End?

.... ✳ ....

Fact: Few single species can shape life on Earth.[1] But in spots like Kansas, Argentina, and the Amazon, hyperproductive farms cover dozens of square miles where almost anything that lives or dies is due to human decisions.[2] Increasingly a few individuals reshape large swaths of this planet, the animals that graze on it, the crops that grow on it, the microbes and viruses that thrive on it; we are pushing harder and harder into more and more realms, establishing more and more control over greater arenas. It is just as important to preserve and protect "all-natural" areas where Darwin and his evolutionary rules reign.

As our technologies advance, as there are more of us, and as more of us become technologically enabled, we become an ever more dominant species. But our ability to shape today's planet does not guarantee our long-term survival. Often there are unintended, ignored, and unknown consequences to what we do and choose.

To ensure our long-term survival we may want to be a little less arrogant in how we treat, colonize, guide, manipulate, and shape other species. We should be far more careful about how many of us there are, how far we spread, how much we consume. But even if we were to do everything right, "go green," be more conscious of our environment and our effect on it, might it be that nature wins?

A brute-force, single-species, ever-more-dominant-on-a-single-planet path likely leads to a dead end. Our hubris in how much we have accomplished sometimes leads to a conviction that we are already smarter and better than we will ever be. The temptation is to ossify the species, protect it at all costs. But if we limit hominin diversity, if we attempt to monoculture our cows, corn, and people, we become more vulnerable. Time and again this strategy eventually leads to a disastrous challenge by an emerging pathogen, a calamitous environmental change, or a new predator.

The key is to strike a balance, to permit nature and Darwinian evolution to survive and thrive alongside our brave new engineered, shaped, manicured, curated world. Perhaps allowing even half the Earth's surface and half of its oceans and lakes to evolve naturally. If we are careless enough to use our extraordinary powers without regard for what other species need, how they naturally evolve, then likely we too will become just another set of interesting fossils. In which case an enormous ecological niche will open in our absence. New entrants will thrive. And perhaps Darwin's rules will reestablish themselves again to guide all life decisions for a few more millennia until another intelligent life form attempts to guide evolution.

On the other hand, if we allow continued, smart, balanced, and rapid evolution, then maybe we ourselves will continue to thrive, and not just on Earth. . . .

Leaving Earth?

.... ✳ ....

The following questions are another good litmus test for how individuals feel about humanity and its future: "What would it take for humans to truly leave this planet?" "Would you advocate doing whatever it takes?"

In March 2014, the Genetics Department at Harvard Medical School organized a symposium with the grand title "Genetics, Biomedicine, and the Human Experience in Space." Juan opened the event with a slide titled "The Morality of Space Colonization." And then, to provoke just a touch of debate, he put forth nine principles to consider should we begin to contemplate leaving our solar system:

1. It's hard to destroy Earth. (But not impossible.) We exist, according to philosopher Jim Holt, in a universe that's 100 percent malevolent, but only 80 percent effective.[1] Almost all of space is completely inauspicious for life, but there are some pockets where organized life can evolve, until annihilated by equivalent antimatter, fission, black holes, solar ovens, overspin, declining orbits, stellar collisions . . .

2. Humanity is much easier to destroy. There have been at least five major species extinctions. The long and varied menu of how things can go horribly wrong includes large asteroids, supervolcanoes, nuclear winter, global warming, supernovae, massive solar flares ... One of the greatest killers was a minute microbe, *Methanosarcina*, which covered the Earth with methane and killed more than 90 percent of all creatures during the Permian extinction 252 million years ago.[2]

3. *If we don't get off Earth, humanity will go extinct.* This is not scaremongering or doom saying. It's fact. Earth experiences periodic extinctions that wipe out almost all life. "Erase the board and start over"–type extinctions. So the math over the millennia is really simple. If we just stay here, we certainly die. The key question is when. Likely it will be far, far before the sun begins to burn out and incinerates the Earth in its expanding corona. Likely it will also be before yet another asteroid, supervolcano, or even the galaxy Andromeda collides with our Milky Way. So leaving Earth is really risky, but it's the only way to hedge against complete human extinction. And if you believe in human rights, then it makes sense to try to protect humanity, and its successors, by minimizing the chances that one catastrophic event takes us all out. It's therefore a moral imperative to diversify our species (including colonies outside the solar system). Martin Rees, Britain's Astronomer Royal, makes the argument most succinctly. Imagine two catastrophic scenarios: "In scenario A 90 percent of humanity perishes. In scenario B 100 percent perishes. Even though there's nominally only a 10 percent difference, one is immeasurably worse ..." Leaving the planet, not in this generation or the next but in the future, should be a universal goal for humanity.

4. *Natural selection does not get you off this planet.* Our current bodies are not designed for other environments. We have to redesign ourselves, as it appears that no amount of random mutation and natural selection is likely to prepare us to survive and reproduce in non-Earth environments.

5. *Traveling to and living on nearby planets will likely require a deliberate reengineering of our bodies.* We are not built for, nor have we evolved to adapt to other planets or atmospheres. Space travel and colonization will likely require a far greater life span and even fundamental changes in *H. sapiens*. If we truly want to adapt and survive, to have

kids in a different gravity, breathe different atmospheres, we need to engineer our bodies, which implies that if humans are to have a long-term future they have to practice unnatural selection and nonrandom mutation. As long as consciousness, free will, fully informed consent, and liberty are preserved, everything should eventually be on the table, from engineering radiation-resistant genes into our bodies to cloning and/or mirroring our brains.

6. *Life can thrive in unlikely places.* Many of the technologies required to survive on other planets, as well as some of the needed medical breakthroughs, are just becoming visible. (And none of them contradict the basic rules of physics or nature.) Life can thrive in the most unlikely of places; there are organisms with millennial life spans and creatures that thrive in extreme environments such as boiling battery acid or tar lakes where they eat oil and breathe metals.[3] It is conceivable that we will find whole zoos living in liquid methane under the ice on a Neptunian moon.

7. *If we do get off the planet, we should take many other species with us.* (Perhaps in frozen zoos or reproducible gene sequences?) Ours is not the only life form worth saving. A series of Noah's Arks, and/or a database of DNA sequences of various species, should preserve at least a part of nature's 4+ billion years of shaping life itself.

8. *Colonize intelligently.* "Colonization" has become a dirty word. But other than a handful of Africans who stayed put in their ancestral lands, all of our ancestors begat and supported colonizers. While there were horrible excesses and abuses, colonization also spread many positive things, like trade, learning, human rights, democracy, and civil service, and sometimes put an end to all-powerful, murderous kings, maharajas, and zealots. Next time around, let's get it right. Preserve and protect the ecosystems we explore and colonize.

9. *Getting off the planet will be neither easy nor safe.* As with many of the earliest voyages of exploration and human expansion, the unknowns and risks are extraordinary. Even getting humans to Mars is far from a trivial hike. But individuals and groups of individuals should be free to knowingly volunteer to face these risks and challenges even against

daunting odds. Rock-star engineer Adam Steltzner (NASA/Jet Propulsion Lab—JPL) is a slightly Elvis-esque character who wears snakeskin boots, can pull off the pompadour haircut, stylishly wears his dad's Navajo turquoise ring, and speaks in a slow drawl. Famous for heading up a team that landed a rover the size of a Mini Cooper on Mars, Steltzner stars in a video that captures the tension and elation of the final descent of this lander to the planet's surface. It showcases the engineering involved in writing 500,000 perfect lines of computer code to accurately fire seventy-six pyrotechnic devices and slow a package the size of a car from 13,000 mph to 1 mph, in an atmosphere that's 1/44th as thick as that of Earth. The last step involved engineering the largest supersonic parachute ever built, plus a "trapeze" to gently lower the lander the last twenty-five feet without any human control whatsoever once the landing sequence was initiated.[4]

Landing is far from the only issue. Imagine living for years locked up in a small tin can with a bunch of *Right Stuff* cowboys; the Russians simulated such a mission by locking six guys—three Russians, two Europeans, and one Chinese—into a very small space for 520 days. As boredom sets in, all kinds of behaviors emerge—for example, hypokinesis, aka *very* slow movement, a lot of sleep . . .[5] It's almost as if lack of activity and stimulation led to a partial hibernation. However, a few folks slept on different cycles—say every twenty-five hours—leading them to wander around like virtual hermits inside a small enclosed space while all others snoozed.

Seriously contemplating leaving this planet requires rethinking and reengineering the human body. The deputy chief scientist at the National Space Biomedical Research Institute, Dorit Donoviel, points out that even short jaunts in space, a few weeks, can seriously alter our visual, cardiovascular, skeletal, muscular, and brain functions. Headaches are a constant. Lack of gravity means you rapidly begin losing muscle and bone mass. Eyesight can change so radically and quickly that many ex-astronauts carry adjustable-lens frames. Some gradually recover postflight, but some suffer optic-disk edemas. Altered circadian rhythms—the constant cycle of sunrises and sunsets when in orbit, or endless darkness after leaving Earth's orbit—drive at least one-third of astronauts partly crazy (much the same as what happens to most of us after several nights of little sleep).[6]

Food is a big issue. As happens with high-altitude mountaineers, humans in space report dulled taste and lack of appetite, which can lead to extreme weight loss. Astronauts continually demand spicier and spicier food. (Just imagine eating stuff that's a hundred times blander and yuckier than airline food for a few years.) Meanwhile, you also have to consume extraordinary amounts of salt; otherwise you lose 30 percent of your body fluid volume. And speaking of food, there's the issue of the calories and resources required to raise children in space; an average male needs about 10 million kcal to get from birth to a useful eighteen years old.[7]

Nevertheless, NASA is not the only one dreaming of Mars and beyond. Elon Musk, creator of the Tesla and PayPal, wants to be a modern-day discoverer and is prepared to travel, colonize, and die there ("just not on impact"). He has already built massive rockets that can take off and land vertically. Now he and his SpaceX team are busily designing interlocking modules to build a first colony. But getting there's one thing; aside from surviving even a few spaceflights and their side effects, getting enough resources to colonize and stay is a problem of a different order of magnitude.

So who gets to go? We will likely begin genotyping and selecting and modifying people long before sending them out for extended missions, and we may end up sending very specific groups of people—a population that perhaps does not, genetically at least, look like the average Earth population. Would you send someone into space with a short 5-HTT gene variant involved in serotonin signaling in the brain? If you have this variant and then experience a lot of stress, something that conceivably could happen during a long, cramped, uncertain space flight, you will be more likely to descend into depression.[8] Would you instead want to load up the spaceship only with those who have long 5-HTTPLR variants, who tend to be happy, cheerful, and optimistic?[9] You may even want to distinguish those who are pre-disposed toward major depression as opposed to bipolar behavior.[10]

How about women? The U.S. government sets limits on radiation exposure based on the equivalent of a 3 percent increase in an average person's overall chance of getting cancer, as compared with an unex-posed population. (A pilot exposed to 0.5 mrem per hour of flight can work far longer than an unshielded X-ray technician, given that an average whole-body CAT scan lets loose 1,000 mrem.)[11] But the inter-esting twist is that men and women react differently to radiation; women absorb more radiation, and suffer more damage.[12] Because astronauts are federal workers, and must abide by federal regulations, no woman can fly to Mars with the government's blessing given these regulations. Men? They're just barely able to meet federal radiation-exposure limits on one round trip. Without better radiation shielding for women, a Mars trip could be quite a dull and nasty place. (Imagine an entire planet as a massive bachelor pad with no women, a big-screen TV, and pizza boxes and old socks lying around.) And did we mention the problem of weak-kneed men? Women astronauts may be more vulnerable to radiation, but they lose bone mass at half the rate of their male colleagues; male astronauts' bone-loss rate is 1 to 1.5 percent of bone mass per month—more than tenfold the rate at which post-menopausal women lose bone.

Long-term space travel, beyond Mars, would require a far more extensive reengineering of the human body. The good news is that astronomers have found thousands of planets in other solar systems. Star Tau Ceti is a mere twelve light-years away, a close neighbor in astronomical terms. Its system has a planet that may be within a human-inhabitable zone and that is only 4.4 times the Earth's mass. (Still, without unnatural evolution, even Schwarzenegger could not do a single push-up there because of higher gravity.) And yet at today's fastest space-travel speed, it would take 90,000 years to get there; that's 4,500 generations of current humans.[13] It took *Voyager I* thirty-six years to become the first human object ever to leave our solar system.[14] But despite traveling at 17 km/sec it is still less than 20 billion miles from Earth.[15] It will likely be 40,000 years before the probe gets close to the next star on its route.

Perhaps someday we will add *Deinococcus radiodurans* genes to our bodies; this microbe can withstand 10,000 times as much ionizing radiation as can normal microbes. It can also live desiccated, with no food, perhaps for centuries. While ubiquitous on Earth, the creature seems particularly well adapted to space travel, which leads to all types of interesting questions . . . Why would such a creature evolve on Earth and be common in so many places? Is it common somewhere else?

There is even hope that we could adapt to some radiation. Many Earth-bound organisms exposed to something truly nasty, like radioactive cobalt-60, can evolve rapidly. In *E. coli*, 99 percent of the first generation exposed to toxic radiation dies, but by the twentieth generation, the surviving cells have figured out how to get rid of reactive oxygen molecules and survive in environments just as toxic as those of *radiodurans*.[16] Even in the most unlikely environments, some life does adapt.

Current diseases are driving knowledge fast; engineering the human body to restitch double-stranded breaks in DNA could benefit fetuses of moms who smoke, cure some cancers, and repair radiation exposure.[17] Someday we may consider using these new techniques to engineer humans. What might you think of adding or subtracting? How about oxygen? On Mars, and all known planets, if you go out with no space suit, you die. Not good after being cooped up in a tin can for a few months. Might you want to stretch your legs, maybe even without a bulky space suit, particularly around the Martian equator when it is a balmy 70 degrees Fahrenheit? Fortunately, the nice folks at Boston Children's Hospital began working on a system that directly infuses oxygen into your blood without its having to go through the lungs.[18] The little particles of pure oxygen are encapsulated inside lipids and injected. And as the lipids circulate they release at different rates and keep a body oxygenated. (You would still need a good radiation block.)

How about tinkering with our genes to get a leg up on space travel? George Church thinks there are about two hundred rare gene variants that already exist in humans that may be useful for future space travelers. One of the clues as to how to address both cancer and space travel may come out of a remote community of Jewish descendants, one exiled to the farthest reaches of the Earth by the Spanish Inquisition. The Laron community, in southern Ecuador, includes about one hundred really short people.[19] While this would not normally be a desirable trait given how much we favor height on Earth, smaller people may do better and stay healthier in confined spaces. Furthermore, the growth factor that is mutated in the Laron community, IGF-1, also lessens the incidence of diabetes and cancer in mice.[20] There are several other traits one might want to engineer into a tightly confined, very long lived species; for instance, Church would like to engineer astronauts with almost no sense of pain, just in case you have to perform surgery and do not have an anesthesiologist.

Maybe we should also send germ-free folks into space? Today's astronauts have to deal with a double threat: The human immune system seems to weaken in space. And radiation and microgravity can lead to rapid mutation and speciation of microbes, sometimes increasing their virulence while at the same time they confuse or evade our immune system.[21] Not a good combo. So while our microbiome performs many essential services for us today, it may turn against us in a different, high-radiation, high-mutation, weightless environment. How antiseptic do you want to be? If you sterilize everything and kill off all the good and bad bugs, what would happen to people who come back to Earth and have no immunity? Or do you perhaps eliminate particular bad-actor microbes and fortify microbial allies and prune a constantly mutating microbiome?

How about reproduction? Sex in a zero-G environment might sound like fun, but it's not reproductively efficient. Microgravity and ionizing radiation can lead to distorted sperm, cancerous mutations, and lower reproductive fitness.[22] It's just one more thing you also have to address before you are able to travel to some perfect beaches somewhere on the estimated 40 billion or so planets that exist within just our galaxy.

Eventually the physical problems of space travel may be far easier to solve than the neurobehavioral ones. For example, should a new planetary colony be free of (or have minimal) mental disorders, at least when they arrive? How might art, painting, theater, literature, poetry, and other creative endeavors differ in a space colony whose artists are descended from pioneers with genomes optimized for logic, space travel, and colonization? Might there be unforeseen consequences in choosing to build such a world? As with all things epigenetic, some human studies seem to show correlation between certain mental states and creativity, others claim direct causation, and still others find nothing.[23]

Perhaps the practical answer is to avoid sending full-living bodies across vast distances; one option, suggested by Francis Crick, is not to send humans but information. Break the human genome up into small pieces, implant them into different bacteria, send the bacteria to another planet, and then reassemble the human genome once there. In a sense, this is similar to how packets of information are sent over the Internet and are reassembled into a whole at their destination. Yet another option would be to send a DNA printer, which would serially spit out G, A, T, and C chemicals to make long strands of DNA to build designer species before the spacecraft arrives.[24] Both of these methods obviate the logistical problems of keeping fully formed bodies shielded from radiation, fed, happy, and healthy for centuries.

The DNA-printing option may be closer to reality than it seems. In November 2013, in a scene reminiscent of *Breaking Bad*, a series of trucks, RVs, and cars gathered at the far reaches of the Mojave Desert. Folks from Synthetic Genomics, the Venter Institute, and NASA had convened on a dusty, barren, dry, desolate spot to simulate being on Mars. They were there to test a futuristic concept—a digital-biological converter.

What drove the experiment was the inefficiency of vaccine production. Using today's methods, you need a biological sample of the organism itself; bring something deadly into your lab; create an antigen cocktail from the organism; inject it into many, many chicken eggs; and harvest enough vaccine to cover everyone. Typically this is a six-month to yearlong process. Not good if millions of people are getting sick in the morning and dying that night. With a digital-biological converter, you can sequence the bug at the same location where the epidemic is occurring; upload the DNA sequence data into the cloud; run it against a database of all known bird, swine, and human flus; design an effective vaccine; and send the file to a DNA printer on an airplane in flight. By the time that airplane lands, it will have enough vaccine on board to treat all the first responders and then keep producing vaccine to distribute.

The first part of this plan has already been proven. Novartis, SGI, and the Venter Institute have taken a sample of H7N9 and made a vaccine without actually possessing or touching any of the live virus from China. They are currently headed toward scaling the process to be able to print the entire U.S. flu vaccine supply in less than a week, which brings us back to the Mojave. If you can print vaccines on Earth, in the future you should also be able to print DNA sequences that can form living microbes. And if you can print microbes, you can do so across very long distances. So the NASA test was a proof of concept as to whether you could send a printer to another planet and then have it print a living organism—say, one that would begin generating oxygen, or algae for food and fuel. Unlike sending seeds or frozen cells, sending a printer provides greater flexibility for adapting to the local environment by programming functions into the DNA as needed; you could have real-time prototyping and rapid evolution. As you have seen throughout this book, humans are already engineering their bodies for sport, beauty, and medicine to such an extraordinary extent that we are well on our way to becoming something different. And the instruments we now possess are so powerful that in the not-too-distant future we are likely to beget humans quite different from ourselves.

As we redesign ourselves for very long-distance travel we won't just generate a successor species; after we leave Earth we will almost certainly continue to speciate, both because of separation and because of a continual need to reengineer our bodies to very different environments. Maybe the right answer is, if we're going to space, it's best not to bet everything either on the equivalent of 17 black on the roulette wheel, or on any one gene or any one "right people." If we have learned anything from bacteria, plants, and animals, it's the value of constant mutation, variation, adaptation, and hybridization. We must respect what we learned from Darwin on Earth and in space.

Space still seems a distant prospect. But we have accomplished an extraordinary number of things in the very short time that our species has been sentient, and our power has trebled as we have mechanized and digitized. Were the history of life on Earth compressed into one year, civilization has been around for the last 1.5 minutes.[25] At this pace, while it seems the stuff of science fiction today, space travel and colonization of other planets is likely a few seconds from now.

Meanwhile, long before we begin engineering astronauts, one might see wholesale gene swaps designed to foil genes like BRCA-1, which leads to early breast or ovarian cancer (particularly given that one current preventive alternative is a pre-cancer double mastectomy). How big dare we dream? CRISPR will soon let us engineer various traits into human cells, including sperm and eggs, as long as we get societal buy-in for large-scale, complex, deliberate human-genome engineering.

New Evolutionary Trees

.... :✳: ....

Chemical biologist Floyd Romesberg's new baby was far from an easy birth; gestation took fifteen years. But it was worth the wait; he ended up with a beautiful newborn, mutant *E. coli*, with DNA unlike that of any other species on Earth: "The resulting bacterium is the first organism to propagate stably an expanded genetic alphabet."[1]

Because every living thing is thought to derive from one universal common ancestor, through November 27, 2013, all known life forms living on Earth were based on one, and only one, genetic code—DNA based on four nucleotides (A, T, C, G).[2] What Romesberg and his team at the Scripps Institute did, in a chemistry tour de force, was to add two new base pairs (with the sexy and easily remembered names d5SICS and dNaM) to DNA. This turned a four-letter code of life into a six-letter code. (Fortunately, the new base pairs are now nicknamed X and Y.)

Federal funders, and most peer reviewers, had thought this new chemistry of life impossible and had refused to support it. So when Romesberg finally got something living by using a different life code, the very smart and colorful founder of the DNA rewrite field, Steven Benner, simply summarized the breakthrough by stating, "That's why Floyd's paper is so friggin' important."[3] What nature could not do in 4 billion years—alter and vary the fundamental code of life—humans achieved in a couple of decades. For the first time in history a six-letter DNA code (A, T, C, G, X, Y) operates and reproduces life. At the very least, in the short term, this discovery plants an entirely new evolutionary tree. We can watch the story of evolution unfold again, in parallel to all existing life on Earth, using a different life chemistry. The new organism's gene code would likely not breed with any living thing, and it would be resistant to all viruses that lack its code, since no parasite has adapted to coexist with the new code.

Medium-term, instead of just having a basic tool kit of twenty amino acids with which to build life code, the new life forms that emerge using this new chemistry will begin with 172 basic amino acids of different sizes and shapes as building blocks for proteins. When the biblical Tower of Babel broke down, each culture's language speciated until it became unintelligible to others. Perhaps similarly, after 4 billion years of a common life language, humans have rewritten the code of life and will grow many, many species that are genetically isolated from one another.

The rules and letters of life code may alter fast, because Romesberg is not the only one involved in rewriting it. Steve Benner busily builds "artificially expanded genetic information systems" (AEGISes)—hundreds of different ways to merge existing DNA with new chemistries in attempts to attack diseases like cancer or Alzheimer's.[4] In 2011, Philippe Marlière replaced one of the basic letters of the four letter gene code, thymine, with 5-chlorouracil; eventually he wants to swap out all four base pairs and have his own alien life garden without G, A, T, or C.[5] And in Japan, Ichiro Hirao, who would likely be a game-board designer or classical guitarist had he not read James Watson's *The Double Helix*, is trying to standardize and industrialize the emerging field.[6]

Long-term, now that we know we can rewrite life, we are very unlikely to stop; using various chemicals, we can start to think about how to "seed" life in very different ways under different conditions. Benner argues that while DNA "might not be the best possible solution for supporting life, it might be the best solution that could emerge from prebiotic Earth."[7] This was a toxic environment where things like hydrogen cyanide were floating around, so some of our early ancestors co-opted these chemicals to make more complex molecules, like adenine, and other basic building blocks of life. Plus there is the water problem. . . .

Anyone who has ever been in a boat, a house after a storm, an office building after a flood, or just let a bathtub overflow knows two things: Water is essential, and it can be really destructive and corrosive. The same principles apply to your genes. Water allows your cells to survive and your gene code to divide, reproduce, evolve. But Benner says, "You're trusting your valuable genetic inheritance that you're sending on to your children to hydrogen bonds in water? If you were a chemist setting out to design this thing you wouldn't do it this way at all."[8] If we truly want to live a long, long time, avoid errors, reliably preserve and pass on our genetic inheritance, we need to deal with electron-rich oxygen tearing apart our molecules: "In your body right now, the DNA in your cells is subtly breaking down many times a second because of the action of water."[9] Because of this continuous decay in our existing biological processes, we use an enormous amount of energy repairing our gene code; it is one of the core processes that drives aging.

Eventually chemistry will progress to the point where we will be able to code self-replicating organisms under what today look like intolerable conditions. Benner and several other scientists are now attempting to build "dry life," using organic solvents like formaldehyde to add more carbon, less oxygen, no water.[10]

If you think about genetic code, and life, as a universal, self-organizing system that can pass on information of increasing complexity, there is no reason it has to be DNA-based. Now that Benner, Romesberg, and others have shown the way, some people are realizing that life may be an inevitable consequence to particular chemical disequilibria. And these chemical disequilibria can occur under various circumstances as long as you have hydrogen plus methane and a few other chemicals to complete a circuit, as well as a pH gradient. This implies that one might find self-replicating life forms on completely different planets, including planets that are very cold and have little or no water.[11] Life on Earth may have evolved without water first and then had to adapt to the presence of water and oxygen.[12] In fact, a hot, dry environment with little water is not incompatible with life, as the complex biomolecules that are thought to have preceded life on Earth can form in the presence of borate, an abundant mineral in arid Death Valley.[13] In the measure that we realize DNA is not the only answer, every one of these discoveries increases the likelihood of life elsewhere and of very different genetics. (Mars, for one, could have supported early boron-chemistry life.)[14]

But coming back to Earth, to the present stage of evolution, NASA astrobiologist Peter Ward put forth the core question we face today most succinctly: "We have directed the evolution of so many animals and plant species. Why not our own? Why wait for natural selection to do the job when we can do it faster and in ways beneficial to ourselves?"[15]

Even before we began deliberately altering, rewriting, inserting, silencing, and deleting DNA code, human evolution was already on a tear; as natural selection gave way to human selection, we adapted so quickly to our changing environment that a full 7 percent of our genes underwent rapid evolution within the last 5,000 years, a blink of an eye on a Darwinian scale, but quite slow compared to what we are capable of now because we can engineer mutations nonrandomly.[16] We have transitioned from observers and actors in the soap opera known as *Life on Earth* to producers and directors of it. (We don't control every actor or the whole audience, but we have a big say as to who makes it onscreen, for how long, and for what purpose.)

As we understand our own gene code in detail, and as we develop instrument after instrument to alter that gene code, we will be able not only to change the code nature gave us but eventually to design it ourselves. This truly is a step that would have surprised Darwin.

And once we alter our gene code in beneficial ways, we will propagate the changes throughout our body and probably those of our descendants. Until recently, synthetic biologists created only a bacterial genome, not higher organisms with larger genomes and multiple chromosomes. In December 2013, scientists at Johns Hopkins built a full eukaryotic chromosome, putting them one large step closer to engineering entirely new plant and animal genomes.[17] Then in May 2014 an Italian team showed how to use CRISPR for "targeted genome editing in human haematopoietic stem cells."[18] In civilianspeak, this means one could take a person's own stem cells, make deliberate changes to one or more specific genes, and transfer these cells with healthy new instructions back into the body. The Italians are far from the only ones playing in this evolutionary sandbox; a team at Memorial Sloan Kettering in New York just announced the introduction of three major changes into human stem-cell lines.[19]

In sum, we can design, build, and transfer new genes into humans within months. We can design, build, and transfer whole genomes into bacteria within months. We can make new chromosomes within months. Millions of years' worth of evolution is being reformulated, redirected by humanity in just a few years.

What might we do with all this rapidly accumulating power? We will accelerate what we have already been doing with plants and animals, modifying and breeding them to suit our agricultural purposes, needs, and tastes. But we are also close to fundamentally modifying the bodies of animals for other reasons; the old sci-fi horror movies predicted human organ farms where we would grow clones to provide body parts for the rich. No need. We are starting to humanize animals to grow individualized organs that can be transplanted without rejection into our bodies.[20]

Soon we may wish to have periodic genetic tune-ups; the Resilience Project identifies not disease-causing genes but those that provide resistance to diseases.[21] In any given environment where plague, malaria, or HIV is rampant, about one person in 20,000 is immune. By isolating the genetics of these rare individuals and rapidly spreading the beneficial mutations to others, we may be able to create the equivalent of disease resistance. And germ line interventions could potentially make subsequent generations resistant to the various diseases. This is precisely what University of California–San Francisco's Yuet Kan is trying to do, using CRISPR to engineer HIV resistance into human blood stem cells.[22]

Would we want to develop "a general mechanism for resistance to all natural viruses"?[23] Yep—that could be a big deal. If you change how DNA reads and operates the viral expression systems, any viruses that have hijacked your genetic machinery for their own purposes may have no place to latch on to and execute their own code. A change of this kind could make us, animals, and plants resistant to many infectious diseases. (But as we develop the coding power to block all viruses, we might best remember that viruses, like our symbiotic bacteria, likely provide us benefits and have a way of playing a key role in evolution.)

Even before we begin intervening directly on a large scale in our own genomes, we are already starting "predictive genomics." GenePeeks, a service launched in 2014 by Princeton professor Lee Silver, claims it can virtually compare two prospective parents' DNA before conception and determine whether the baby would be healthy. This could be useful and important for 100,000 variants of single-gene diseases like cystic fibrosis.[24] But the new company also sought broad patent protection on hundreds of complex, intermingled gene traits, including aggression, blood pressure, body mass index, breast size/shape, cleft chin, conditioned emotional response, dimples, discrimination learning, drinking behavior, drug abuse, ear size/shape including earlobe attachment, eating behavior, ejaculation function, grip strength, grooming behavior, hair color . . .[25] We do not know enough about how genes function and their complex network of interactions to accurately assess most of these traits, but even if we did, should we draw a line between selecting for disease and selecting for looks? Or should parents simply have access to all we know and make their own decisions?

On his way to lecture halls, Nobel Prize–winning economist and Stanford professor Ken Arrow likes to think about what he is going to teach that day not so much in terms of how wise he is and all he knows, but in terms of the questions he would ask.[26] Rather than preach, he prefers to ask questions about things partly known and then explore them together with some very smart students. As new arguments or discoveries come to light he freely switches topics as more interesting things come up. Do we know all the implications of a world increasingly driven by unnatural selection and nonrandom mutation? Far from it. But inspired by Arrow's teaching style, what we wanted to do in this book was to open a dialogue with you. We have asked you to consider a range of topics, including chemistry, ethics, genomics, religion, robots, politics, behavior, space travel, and the future of humanity. In that spirit, how would you answer the question Juan asked at a recent TED conference: "If you could redesign humans, what changes would you want archeologists to find in 100,000 years?"

We face extraordinary challenges. But we also have an array of opportunities never previously available to any other life form on this planet. It's our responsibility to choose wisely, to continue to build, to prepare ourselves for very different environments, to eventually leave Earth. After dozens of versions of hominins, it would be bullheaded to think our species does not continuously change, that we cannot improve the species, that we will not beget other species. We cannot improve ourselves if we do not recognize what is happening and how quickly, and then forthrightly create a world we would wish for the variety of humans who will surely follow us. Darwin would have been a better guide. But he left us enough to build on as we embark upon the greatest of all human adventures: the creation of our own successors. For better and worse, we are increasingly in charge. We are the primary drivers of change. We will directly and indirectly determine what lives, what dies, where, and when. We are in a different phase of evolution; the future of life is now in our hands.

EPILOGUE
Eppur Si Muove

····················· ✳ ·····················

A lot has happened in the brief period between the hardcover and paperback printings of this book. Floyd Romesberg now has a spectacular office, top floor, overlooking storied Torrey Pines Golf Course and the Pacific Ocean. Paragliders float past. Still bent on modifying the chemistry of life on the planet, he hardly notices. Now that he knows DNA is not a unique solution for life forms, he continues to tweak, alter, and augment that which can reproduce. And strange menageries may emerge, based on alternate chemistries which create new genetic codes that include unnatural amino acids, and which might help us understand both the origins of life on Earth as well as alternate ways to build life elsewhere. After all, Romesberg argues, "The fact that you can augment DNA means there is nothing unique to DNA."[1] Perhaps one day he will substitute all four bases of DNA in a self-replicating system. Meanwhile, others are in hot pursuit, and a second team expanded the genetic code to six letters.[2]

Many of the trends and emerging experiments we glimpsed, or predicted, are now common discourse. Once obscure concepts, like CRISPR, are part of popular ethical, business, and political discourse. And yet there are still a lot of dyed in the wool traditional Darwinists out there; folks who having strayed too far from the research bench, and having failed to keep up with the myriad life sciences startups, remain driven by past ideologies and truths, arguing that anything other than "cultural" evolution only happens over centuries.

Even given overwhelming evidence of a new paradigm, many still don't get it. But as Galileo supposedly said: *Eppur si muove* (And yet it moves). And just as the Earth moves, so too does evolution, both naturally and unnaturally. The evidence accumulates daily; we see it in mosquitoes that wandered into the London underground in the 1860s and no longer mate with their above ground brethren.[3] Florida's green anole iguanas developed larger toepads in 20 generations.[4] David Reznick showed that Trinidad guppies evolve quickly and, in doing so, change not just themselves but also drive significant evolution of their surrounding ecosystem.[5] Or apple maggots can evolve from feeding on hawthorn fruit to eating apples in one generation.[6] In the unnatural world of laboratory science, Harvard and Yale researchers altered the genes that create bird beaks such that chickens develop dinosaur-like snouts.[7] And gene editing has been used to create bulls that lack horns (to avoid painful surgery) as well as mosquitoes that cannot transmit malaria.[8] The pace is rapid and accelerating.

In not "tuning in" or "looking to find common ground" the ideologues and fundamentalists of both religion and of past science do a disservice to their students, to science, to their regional economies, and to truth. This has consequences. The longer we take to recognize that we are major drivers of evolution, and that evolution is happening very fast, along different paths, the longer we postpone debates on an absolutely fundamental question: "Now what?"

Polymath Danny Hillis describes our new era as one in which "we have outgrown the distinction between the natural and the artificial. We are what we make."[9] It is an era of entanglement between ourselves and nature, chemistry, artificial intelligence, robotics, chemistry, genetics, and a host of other forces driving rapid evolution. We are a species that now conceives in test tubes, grows new organs, alters its own genes, and hyper-connects brains. And, as Hillis puts it, "Instead of classifying organisms, we construct them. Instead of discovering new worlds, we create them."

Humans are on the way to becoming something else. Something of their own design. We are trying out various temporary solutions and options before making permanent alterations. But the direction is clear; Hugh Herr's bionics evolved such that bombing victim Adrianne Haslet-Davis returned to ballroom dancing and then completed the 2016 Boston Marathon.[10] Pop star Viktoria Modesta, with her spiky, sexy, black metal leg, shows off the extraordinary range of wardrobe/appendages enabled by 3-D printers.[11]

Prosthetics are but a first step. During snowy April 2016 an extraordinary group launched a new MIT initiative, the Center for Extreme Bionics. Ed Boyden and Hugh Herr joined MIT's most productive inventor Bob Langer in a quest to merge robotics, tissue and cell bioengineering, and optical brain control. The dream is to eliminate human disabilities by integrating flesh and metal, with direct brain control of new limbs and organs. Were any other group leading such an initiative it would seem quite farfetched, but these three have a record of achieving extraordinary breakthroughs in short time scales. Boyden and Herr you already know from previous chapters. Langer is author and coauthor of over 1,100 patents, and his office is wallpapered, floor to ceiling, with honorary PhDs and major prizes.

What if, eventually, we are awed, intimidated, and even desirous of some of these artificial appendages/enhancements?[12] We may see hyper-strong legs, hypersensitive hands, and extraordinary reaction times. In Herr's world we may be able to rewire nerves and change their diameters, enabling us to execute ever more complex and sensitive tasks; eventually touching an artificial appendage may feel just like touching your real leg or arm. This direct brain wiring, according to Herr, might accelerate reaction times to the point where we could even dodge bullets fired from a long ways away.[13]

In Herr's mind a century from now humans will be unrecognizable. Our concepts of diversity and beauty will fundamentally change. Some may even end up looking like ancient Hindu Goddess statues, with multiple limbs for multiple purposes. But flashy, multi-limbed humans are not the fundamental change; the core alteration will take place in the brain. One of the stated objectives of the Center for Extreme Bionics, and of Boyden's new research, is to "devise new strategies to grow 3-D brains from scratch, in a dish." And, ho hum, "in the long term, of course, these mini-brains may provide new replacement parts for the brain . . . interface with the natural brain, and compute similarly to the human brain with low power and high parallelism."[14]

Even before we get anywhere near brain implants scientists are experimenting with ways of repairing, enhancing, and posting basic upgrades to that most human of organs. Experiments like those that use genes like Dnmt3a2 can boost memories in mice and erase bad memories and traumas.[15] And even if we can't fix your brain today, neuroscientist Kenneth Hayworth's work may allow a fix in a few decades. Until 2016 most cryopreservation techniques crushed neuronal connections and dehydrated the organ. But now a rabbit brain frozen to 135°C, which is something that looks and acts like solid glass, can be thawed leaving the cell membranes, synapses, and intracellular structures intact. So, in theory, this means brains can be recovered in centuries in near perfect condition.[16]

Some think our brains, especially if they do not adapt/augment quickly, could breed our own demise. Artificial intelligence (AI), and whether it will overwhelm us, is a hot topic. Gates, Hawking, and Musk all worry about how humans might control and outthink emerging intelligences or logic devices orders of magnitude more powerful than our brains. Brilliant tech historian Steven Johnson and philosopher Nick Bostrom explore this theme highlighting a specific experiment as a paradigm; a few decades ago, at the beginning of the AI field, scientists tried to get a computer to build an electronic oscillator.[17] They asked the machine to "evolve" approximations to a set of curves in a Darwinian fashion. And sure enough, the screen created ever better sine waves. But there is a weird twist; when they deconstructed the details of the program the transistors iterated to create the oscillator, they found the computer had not designed an oscillator at all. Instead it virtually rewired the circuit board such that it functioned as an antenna and could detect the very faint emissions from a computer across the room. It was simply detecting the other machine's oscillating electronic transmissions instead of building its own oscillator. In other words, the machine built a radio, not an oscillator. The lesson, as far as Johnson is concerned, is that ever more intelligent machines could develop an alternative consciousness, one completely different from ours. And that such a leap could come out of left field, which makes it very hard to predict or control. If we programmed/birthed such evolving AI, small initial mistakes in programing instructions could lead to catastrophic consequences, e.g., an instruction to maximize human happiness could lead to nanobots that continuously stimulate our brain's pleasure centers.

We cannot have a sensible conversation on AI or any other implication of rapid unnatural evolution if we do not recognize just how powerful we are and what it means to control and drive life code. We need an ongoing ethics and permitted uses conversation. You now understand what life code is and how we control ever larger evolutionary streams; help us develop a set of guiding principles. Here are six initial suggestions, which are not meant to be definitive.[18] These are intended to trigger debate that leads to a set of minimum collective agreements over life, going forward.

First, we have to take responsibility. As we face the extraordinary opportunity and peril of taking control of life, of iterating new life, we can no longer just blame change on random mutations, on "stuff happens," on the will of some divine being. So long as we work toward re-engineering our biological world, and certainly for the foreseeable future, scientists need to continue efforts to incorporate "safety switches" into genetic technologies so they can be halted or reversed. We are increasingly in charge; the consequences are ours to enjoy or to bear.

Second, the technologies we are letting loose will allow us to fundamentally modify our bodies and brains going forward, so we must recognize and celebrate diversity. It is normal and natural for there to be various humanoids walking around. Our ancestors mated with several types. Going back to this multi-hominin state would be natural. Having one single species forever forward would not.

Third, respect other's choices. Some will never alter, no matter what. Others will only alter if faced with deadly disease. Others will seek to enhance beauty, brains, brawn, or other traits. And some will push the envelope far further than most ever would. Your body, your choice, especially if alterations do not involve germ line engineering. Don't force people into anything, but, if it is not harmful to others and an individual fully consents, let that person choose, provided they do not impose undue consequences on others' rights.

Fourth, let's set aside at least a quarter of the Earth and seas (does not need to be contiguous) and let Darwin, and only Darwin, run the show with natural selection and random mutation. It is a really bad idea to bet all evolution on human choice alone. This preserves the natural and provides a backup in case our choices become disastrous. (Of note, species in designated "natural environments" will still need to adapt to global consequences of natural or man-made disruptions such as global warming.)

Fifth, educate yourself and others. Having read this book you already know a lot. But so many fields are moving so quickly you have to continuously update your knowledge. Even more important, you have to help the next generations understand the wonderful possibilities and serious pitfalls of controlling life. Nothing humans have ever done matters more, or has a greater long-term impact.

Finally, remain optimistic. When faced with great power, it is too easy to scare oneself and others. We had best heed Stewart Brand's wise admonishment, "Defining potential long term problems is a great public service. Over defining solutions early is not. Some problems just go away on their own. For others, eventual solutions that emerge are not at all imaginable from the start."[19] We have to be patient, creative, vigilant, tolerant, and a little afraid. We have a superpower that no previous generation of humans even dreamed of: the ability to redesign and shape life on Earth.

ACKNOWLEDGMENTS

.... :✳: ....

The scientific community, which we know well and greatly admire, is extremely dedicated, hardworking, and committed to unearthing truths. We applaud and thank this community for providing insights that have vastly improved the human lot and allowed us to take charge of a part of our evolution. Every day we see new articles updating and advancing our understanding of biology; they support, refine, and sometimes refute some of our thoughts and theses. Our priority is to use what we have learned to help present an overarching theory and context; we hope it's one that helps you better understand the world we live in.

Our research was a labor of love over decades of observing, thinking, and testing. Nevertheless, this book is already out of date. Observations change, theories pivot, concepts arise, technologies emerge. We realize that as research advances, many of the facts presented in this book will turn out to be obsolete, but this is characteristic of the dynamic changes that underlie scientific research—two steps forward, one step back.

While we tried to highlight the relevant science, inevitably there are scientists, ideas, and observations that we overlooked or missed.

We apologize for any omissions or misunderstandings—they were not purposeful. We realize that, like any concept, our theory is built on the ideas of others, though we are unaware of anyone bringing all these ideas together in one place, into one new theory.

We thank the hundreds of people who contributed to our thinking in creating this book. We especially thank Chris Anderson, Dianne Barratt, Rick Blume, Gaye Bok, Joan Bozek, Louis and Mabel Cabot, Jimena Canales, Marc Cecere, Casey Cunningham, Antonio Enriquez, Nico Enriquez, Eric Gordon, Graham Gullans, Jeanne Henry, Malcolm Kottler, Matthew Lawrence, Karen Leese, Kevin Liang, Tara Lemmey, Rodrigo Martinez, Jeff Novins, Eileen O'Connell, Steve Petranek, Rob Reid, Edith Shi, Alan Stoga, Sheldon Wagner, Fred Wang, Phyllis Wyeth, and Bruce Zetter for carefully reading early versions of the manuscript to let us know where we were off base, missed the point, or were just plain ignorant or wrong. Your contributions, in the early stages of making the sausage, are evident and greatly appreciated. We also thank the readers who alerted us to several errors in the hardcover edition, which we have corrected.

There have been innumerable scientists who have generously shared their discoveries, ideas, and criticisms. Among the many who were generous with their time and brilliant thoughts, we especially thank Ed Boyden, Stewart Brand, George Church, James Collins, Monica Colaiácovo, Daniel Dennett, Susan Dymecki, Hugh Herr, Danny Hillis, Michael McCullough, Jessica Richman, Floyd Romesberg, Andrey Rzhetsky, Pardis Sabeti, Karl Skorecki, Hamilton Smith, Cliff Tabin, Craig Venter, and Ting Wu. For those who appear in the text, we thank you for being candid and sharing your experiences and stories with us (sometimes in person, sometimes unknowingly, from your public pulpit). We look forward to learning about your future discoveries and continuing the dialogue. We thank Nico Enriquez, Jennifer Stone, Jeff Szeszko, and Alex Trautman for research assistance. We are grateful to Lorna Prescott for helping us with countless endnote citations and general edits. We are truly indebted to Lisa Goldberg, who facilitated endless tasks and spent months researching, fact checking, and challenging us; she is truly an integral part of this book. Suzanne LaFleur also deserves special acknowledgment for assisting with edits and helping make science jargon more user-friendly. Thanks to Wesley Neff and Emily Loose, who were particularly helpful in early stages of launching this project. We are grateful to the team at Current/Penguin Random House, led first and foremost by the great editor Niki Papadopoulos, plus her colleagues publisher Adrian Zackheim, associate publisher Will Weisser, editorial assistants Kary Perez and Leah Trouwborst, and senior production editor Bruce Giffords, as well as Kathy Damen and Taylor Fleming, who headed up marketing. We also thank our many valuable literature sources and the people who make them possible, especially *MIT Tech Review*, *Nature*, *New Scientist*, the *New York Times*, *PLoS*, PubMed, *Science*, *Science Daily*, *Science News*, *Wikipedia*, and *Wired*. We would, obviously, not be here but for our ancestors, Marjorie and Tony, Mary Lou and Rolf, who taught us to respect and learn from the past but always focus on building the future. We hope with this work we have left our descendants a record of some of the choices we faced and why we, as a species, did some of the things that we did to make them so very different . . .

GLOSSARY

.... :✳: ....

Asperger's—Asperger's syndrome is viewed as a high-functioning form of autism. It is a developmental disorder in which individuals have difficulty interacting socially, show restricted interests, and display clumsiness and repetitive behaviors. Language and cognitive skills are typically normal, but those with Asperger's can possess superior performance in a specific field.

Attention Deficit Hyperactivity Disorder (ADHD)—ADHD is a behavioral condition that arises in childhood and is characterized by having problems focusing, paying attention, and concentrating. Individuals are constantly active, moving, and frequently impulsive.

Autism and Autism Spectrum Disorder (ASD)—A developmental brain disorder evident by age three that is characterized by difficulties in social interaction, verbal and nonverbal communication, and repetitive behaviors. In May 2013, in the DSM-5 diagnostic manual, the many subtypes of autism, including Asperger's, were collectively grouped into one diagnosis termed ASD. The spectrum encompasses the varying degrees of behaviors, skills, and functioning.

CRISPR—CRISPR is an acronym for "clustered regularly interspaced short palindromic repeats." CRISPR familiarly refers to CRISPR-associated systems (Cas) that are genetic-engineering tools used to edit the genome of many species, including humans. Discovered in bacteria as a natural system for removing harmful viruses from the genome, CRISPR/Cas9 has been engineered by scientists to enable a DNA sequence in any genome to be removed and replaced with a desired sequence. This has many potential applications, including repair of disease mutations or modification of normal genes to alter or enhance function. It could also be used to fundamentally alter a species or create a new one.

DarWa—A term used herein to describe the evolutionary theory espoused by Charles Darwin and Alfred Russel Wallace. "DarWa 1.0" refers to the theory as proposed by Darwin and Wallace in the nineteenth century, plus updates to their theories during the early twentieth century with the modern synthesis, which incorporated Mendel's genetic concepts. In sum, DarWa 1.0 is the traditional theory that evolution is driven by natural selection and random mutation. "DarWa 2.0" is a derivative of Darwinian theory indicating that human-driven activities enable evolution to also operate under the forces of unnatural selection and nonrandom mutation. Evolution in DarWa 2.0 is a balance between natural selection/random mutation and unnatural selection/nonrandom mutation, which can vary in different environmental niches and over time.

Darwinism—The theory espoused by Charles Darwin that species arise by descent with modification through a process of natural selection favoring those individuals with variations or traits best adapted for survival and reproductive success.

DESTINY—The environmental stimuli that promote changes in the hologenome, particularly the epigenome, of you and your progeny. The acronym D-E-S-T-I-N-Y stands for: *Diet, Enriched Environment, Stress, Toxins, Infections,* and *Nurture,* which affect *You* and your progeny. DESTINY operates minute by minute, day by day, across a lifetime, and across multiple generations. The human body has at least three organ systems that signal these DESTINY inputs to each cell in the body: the endocrine system, the nervous system (including the brain), and the immune system.

Domesticated Humans—Unlike our prehistoric ancestors, modern humans display traits similar to those one sees in domesticated animals, such as cats and dogs, and plants, such as many crops. Basically, modern humans are tame, live well with others, and would struggle to survive in the wild. Humanity has less genetic diversity in the population and can breed in "captivity" (for example, urban dwellings). Like other domesticated species, we are overcrowded and may be more prone to epidemics. Humans are adapted to living in a secure, comfortable environment. Behaviors such as overt aggression and violence have also lessened, as has been seen with domesticated animals. Finally, because of the increasing uniformity of lifestyles and environments across the globe, there is likely increasing similarity in the four genomes across the population, which begets greater uniformity of traits.

At the highest level, a domesticated species is one "in which the evolutionary process has been influenced by humans to meet their needs."[1] We certainly meet this metric ourselves, since humans influence, if not generally control, their own survival and reproduction, which are the basic elements of the evolutionary process. Like those of dogs and cats, our behaviors and genomes (for instance, the epigenome and microbiome) are adapted to fit the needs of modern living in a crowded urban society. Survival in the wild is a problem for domesticated breeds, which goes without saying for those people who frequent spas and red-carpet events. On occasion, a domesticated animal will escape into the wild, becoming feral, kind of like our survivalists. As for breeding in "captivity," very crowded apartment buildings don't seem to inhibit anyone's sex life. While some may disagree, it appears that humans certainly resemble their pet dogs—unwilling to live outdoors, eating food from the grocery, readily making new friends, protecting loved ones from danger, playing with toys, and typically having offspring under highly controlled circumstances.

Epigenetics/Epigenomics—The study of the heritable changes in the activity of specific genes in an organism that do not alter the basic DNA code (GATC) itself. These changes occur in multiple ways through biochemical mechanisms that typically switch a gene or collection of genes "on" or "off," leading to modifications in an organism's development, behavior, or countless other functions. Some epigenetic modification with respect to whether a gene is active or inactive can be passed to future generations. Moreover, these changes are reversible. Epigenetic gene switches are controlled by evolutionarily conserved mechanisms and are triggered by environmental influences, particularly diet, enriched environment, stress, toxins, infections, and nurturing. Epigenetic switches are a key way in which lifestyle choices and experiences at an early age or in utero lead to diseases like cancer, as well as other good and bad outcomes later in life or in our progeny. Technically multiple systems underlie epigenetics, including DNA methylation, histone modification, and noncoding RNAs. Epigenomics is the study of the biochemical changes to the DNA genome that modify the activity of individual genes. Every individual has a unique epigenome.

Evolution—A process that results in heritable changes in a population across multiple generations.[2]

Genome/Genomics—A genome is all the core genetic material that constitutes an organism, from viruses to humans. It contains all the basic genetic instructions to enable a species to survive, thrive, and reproduce. The human genome is 3.2 billion bases (GATC) long, and every person has two copies, one from each parent. There is individual variation in the genome sequence of individuals so any two humans differ by 0.1 percent in the genome sequence, and this difference underlies many individual traits such as eye color or many diseases. Virus genomes can be as small as several thousand bases and fewer than ten genes. Genomics is the study of genomes.

Hologenome—The total integrated information contained within an organism's four genomes: core DNA genome, epigenome, microbiome DNA, and virome DNA. This concept extends and amplifies that described initially using the same term by two Israeli ecologists, Eugene Rosenberg and Ilana Zilber-Rosenberg, in their description of how microbes and coral mutually interact.[3] Our expanded definition of the hologenome has three basic aspects: (1) Plants and animals utilize the genetic information of their composite four genomes to survive, thrive, adapt, and reproduce. (2) The four genomes are highly interactive and change dynamically over time through a variety of mechanisms, particularly in response to external stimuli and to one another. (3) The hologenome is passed from one generation to the next. The hologenome provides more stability to the adaptive process for an organism and species, while providing a wider range of traits that can be accessed rapidly.

Hominin (Hominid)—In this book the scientifically correct term "hominin" is used to refer to modern humans and close extinct relatives. Hominin excludes great apes, which includes gorillas and chimpanzees. "Hominids" is the more familiar term that is often used in the nonscience literature, but it includes hominins plus great apes and their ancestors.

Horizontal Gene Transfer—The direct transfer of genes from one organism to another without reproduction. Contrast with vertical gene transfer, the transmission of genes from the parents to offspring. Horizontal transfer can occur between individuals of the same species or different species. Viruses are often involved in the mechanism of horizontal gene transfer. This process is important in evolution and is demonstrated in the rapid and widespread appearance of antibiotic resistance in many pathogens.

Hygiene Hypothesis—A hypothesis that arose in the 1980s that a child's environment can be "too clean" and the immune system will fail to mature properly, rendering the child allergic to many harmless environmental stimuli, including some foods. In early life, a child's immune system is educated about natural substances that are safe, like food, and those that are not, such as pathogens. The hypothesis suggests that a child exposed to foods, pollens, dander, germs, and other potential allergens in early life will not develop allergies. Hence, growing up in a modern urban environment devoid of farm animals, pollens, soil bacteria, and so on predisposes a child to allergies.

Microbial Genome—The microbial genome is the collective DNA from all the bacterial microorganisms in an environmental niche, which could range from one species to a multitude of highly interactive cohabiting species.

Microbiome—The composite mixture of all the microbes that live together within an environmental niche such as the gut, skin, placenta, or other areas of the body. It is also a key component in understanding animals and plants, as well as a wide variety of habitats such as soil, air, seawater, hospitals, kitchens, farms, and so on. The microbiome does not include the virome.

Mutation—A change in the DNA sequence of an organism. If the change occurs in germ-line cells (sperm and egg), it can be passed on to future generations. Some mutations cause changes in biological traits, such as a change in a gene, while others are biologically silent.

Natural Selection—The process by which living organisms, from viruses to humans and plants, adapt to specific environmental pressures such as diet, predation, stress, and climate. Natural selection favors individuals with traits that are beneficial for survival and reproduction, particularly when propagated across a population. Closely associated with the phrase "survival of the fittest."

Nonrandom Mutation—Changes in any of the four genomes— DNA genome, epigenome, microbial genome, viral genome—that do not arise by complete chance, and are usually instigated by human actions. Specifically, this includes directed changes in the genome sequence or related properties by genetic engineering, transfer, and editing technologies. It could potentially also include nonrandom, highly predictable changes in any of the four genomes that result from selective pressure applied by specific human actions, whether purposeful or serendipitous. With the advent of new gene engineering techniques such as CRISPR, humans have new ways to explicitly modify an organism's genetic code in a specific, nonrandom manner.

Optogenetics—The use of genes engineered to make neurons light-sensitive. Typically a light-sensitive opsin gene containing an electrically sensitive ion channel, together known as a channelrhodopsin, is introduced into many individual neurons using a virus carrying the gene. An infected cell becomes electrically active when a flash of light is shone upon it. Alternatively, a channelrhodopsin-infected neuron will emit a flash of light when it is active, enabling a scientist to record visually and noninvasively the activity of neurons in real time. Optogenetics is used to record brain activity at the level of individual cells or to control specific cells within the brain.

Random Mutation—A change in the DNA sequence of an organism that occurs by chance without regard to altering a specific function or trait in an organism. It can occur naturally. If the change occurs in germ-line cells (sperm and egg) it can be passed on to future generations. Some mutations cause changes in biological traits, and others are biologically silent.

Species—High school biology courses teach that two living entities are different species when they cannot create viable offspring capable of procreating themselves. This definition is inadequate because of numerous counterexamples. For this reason, Darwin himself was unable to define the word "species," which is ironic given the title of his seminal book, *On the Origin of Species*. Biologists have provided at least twenty-six definitions of species, which underscores the difficulty of pinpointing this elusive concept.[4]

In the present context, humans and other organisms are more rapidly acquiring new natural and unnatural traits, so at what point is an individual or a population a new species? The reader is advised to make his or her own decision about what constitutes a new or different species, since, as Justice Potter Stewart taught us about the nature of the indefinable, "I know it when I see it."[5] Or, to be more direct, as more than one person has commented, "I know a different species when I see one, because it is something I certainly would not mate with."

Unnatural Selection—An evolutionary process where survival, reproduction, or acquisition of new traits is influenced or caused by events or pressures that would not normally occur without human actions. Human-driven unnatural selection can be either purposeful or entirely accidental and can lead to beneficial or detrimental outcomes for any particular species, including humans themselves. Note that Darwin used several terms to describe human-driven selection, such as "artificial selection," "methodical selection," and "unconscious selection."

Virome—The genomes of all the viruses that live within or on an organism or in a specific environment. The human virome comprises quadrillions of individual viruses whose functions are poorly understood. The virome changes with diet, geography, season, weather, and many other environmental conditions. On occasion, a pathogenic virus enters our bodies, making us sick until our immune system can destroy it.

NOTES

····························· ·**米**· ·····························

What Would Darwin Write Today?

1. Just to point out how radically the world has changed, the word "allergies" was not even around when Darwin was writing. It was not until 1906 when a Viennese MD noticed the reactions some of his patients had to certain foods or seasonings that he coined the word and medical concept. See Richard Wagner, "Clemens Von Pirquet, Discoverer of the Concept of Allergy," *Bulletin of the New York Academy of Medicine*, 40, no. 3 (1964): 229–35.

2. Because this is not an academic treatise on the origins and development of the theory of evolution, throughout this book we will use "Darwin" as a shorthand for a far more complex and varied cast of scholars and theorists who helped flesh out the concepts and drivers of how life evolves. But first and foremost we do want to recognize that alongside the far more famous Charles Darwin, one underrecognized and underappreciated mind stands out. Alfred Russel Wallace should be described as a co-discoverer of the theory of evolution. (Not that Darwin and Wallace agreed on everything; see, for instance, M. J. Kottler, "Darwin, Wallace, and the Origin of Sexual Dimorphism," *Proceedings of the American Philosophical Society* 124, no. 3 (1980): 203–26, accessed August 28, 2014, http://www.jstor.org/discover/10.2307/986370?.

3. Though Darwin and Wallace did not speak specifically of random mutation, the basic principles of Darwinian evolution were merged with Mendelian genetics in the early twentieth century to create the modern synthesis theory of evolution, which continues to be taught today, and which prescribes that evolution involves the combination of natural selection and random mutation.

4. Were Darwin writing an updated version of *On the Origin of Species* or *The Descent of Man* today, many of the original chapters, the deep historical ones—on the natural origin of species, on the history and evolution of life on Earth, of humans—would likely be quite similar, though he would be able to flesh them out with overwhelming scientific evidence. But the book's new chapters, those covering our time, current evolution, and the future of life would likely be quite different.

5. Max Roser, "Life Expectancy," OurWorldinData.org, accessed August 29, 2014, http://www.ourworldindata.org/data/population-growth-vital-statistics/life-expectancy.

6. James Melkie, "Sir David Attenborough Warns Against Large Families and Predicts Things Will Only Get Worse," *Guardian*, September 9, 2014, accessed August 29, 2014, http://www.theguardian.com/tv-and-radio/2013/sep/10/david-attenborough-human-evolution-stopped.

Is Autism a Harbinger of Our Changing Brains?

1. L. Torian, M. Chen, P. Rhodes, and H. I. Hall, "HIV Surveillance—United States: 1981–2008," *Morbidity and Mortality Weekly Report (MMWR)* 60, no. 21 (2011), accessed August 29, 2014, http://www.cdc.gov/mmwr/pdf/wk/mm6021.pdf.

2. For a gene to spread really quickly across a population, there needs to be an extraordinary selection bias favoring a beneficial trait or disfavoring a deleterious feature. For example, a gene mutation that confers resistance to a new infectious disease or a mutation that slows metabolism during times of famine. Those people who lack the beneficial genetic variant die without reproducing, or reproduce at a diminished rate. Deleterious genetic variants are generally lost from a population and thus difficult to observe in humans today. Studies of genetic adaptation in humans suggest there are hundreds of mutations that have been positively selected for over the past 80,000 years since humans departed Africa. The newly acquired traits appear to enhance or modify resistance to infections, reproduction, olfaction, external body appearance, and nutrition, among other things. The rate at which a mutation propagates varies according to its relative benefit and the rate of reproduction. As an example, it has taken fewer than 10,000 years for mutations encoding lightened skin, which enhances vitamin D synthesis, to become highly prevalent in humans in northern latitudes where sunlight is diminished. See B. F. Voight, S. Kudaravalli, X. Wen, and J. K. Pritchard, "A Map of Recent Positive Selection in the Human Genome," *PLoS Biology* 4, no. 3 (2006): 446–58; P. C. Sabeti, S. F. Schaffner, B. Fry, J. Lohmueller et al., "Positive Natural Selection in the Human Lineage," *Science* 312, no. 5780 (2006): 1614–20. Also an overview of adaptive evolution is available at "Adaptive Evolution in the Human Genome," *Wikipedia*, accessed October 31, 2014, http://en.wikipedia.org/wiki/Adaptive_evolution_in_the_human_genome.

3. Scott Grosse et al., "Centers for Disease Control and Prevention. Newborn Screening for Cystic Fibrosis," *Morbidity and Mortality Weekly Report* 53, no. RR13 (2004): 1–36.

4. Centers for Disease Control and Prevention, "Sickle Cell Disease (SCD)," CDC.gov, accessed September 7, 2014, http://www.cdc.gov/ncbddd/sicklecell/data.html.

5. Centers for Disease Control and Prevention, "Autism Spectrum Disorder (ASD)," CDC.gov, accessed October 31, 2014, http://www.cdc.gov/ncbddd/autism/data.html.

6. Dr. Gerald Fink, Whitehead Institute seminar, November 2, 2012, attended by one of the authors (Juan Enriquez). These numbers are way too specific, given what we currently know about autism and related spectrum disorders. There are likely many underlying and overlapping conditions with the same symptoms but different causes. But these estimates do provide an order of magnitude of likelihood.

7. X. Liu and T. Takumi, "Genomic and Genetic Aspects of Autism Spectrum Disorder," *Biochemical & Biophysical Research Communications* 452, no. 2 (2014): 244–53; A. M. Persico and V. Napolioni, "Autism Genetics," *Behavioural Brain Research* 251 (2013): 95–112; I. Iossifov, B. J. O'Roak, S. J. Sanders, M. Ronemus et al., "The Contribution of *De Novo* Coding Mutations to Autism Spectrum Disorder," *Nature* 515, no. 7526 (2014): 216–21; doi:10.1038/nature13908. S. De Rubeis, X. He, A. P. Goldberg,

C. S. Poultney et al., "Synaptic, Transcriptional and Chromatin Genes Disrupted in Autism," *Nature* 515, no. 7526 (2014): 209–15; doi:10.1038/nature13772.

8. J. Baio, "Prevalence of Autism Spectrum Disorders—Autism and Developmental Disabilities Monitoring Network, 14 Sites, United States, 2008," *Morbidity and Mortality Weekly Report (MMWR)* 61, no. SSO3 (2012), accessed August 21, 2014, http://www.cdc.gov/mmwr/preview/mmwrhtml/ss6103a1.htm.

9. J. Baio, "Prevalence of Autism Spectrum Disorder Among Children Aged 8 Years—Autism and Developmental Disabilities Monitoring Network, 11 Sites, United States, 2010," *Morbidity and Mortality Weekly Report (MMWR)* 63, no. SS02 (2014): 1–21, accessed August 28, 2014, http://www.cdc.gov/mmwr/preview/mmwrhtml/ss6103a1.htm.

10. American Psychiatric Association, *Diagnostic and Statistical Manual of Mental Disorders, Fifth Edition* (DSM-5) (Arlington, VA: American Psychiatric Publishing, 2013). This is considered the textbook "bible" on all mental disorders and conditions. In May 2013 a controversy exploded around the publication of this new edition, the DSM-5. Previously, the expert authors, nicknamed by one blogger "The Gods of the Mind," differentiated Asperger's as a separate condition from autism, but in the updated version Asperger's is part of autism spectrum disorder (ASD). (Perhaps this redefinition might reclassify people like Einstein and Warhol as autistic?) "DSM-5 Diagnostic Criteria," AutismSpeaks.org, accessed August 28, 2014, http://www.autismspeaks.org/what-autism/diagnosis/dsm-5-diagnostic-criteria. For a more non-technical discussion of the topic see Amy S. F. Lutz, "You Do Not Have Asperger's," Slate.com, accessed August 28, 2014, http://www.slate.com/articles/health_and_science/medical_examiner/2013/05/autism_spectrum_diagnoses_the_dsm_5_eliminates_asperger_s_and_pdd_nos.html. It is worth noting that the DSM-5 text, like Darwin's books, is largely descriptive with little regard to underlying biology or disease mechanisms. It is ripe for major revisions in the future as a more mechanistically based understanding of psychiatric conditions is uncovered, a vision promulgated by Dr. Tom Insel, director of the National Institute of Mental Health (NIMH). See Tom Insel, "Director's Blog: Transforming Diagnosis," NIMH.NIH.gov, April 29, 2013, accessed August 28, 2014, http://www.nimh.nih.gov/about/director/2013/transforming-diagnosis.shtml.

11. S. Sandin, P. Lichtenstein, R. Kuja-Halkola, H. Larsson et al., "The Familial Risk of Autism," *Journal of the American Medical Association* 311, no. 17 (2014): 1770–77.

The DarWa Theory Revisited

1. Darwin's grandfather was a polymath, poet, abolitionist, and bohemian, a far-from-uptight gentleman who was far ahead of his time on many subjects, including evolution, something he foreshadowed in two books, *Zoonomia* and *The Botanic Garden*. He was a great mentor, inspirer, and supporter of his young and more conservative grandson.

2. Charles Darwin, *The Descent of Man* and *Selection in Relation to Sex* (London: John Murray, 1871), 2.

3. Darwin preferred the term "variation" instead of "mutation," the latter of which became the dominant term with the advent of the modern synthesis theory of evolution that fused Darwin, Wallace, and Mendel in the early twentieth century.

4. Darwin Correspondence Project, "Darwin, C. R. to Lyell, Charles, 18 [June 1858], Letter 2285," DarwinProject.ac.uk, accessed August 29, 2014, https://www.darwinproject.ac.uk/letter/entry-2285.

5. Two of the best and most readable works on Darwin are by the great historian of science Janet Browne. The description of these dramatic days is based on her volume

Charles Darwin: The Power of Place (Princeton: Princeton University Press, 2003). You might also enjoy reading the preceding volume: Janet Browne, *Charles Darwin: Voyaging* (Princeton: Princeton University Press, 1996).

6. John Offer, ed., *Herbert Spencer, Critical Assessments*, vol. 2 (London: Taylor & Francis, 2000), 3.

7. Gregor Mendel, in his 1866 book, conceived that inheritance involved "discrete units" of heredity that existed as dominant and recessive traits in offspring. Mendel did not use the word "gene." The word, which derives from "pangenesis," was first used by a Danish botanist, Wilhelm Johannsen, in 1909. Johannsen's text was published in German in 1913 as *Elemente der Exakten Erblichkeitslehre* (Jena: Gustav Fischer), and "gene" was described on pages 143–44. The German book is here: http://caliban.mpipz.mpg.de/johannsen/elemente/index.html. A translation to English of the relevant passage is available in Marcel Weber's *Philosophy of Experimental Biology* (Cambridge, UK: Cambridge University Press, 2005), 195.

8. World history was traditionally driven by one overwhelming idea: An all-powerful God, or gods, created all life, all at once, unchangeable thereafter. Sometimes life was seen as a cycle of reincarnation. Others saw life as a journey through pain, suffering, and purgatory, to eventual heaven or hell—a vision that persists for many today. In the United States in June 2014, despite all scientific evidence, about 42 percent of people still believe the creationist version and reject Darwin. See Gallup, "In U.S., 42% Believe Creationist View of Human Origins," Gallup.com, accessed August 29, 2014, http://www.gallup.com/poll/170822/believe-creationist-view-human-origins.aspx. But as the T-shirt says, science is true whether you believe it or not.

9. G. B. Dalrymple, "The Age of the Earth in the Twentieth Century: A Problem (Mostly) Solved," *Special Publications, Geological Society of London* 190, no. 1 (2001): 205–21.

10. Ben Schiller, "A Massive Global Map of Where All the Cattle, Pigs, and Other Livestock Live," *Fast Company*, fastcoexist.com, accessed August 29, 2014, http://www.fastcoexist.com/3031945/a-massive-global-map-of-where-all-the-cattle-pigs-and-other-livestock-live.

11. See, for example, the circular map of the maize genome: P. S. Schnable, D. Ware, R. S. Fulton, J. C. Stein et al., "The B73 Maize Genome: Complexity, Diversity, and Dynamics," *Science* 326, no. 5956 (2009): 1112–115.

12. D. A. Jackson, R. H. Symons, and P. Berg, "Biochemical Method for Inserting New Genetic Information into DNA of Simian Virus 40: Circular SV40 DNA Molecules Containing Lambda Phage Genes and the Galactose Operon of *Escherichia coli*," *Proceedings of the National Academy of Sciences USA* 69, no. 10 (1972): 2904–9.

13. High school students can now easily apply these techniques thanks to the restriction enzymes discovered by Hamilton Smith. See H. O. Smith and K. W. Welcox, "A Restriction Enzyme from *Hemophilus influenzae*. I. Purification and General Properties," *Journal of Molecular Biology* 51, no. 2 (1970): 379–91; T. J. Kelly Jr. and H. O. Smith, "Restriction Enzyme from *Haemophilus influenzae*. II. Base Sequence of the Recognition Site," *Journal of Molecular Biology* 51, no. 2 (1970): 393–409. And because of the follow-up to Paul Berg's work done by Herb Boyer and Stanley Cohen, who later became Nobel laureates as well.

Twenty Generations to Domesticate Humans

1. D. K. Belyaev, "Domestication of Animals," *Science (Russ.)* 5, no. 1 (1969): 47–52. Ongoing research: "Study of the Molecular Basis of Tame and Aggressive Behavior in the Silver Fox Model," CBSU.TC.Cornell.edu, accessed August 29, 2014, http://cbsu.tc.cornell

.edu/ccgr/behaviour/History.htm. Lay summary and videos: "Thoughtful Animal: Monday Pets: The Russian Fox Study," *ScienceBlogs*, June 14, 2010, accessed August 29, 2014, http://scienceblogs.com/thoughtfulanimal/2010/06/14/monday-pets-the-russian-fox-st.

2. B. B. Horswell and C. J. Chahine, "Dog Bites of the Face, Head and Neck in Children," *West Virginia Medical Journal* 107, no. 6 (2011): 24–27.

3. For Darwin, "artificial selection" described the human activity of interbreeding specific plants or animals within a species or across two closely related species to create offspring/hybrids with specific beneficial traits. This has been the approach to creating most domesticated plants and animals on the planet today, such as our major crops, livestock, many pets, and many horticultural varieties. Charles Darwin, *On the Origin of Species by Means of Natural Selection*, 1st ed. (London: John Murray, 1859), 109.

4. Darwin wrote about the role and power of human-driven selection as both a conscious and an unconscious act, predominantly in the context of plant and animal breeding. He did not and could not appreciate the degree to which it has been applied and expanded over the past 150 years.

5. Carl Haub, "How Many People Have Ever Lived on Earth?" *Population Reference Bureau Publications*, October 2011, accessed August 23, 2014, http://www.prb.org/Publications/Articles/2002/HowManyPeopleHaveEverLivedonEarth.aspx.

6. U.S. Census, "World Population," Census.gov, December 2013, accessed August 29, 2014, https://www.census.gov/population/international/data/worldpop/table_history.php.

7. World Health Organization, "Urban Population Growth," WHO.org, accessed August 29, 2014, http://www.who.int/gho/urban_health/situation_trends/urban_population_growth_text/en/#.

8. Ian Johnson, "China's Great Uprooting: Moving 250 Million into Cities," *New York Times*, June 16, 2013, accessed August 23, 2014, http://www.nytimes.com/2013/06/16/world/asia/chinas-great-uprooting-moving-250-million-into-cities.html.

9. E. Trinkaus, "Late Pleistocene Adult Mortality Patterns and Modern Human Establishment," *Proceedings of the National Academy of Sciences USA* 108, no. 4 (2011): 1267–271.

10. But of course not all recognize how far we have come and where important fundamental changes in diet and hygiene have taken us; how else might one explain "Paleo" as the most searched term for diet on Google in 2013? Despite having many followers, the diet—basically meat, occasional wild plants, no dairy, sugar, grain, or veggies—was ranked absolutely the worst among those reviewed by *U.S. News and World Report*. See "Paleo Diet," *US News and World Report*, October 20, 2014, accessed August 29, 2014, http://health.usnews.com/best-diet/paleo-diet/reviews.

11. Want to rubberneck a little? Here is every known, historically confirmed shark attack incident worldwide: Shark Attack File, "Incident File," SharkAttackFile.com, accessed August 29, 2014, http://www.sharkattackfile.net/incidentlog.htm.

12. P. Hunter, "The Human Impact on Biological Diversity. How Species Adapt to Urban Challenges Sheds Light on Evolution and Provides Clues About Conservation," *EMBO Reports* 8, no. 4 (2007): 316–18.

13. Trinkaus, "Late Pleistocene Adult Mortality Patterns."

14. Sharks kill about 5 people per year. Death from war-related violence kills about 55,000 per year. National Geographic Channel, "Human Shark Bait," NatGeoTV.com, accessed August 29, 2014, http://natgeotv.com/ca/human-shark-bait/facts; Andrew Mack et al., "The Human Security Report 2013: The Decline in Global

Violence: Evidence, Explanation, and Contestation," Human Security Report Project, accessed September 1, 2014, http://www.hsrgroup.org/docs/Publications /HSR2013/HSRP_Report_2013_140226_Web.pdf; B. A. Lacina and N. P. Gleditsch, "Monitoring Trends in Global Combat: A New Dataset of Battle Deaths," *European Journal of Population* 21, no. 2 (2005): 145–65.

15. Steven Pinker, *The Better Angels of Our Nature: Why Violence Has Declined* (New York: Viking, 2011). The entire book details, with hundreds of concrete examples and statistics, the collapse in overall societal violence.

16. Mack et al., "Human Security Report 2013." These days when something bad happens to someone almost anywhere, it increasingly becomes world news. In February 2014 a huge round boulder loosened from a picturesque Tuscan hilltop and rolled right over a perfect palazzo. While it made for an impressive photo, and was certainly a shame, the notion that a random building had been partly destroyed, in some distant land, is not something that used to worry an entire continent when it was about to be razed by invading Mongols, Huns, or Vikings.

17. War on Irrational Fear, "So, What's This All About?" WarOnIrrationalFear.org, accessed August 29, 2014, http://waronirrationalfear.com/facts.

18. Michael H. Reggio, "History of the Death Penalty," PBS.org, accessed August 29, 2014, http://www.pbs.org/wgbh/pages/frontline/shows/execution/readings/history.html.

19. The Hobbes comment is attributable to Steven Pinker from this article: Peter Singer, "Is Violence History?" *New York Times*, October 6, 2011, accessed September 1, 2014, http://www.nytimes.com/2011/10/09/books/review/the-better-angels-of-our -nature-by-steven-pinker-book-review.html.

20. "List of Countries by Intentional Homicide Rate," *Wikipedia*, accessed August 29, 2014, http://en.wikipedia.org/wiki/List_of_countries_by_intentional_homicide_rate.

21. Monica Mark, "African Leaders Vote Themselves Immunity from New Human Rights Court," *Guardian*, July 3, 2014, accessed August 29, 2014, http://www .theguardian.com/global-development/2014/jul/03/african-leaders-vote-immunity -human-rights-court.

22. Food and Agriculture Organization of the United Nations, "The State of Food Insecurity in the World 2013: The Multiple Dimensions of Food Security," FAO, International Fund for Agricultural Development (IFAD), and World Food Program (WFP), accessed August 29, 2014, http://www.fao.org/docrep/018/i3434e /i3434e.pdf.

23. J. W. Pryce, M. A. Weber, M. T. Ashworth, S. Roberts et al., "Changing Patterns of Infant Death over the Last 100 Years: Autopsy Experience from a Specialist Children's Hospital," *Journal of the Royal Society of Medicine* 105, no. 3 (2012): 123–30.

24. Over-loving and overdesigning particular breeds of dogs is a common phenomenon. The more popular the breed, the likelier it is to suffer inherited disorders. Just look at bulldogs and then review their overall health records. S. Ghirlanda, A. Acerbi, H. Herzog, and J. A. Serpell, "Fashion vs. Function in Cultural Evolution: The Case of Dog Breed Popularity," *PLoS ONE* 8, no. 9 (2013): e74770.

Side Effects of Nonviolence

1. D. Finkelhor, A. Shattuck, H. A. Turner, and S. L. Hamby. "Trends in Children's Exposure to Violence, 2003 to 2011," *JAMA Pediatrics* 168, no. 6 (2014): 540–46.

2. Max Roser, "Homicides: Time-Series of Homicide Rates Pre 1800," Our World in Data, accessed September 2, 2014, http://www.ourworldindata.org/data/violence -rights/homicides.

3. Liberty Knowledge Reason "The Systematic Idiocy," accessed October 31, 2014, http://www.vency.com/wars.html; this Web site summarizes statistics compiled by George C. Kohn, ed., *Dictionary of Wars*, 3rd ed. (New York: Facts on File, 2006).

4. Joseph Carrol, "Most Americans Approve of Interracial Marriages: Blacks More Likely Than Whites to Approve of Black-White Unions," Gallup, August 16, 2007, accessed September 2, 2014, http://www.gallup.com/poll/28417/most-americans-approve-inter racial-marriages.aspx.

5. Mike Develin, "Love and Religion," Facebook.com (Facebook Data Science), February 10, 2014, accessed September 2, 2014, https://www.facebook.com/notes/facebook-data -science/love-and-religion/10152056520123859.

6. Douglas Belkin, "Blue Eyes Are Increasingly Rare in America," *New York Times*, October 18, 2006, accessed September 2, 2014, http://www.nytimes.com/2006/10 /18/world/americas/18iht-web.1018eyes.3199975.html.

7. D. J. Handelsman, "Global Trends in Testosterone Prescribing, 2000–2011: Expand- ing the Spectrum of Prescription Drug Misuse," *Medical Journal of Australia* 199, no. 8 (2013): 548–51.

8. K. A. Pavlov, D. A. Chistiakov, and V. P. Chekhonin, "Genetic Determinants of Aggres- sion and Impulsivity in Humans," *Journal of Applied Genetics* 53, no. 1 (2012): 61–82.

9. Y. Lood, A. Eklund, M. Garle, and J. Ahlner, "Anabolic Androgenic Steroids in Police Cases in Sweden 1999–2009," *Forensic Science International* 219, nos. 1–3 (2012): 199–204. Of course, not everyone believes the pointy-headed scientists; former bodybuilder and Mr. Universe Lee Priest, for one, vehemently disagrees. In a 2014 press conference, he explained that he had used steroids for two decades and that the real problem was not 'roids but alcohol and the inability of parents to cane their kids: Ben Fordham, "Australia's Lead- ing Body Builder Says 'Roid Rage' Is a Myth and the Real Problem Is Alcohol," *Sunday Telegraph*, February 15, 2014, accessed September 2, 2014, http://www.dailytelegraph.com .au/news/nsw/australias-leading-body-builder-says-roid-rage-is-a-myth-and-the-real -problem-is-alcohol/story-fni0cx12-1226828027848.

10. Garrett Hellenthal et al., "A Genetic Atlas of Human Admixture History," compan- ion Web site for research article by the same title published in *Science*, February 14, 2014, accessed September 2, 2014, http://www.admixturemap.paintmychromosomes .com. You can look at many of the world's populations and who mixed with whom.

11. "Thomas Jefferson and Sally Hemings: A Brief Account," Monticello, accessed September 2, 2014, http://www.monticello.org/site/plantation-and-slavery/thomas -jefferson-and-sally-hemings-brief-account. Many people are still not reconciled to the genetic fact that Jefferson was very likely father of at least one, if not six, children by his slave. See also "Is It True?" *Frontline*, accessed September 2, 2014, http://www .pbs.org/wgbh/pages/frontline/shows/jefferson/true/.

12. I. Al-Gazali, H. Hamamy, and S. Al-Arrayad, "Genetic Disorders in the Arab World," *British Medical Journal* 333, no. 21 (2006): 831–34; G. Tadmouri et al., "Table 1: Consanguinity Rates in Arab Populations," in "Consanguinity and Reproductive Health Among Arabs," *Reproductive Health*, October 8, 2009, accessed September 2, 2014, http://www.ncbi.nlm.nih.gov/pmc/articles/PMC2765422/table/T1/.

13. C. Rivoisy, L. Gérard, D. Boutboul, M. Malphettes et al. (DEFI study group), "Paren- tal Consanguinity Is Associated with a Severe Phenotype in Common Variable

Immunodeficiency," *Journal of Clinical Immunology* 32, no. 1 (2012): 98–105; C. Stoll, Y. Alembik, B. Dott, and J. Feingold, "Parental Consanguinity as a Cause of Increased Incidence of Birth Defects in a Study of 131,760 Consecutive Births," *American Journal of Medical Genetics* 49, no. 1 (1994): 114–17.

14. European Surveillance of Congenital Anomalies, "Prevalence Tables," Eurocat, accessed October 28, 2014, http://www.eurocat-network.eu/accessprevalencedata /prevalencetables.

15. Betsy McKay and Ellen Knickmeyer, "Saudi Researchers Mount Genome-Sequencing Push," *Wall Street Journal*, February 5, 2014, B8, accessed September 2, 2014, http://online.wsj.com/news/articles/SB20001424052702304887104579306831456121354; M. Al-Owain, H. Al-Zaidan, and Z. Al-Hassnan, "Map of Autosomal Recessive Genetic Disorders in Saudi Arabia: Concepts and Future Directions," *American Journal of Medical Genetics A* 15A, no. 10 (2012): 2629–40.

16. A. Bener, R. Hussain, and A. S. Teebi, "Consanguineous Marriages and Their Effects on Common Adult Diseases: Studies from an Endogamous Population," *Medical Principles and Practice* (International Journal of the Kuwait University, Health Science Centre) 16, no. 4 (2007): 262–67.

17. John Siple, "A Brief History of the 'Habsburg Chin,'" *The Society of the Golden Fleece* (blog), accessed September 7, 2014, http://www.antiquesatoz.com/habsburg /habsburg-jaw.htm; Craig Stillwell, "The Hapsburg Lip," Topics in the History of Genetics and Molecular Biology Fall 2000 syllabus, Michigan State University, accessed September 2, 2014, https://www.msu.edu/course/lbs/333/fall/hapsbur glip.html; see also *Wikipedia*'s entry for "Inbreeding" for all kinds of great tales, accessed September 4, 2014, http://en.wikipedia.org/wiki/Haemophilia_in_European _royalty.

Allergies: Another Harbinger of Our Evolving Bodies?

1. Centers for Disease Control and Prevention, "Trends in Allergic Conditions Among Children: United States, 1997–2011," CDC.gov, NCHS Data Brief 121, May 2013, accessed August 29, 2014, http://www.cdc.gov/nchs/data/databriefs/db121.htm.

2. Melanie Thernstrom, "The Allergy Buster: Can a Radical New Treatment Save Children with Severe Food Allergies?" *New York Times Magazine*, March 7, 2013, accessed August 29, 2014, http://www.nytimes.com/2013/03/10/magazine/can-a -radical-new-treatment-save-children-with-severe-allergies.html. Thernstrom's article cites this study: A. M. Barnum and S. L. Lukacs, "Food Allergy Among U.S. Children: Trends in Prevalence and Hospitalizations," *NCHS Data Brief* 10, October 2008, accessed August 29, 2014, http://www.cdc.gov/nchs/data/databriefs/db10.htm #howdoes.

3. AllergyEats, "Dunkin' Donuts Allergy Sign Creates Controversy," *AllergyEats* (blog), accessed August 29, 2014, http://www.allergyeats.com/blog/index.php/dunkin-donuts -allergy-sign-creates-controversy.

4. H. A. Sampson, "Update on Food Allergy," *Journal of Allergy and Clinical Immunology* 113, no. 5 (2004): 805–19; S. H. Sicherer, "Food Allergy," *Lancet* 360, no. 9334 (2002): 701–10.

5. S. H. Sicherer, A. Muñoz-Furlong, J. H. Godbold, and H. A. Sampson, "US Prevalence of Self-Reported Peanut, Tree Nut, and Sesame Allergy: 11-Year Follow-up," *Journal of Allergy and Clinical Immunology* 125, no. 6 (2010): 1322–26.

6. The *New York Times Magazine* had a great overview article on the topic. See Thernstrom, "The Allergy Buster."

7. D. S. Kim and A. B. Drake-Lee, "Infection, Allergy and the Hygiene Hypothesis: Historical Perspective," *Journal of Laryngology and Otology* 117, no. 12 (2003): 946–50; Maggie Moon, "The Hygiene Hypothesis," *Today's Dietician*, July 12, 2000, accessed August 29, 2014, http://www.todaysdietitian.com/newarchives/062909p12.shtml.

8. M. Holbreich, J. Genuneit, J. Weber, C. Braun-Fahrländer et al., "Amish Children Living in Northern Indiana Have a Very Low Prevalence of Allergic Sensitization," *Journal of Allergy and Clinical Immunology* 129, no. 6 (2012): 1671–73.

9. Centers for Disease Control and Prevention, "Trends in Allergic Conditions Among Children." On page 4: "The prevalence of both food allergy and respiratory allergy increased with the increase of income level. Among children with family income less than 100% of the poverty level, 4.4% had a food allergy and 14.9% had a respiratory allergy. Food allergy prevalence among children with family income between 100% and 200% of the poverty level was 5.0%, and respiratory allergy prevalence was 15.8%. Among children with family income above 200% of the poverty level, food allergy prevalence was 5.4%, and respiratory allergy prevalence was 18.3%. There was no significant difference in the prevalence of skin allergy by poverty status."

10. Adam Bible, "What's Going on Inside Your Stomach," *Men's Fitness*, April 2014, 98, accessed September 1, 2014, http://www.mensfitness.com/nutrition/what-to-eat/whats-going-inside-your-stomach. Bible's article is based on research by Jeff Leach, "Human Food Project, Anthropology of Microbes," accessed September 1, 2014, http://humanfoodproject.com/author/jeff-leach.

11. R. E. Ley, D. A. Peterson, and J. I. Gordon, "Ecological and Evolutionary Forces Shaping Microbial Diversity in the Human Intestine," *Cell* 124, no. 4 (2006): 837–48.

12. C. Boschi-Pinto, L. Velebit, and K. Shibuya, "Estimating Child Mortality Due to Diarrhoea in Developing Countries," *Bulletin of the World Health Organization* 86, no. 9 (2008): 710–17.

13. United Nations, "Unsafe Water Kills More People Than War, Ban Says on World Day," UN.org, accessed August 29, 2014, http://www.un.org/apps/news/story.asp?NewsID=34150.

Our Unnatural "All-Natural" World

1. C. Warinner, J. F. Rodrigues, R. Vyas, C. Trachsel et al., "Pathogens and Host Immunity in the Ancient Human Oral Cavity," *Nature Genetics* 46, no. 4 (2014): 336–44; G. H. Sperber, "The Role of Teeth in Human Evolution," *British Dental Journal* 215, no. 6 (2013): 295–97.

2. C. J. Adler, K. Dobney, L. S. Weyrich, J. Kaidonis et al., "Sequencing Ancient Calcified Dental Plaque Shows Changes in Oral Microbiota with Dietary Shifts of the Neolithic and Industrial Revolutions," *Nature Genetics* 45, no. 4 (2013): 450–55.

3. Holly Wagner, "Men from Early Middle Ages Were Nearly as Tall as Modern People," *Ohio State University Research News*, September 1, 2004, accessed September 7, 2014, http://researchnews.osu.edu/archive/medimen.htm.

4. Robert Longley, "Americans Getting Taller, Bigger, Fatter, Says CDC," *AboutNews*, accessed September 7, 2014, http://usgovinfo.about.com/od/healthcare/a/tallbutfat.htm.

5. And perhaps you did not know that fear of vegetables is called "lachanophobia."

6. D. M. Spooner, K. McLean, G. Ramsay, R. Waugh et al., "A Single Domestication for Potato Based on Multilocus Amplified Fragment Length Polymorphism Genotyping," *Proceedings of the National Academy of Sciences USA* 102, no. 41 (2005): 14694–99.

7. M. McMillan and J. C. Thompson, "An Outbreak of Suspected Solanine Poisoning in Schoolboys: Examinations of Criteria of Solanine Poisoning," *Quarterly Journal of Medicine* 48, no. 190 (1979): 227–33.

8. The National Academies Press produced a version of "Everything You Always Wanted to Know About Obscure Peruvian Plants (but were afraid to ask . . .)": Ad Hoc Panel of the Advisory Committee on Technology Innovation, Board on Science and Technology for International Development, National Research Council, *Lost Crops of the Incas: Little-Known Plants of the Andes with Promise for Worldwide Cultivation* (Washington, DC: National Academies Press, 1989), 93ff, accessed August 23, 2014, http://www.nap.edu/catalog.php?record_id=1398. The section on potatoes begins here: http://www.nap.edu/openbook.php?record_id=1398&page=93.

9. *Smithsonian* magazine had a good article on the potato and its importance: Charles C. Mann, "How the Potato Changed the World," *Smithsonian*, November 2011, accessed August 25, 2014, http://www.smithsonianmag.com/history/how-the-potato -changed-the-world-108470605.

10. It took 150 years, but genomics finally allowed scientists to figure out exactly why this blight was so severe. Kentaro Yoshida et al., "The Rise and Fall of the *Phytophthora infestans* Lineage That Triggered the Irish Potato Famine," *eLife* 2 (2013): e00731.

11. We realize you are just dying to read all the statistics, particularly since every 1 percent increase in inbreeding lowers lifetime milk production by an average of 389.4 pounds of milk, so here is the source: Dairy Cattle Reproduction Council, "Implications of Inbreeding on the Dairy Industry," accessed August 23, 2014, http://www.dcrcouncil.org/media /Public/Implications%20of%20inbreeding%20in%20the%20dairy%20industry.pdf.

12. Committee on the Strategic Planning for the Florida Citrus Industry, *Addressing Citrus Greening* (Washington, DC: National Academies Press, 2010), 13, accessed August 23, 2014, http://www.nap.edu/openbook.php?record_id=12880&page=13.

13. Kal Kupferschmidt, "How Tomatoes Lost Their Taste," *Science*, June 28, 2012, accessed August 23, 2014, http://news.sciencemag.org/sciencenow/2012/06/how-tomatoes-lost -their-taste.html.

14. "Nature: Wildlife: Cattle and Aurochs," BBC, accessed September 7, 2014, http: //www.bbc.co.uk/nature/life/Aurochs.

15. Susan Schoenian, "Sheep Breeds," *Sheep101* (blog) accessed September 7, 2014, http: //www.sheep101.info/breedsM-N.html; "Mouflon," *Wikipedia*, accessed September 7, 2014, http://en.wikipedia.org/wiki/Mouflon.

16. "Half Empty or Half Full: Science and Biodiversity," posted by Sharon on April 13, 2014, 9:45 a.m., *A New Century of Forest Planning* (blog), accessed August 23, 2014, http://forestpolicypub.com/2014/04/13/half-empty-or-half-full-science-and -biodiversity.

17. There are hundreds of examples in nature and in the laboratory of acclimation of a species to an acute environmental change in as few as one to three generations— including a wide variety of plants, insects, animals, and even Darwin's finches. Recent studies have drawn attention to the ability of marine species living in coral reefs to adapt to conditions anticipated to occur with global warming. Some recent literature: A. Belgrano and C. W. Fowler, "How Fisheries Affect Evolution," *Science* 342, no. 6163 (2013): 1176–177. J. M. Donelson, P. L. Munday, M. I. McCormick et al., "Rapid Transgenerational Acclimation of a Tropical Reef Fish to Climate Change," *Nature*

Climate Change 2 (2012): 30–2. H. D. Veilleux, T. Ryu, J. M. Donelson, L. van Herwerden et al., "Molecular Processes of Transgenerational Acclimation to a Warming Ocean," *Nature Climate Change* 5 (2015): 1074–78. C. K. Ghalambor, K. L. Hoke, E. W. Ruell, E. K. Fischer et al., "Non-Adaptive Plasticity Potentiates Rapid Adaptive Evolution of Gene Expression in Nature," *Nature* 525, no. 7569 (2015): 372–5. S. P. Egan, G. J. Ragland, L. Assour, T. H. Q. Powell et al., "Experimental Evidence of Genome-Wide Impact of Ecological Selection During Early Stages of Speciation-with-Gene-Flow," *Ecology Letters* 18, no. 8 (2015): 817–25.

18. Duane Jeffery, "Science and Society: Evolution and the Shrinking Fish," *Daily Herald*, (Provo, UT) February 27, 2013, accessed September 1, 2014, http://www.herald extra.com/lifestyles/science-and-society-evolution-and-the-shrinking-fish/article _d5d831aa-8ba0-5486-9b09-f990d13ae454.html; David Malakoff, "Ocean Fishing May Spread 'Runt' Genes," NPR, May 27, 2006, accessed September 1, 2014, http: //www.npr.org/templates/story/story.php?storyId=5434698.

19. Jeremy Hobson, "Glut of Lobster Brings Price to a 20-Year Low in Maine," *Here & Now*, August 12, 2013, accessed August 25, 2014, http://hereandnow.wbur.org/2013 /08/12/lobster-price-low.

20. Paul Recer, "Studies: Human Hunting Led to Extinctions," ABC News, accessed September 6, 2014, http://abcnews.go.com/Technology/story?id=98510; J. Alroy, "A Multi-Species Overkill Simulation of the End-Pleistocene Megafaunal Mass Extinction," *Science* 292 (2001): 1893–96, accessed September 6, 2014, https://www.sciencemag .org/content/292/5523/1893.

21. "The Extinction Crisis," The Center for Biological Diversity, accessed August 25, 2014, http://www.biologicaldiversity.org/programs/biodiversity/elements_of_biodiversity /extinction_crisis.

22. C. Soto-Azat, A. Valenzuela-Sánchez, B. T. Clarke, K. Busse et al., "Is Chytridiomy-cosis Driving Darwin's Frogs to Extinction?" *PLoS ONE* 8, no. 11 (2013): e79862. This odd creature, discovered by Darwin in 1834 on Lemuy Island, Chile, was one of two frogs that incubates eggs in its mouth.

23. S. L. Pimm, C. N. Jenkins, R. Abell, T. M. Brooks et al., "The Biodiversity of Species and Their Rates of Extinction, Distribution, and Protection," *Science* 344, no. 6187 (2014): 1246752.

24. Natalie Angier, "New Creatures in an Age of Extinction," *New York Times*, July 26, 2009, accessed September 6, 2014, http://www.nytimes.com/2009/07/26/weekin review/26angier.html; "New Species Appear to Arise from Sudden Changes," Phys.org, February 19, 2013, accessed September 6, 2014, http://phys.org/news/2013-02-species -sudden.html.

25. Fred Pearce, "Human Meddling Will Spur the Evolution of New Species," *New Scientist*, January 17, 2014, accessed August 25, 2014, http://www.newscientist.com/article /mg22129510.400-human-meddling-will-spur-the-evolution-of-new-species.html.

26. Madhusudan Katti, "Biodiversity Can Flourish on an Urban Planet," *The Conversation* (blog), January 22, 2014, accessed September 6, 2014, http://theconversation. com/biodiversity-can-flourish-on-an-urban-planet-18723; M. F. Aronson, F. A. La Sorte, C. H. Nilon, M. Katti et al., "A Global Analysis of the Impacts of Urbanization on Bird and Plant Diversity Reveals Key Anthropogenic Drivers," *Proceedings of the Royal Society B* 281, no. 1780 (2014): 20133330.

27. E. C. Snell-Rood and N. Wick, "Anthropogenic Environments Exert Variable Selection on Cranial Capacity in Mammals," *Proceedings of the Royal Society B* 280, no. 1769 (2013): 20131384.

28. *The Onion* has a wonderful sight gag on the wolf-to-dog transition: "Study: This Descended from Wolves," *The Onion*, January 8, 2014, accessed September 6, 2014, http://www.theonion.com/articles/study-this-descended-from-wolves,34898.

29. P. Savolainen, Y. P. Zhang, J. Luo, J. Lundeberg et al., "Genetic Evidence for an East Asian Origin of Domestic Dogs," *Science* 298, no. 5598 (2002): 1610–13.

30. We love our newly designed, genetically modified pets. Acquiring a "purebred" is ever more expensive and complex. So too is feeding it. In March 2014, near the verdant pastures of Greenwich, Connecticut, Bob Vetere, president and CEO of the American Pet Products Association (APPA), was "pleased to announce overall spending in the pet industry for 2013 exceeded early estimates coming in at an all-time high of more than $55.7 billion." See "Pet Spending Higher Than Ever with an Estimated $58.5 Billion in Spending in 2014," Insurance News Net, March 15, 2014, accessed August 25, 2014, http://insurance newsnet.com/oarticle/2014/03/15/pet-spending-higher-than-ever-with-an-estimated -%24585-billion-in-spending-in-201-a-474886.html.

Fat Humans, Fat Animals: Another Symptom?

1. Harvard School of Public Health, "Obesity Has Doubled Since 1980, Major Global Analysis of Risk Factors Reveals," February 3, 2011, accessed August 30, 2014, http://www .hsph.harvard.edu/news/press-releases/worldwide-obesity; World Health Organization, "Obesity and Overweight," WHO.org, accessed September 7, 2014, http://www.who.int /mediacentre/factsheets/fs311/en; M. M. Finucane, G. A. Stevens, M. J. Cowan, D. Goodarz et al., "National, Regional, and Global Trends in Body-Mass Index Since 1980: Systematic Analysis of Health Examination Surveys and Epidemiological Studies with 960 Country-Years and 9.1 Million Participants," *Lancet* 377, no. 9765 (2011): 557–67.

2. Michael Moss, "The Extraordinary Science of Addictive Junk Food," *New York Times Magazine*, February 20, 2013, accessed August 28, 2014, http://www.nytimes.com /2013/02/24/magazine/the-extraordinary-science-of-junk-food.html. Want to know how clinically obese BMI is calculated? It is a combination of height and weight. There are, of course, far more expensive, detailed, and complex ways to parse a large body: Centers for Disease Control and Prevention, "Healthy Weight—It's Not a Diet, It's a Lifestyle!" CDC.gov, accessed August 28, 2014, http://www.cdc.gov /healthyweight/assessing/bmi/adult_bmi/index.html.

3. World Health Organization, "Obesity and Overweight." Of note, the trend in childhood obesity may be starting to plateau or even reverse in the United States: Centers for Disease Control and Prevention, "New CDC Data Show Encouraging Development in Obesity Rates Among 2 to 5 Year Olds," CDC.gov, February 25, 2014, accessed August 30, 2014, http://www.cdc.gov/media/releases/2014/p0225-child-obesity.html.

4. "Mexico Obesity Rate Higher than U.S., Says U.N. Report," *Huffington Post*, July 9, 2013, accessed August 30, 2014, http://www.huffingtonpost.com/2013/07/09/mexico -obesity-rate-united-states_n_3568537.html.

5. Lisa Young, "Bloomberg's Cap on Supersize Soda May Be Contagious," *Huffington Post*, December 26, 2013, accessed August 30, 2014, http://www.huffingtonpost.com /dr-lisa-young/bloomberg-soda_b_4494788.html. If you want to see the graphic equivalent of sugar and your food, go to: http://www.sugarstacks.com/beverages.htm.

6. As you might imagine, the soft-drink/junk-food industrial complex would love to find alternative explanations/causality to partly reduce their share of the blame and perhaps help confuse looming courtroom battles. Our objective is not to minimize their enormous contribution to the pandemic but to also examine additional inputs and accelerators.

7. C. A. Befort, N. Nazir, and M. G. Perri, "Prevalence of Obesity Among Adults from Rural and Urban Areas of the United States: Findings from NHANES (2005–2008)," *Journal of Rural Health* 28, no. 4 (2012): 392–97.

8. T. S. Church, D. M. Thomas, C. Tudor-Locke, P. T. Katzmarzyk et al., "Trends over 5 Decades in U.S. Occupation-Related Physical Activity and Their Associations with Obesity," *PLoS ONE* 6, no. 5 (2011): e19657.

9. A. M. Linabery, R. W. Nahhas, W. Johnson, A. C. Choh et al., "Stronger Influence of Maternal Than Paternal Obesity on Infant and Early Childhood Body Mass Index: The Fels Longitudinal Study," *Pediatric Obesity* 8, no. 3 (2013): 159–69.

10. D. P. McCormick, K. Sarpong, L. Jordan, L. A. Ray et al., "Infant Obesity: Are We Ready to Make This Diagnosis?" *Journal of Pediatrics* 57, no. 1 (2010): 15–19. And if you want to bring dollars and cents into the discussion, each obese baby costs the health-care system about $19,000 more over a lifetime than a normal-weight child. See E. A. Finkelstein, W. C. Graham, and R. Malhotra, "Lifetime Direct Medical Costs of Childhood Obesity," *Pediatrics* 133, no. 5 (2014): 854–62.

11. M. M. Kelsey, A. Zaepfel, P. Bjornstad, and K. J. Nadeau, "Age-Related Consequences of Childhood Obesity," *Gerontology* 60, no. 3 (2014): 222–28.

12. American Heart Association, "Kids Less Fit Today Than Those in 1970s," American Heart Association, November 19, 2013, accessed August 30, 2014, http://www.heart.org/HEARTORG/News/Global/SimpleScience/Kids-less-fit-today-than-those-in-1970s_UCM_458398_Article.jsp.

13. PetFirst, "PetFirst: Pet Insurance to Be More Popular in 2008," PetFirst Pet Insurance Blog, February 7, 2008, accessed August 30, 2014, http://blog.petfirst.com/PetfirstBlog/index.php/2008/02/07/petfirst-pet-insurance-popular-2008.

14. PetFirst, "Pet Health Crisis: Americans Skimp on Preventative Care," PetFirst Pet Insurance Blog, January 12, 2014, accessed August 30, 2014, http://blog.petfirst.com/PetfirstBlog/index.php/2014/01/12/pet-health-crisis-americans-skimp-preventative-care.

15. Meanwhile, back in Washington, DC, a bipartisan "compromise" cut $9 billion out of the Supplemental Nutrition Assistance Bill, otherwise known as food stamps, which means the 47,636,000 poorest people in the country will receive a wildly generous $1.40 per person per meal. As animal spending relentlessly increases, and as we cut programs for the poor, we should soon be able to announce that we spend more on our furry friends than on the poor. We are almost 90 percent of the way there already. See Stacy Dean and Dottie Rosebaum, "SNAP Benefits Will Be Cut for Nearly All Participants in November 2013," Center on Budget and Policy Priorities, August 2, 2013, accessed August 30, 2014, http://www.cbpp.org/cms/?fa=view&id=3899.

16. Y. C. Klimentidis, T. M. Beasley, H. Y. Lin, G. Murati et al., "Canaries in the Coal Mine: A Cross-Species Analysis of the Plurality of Obesity Epidemics," *Proceedings of the Royal Society B* 278, no. 1712 (2011): 1626–632.

17. G. Cizza and K. I. Rother, "Beyond Fast Food and Slow Motion: Weighty Contributors to the Obesity Epidemic," *Journal of Endocrinological Investigation* 35, no. 2 (2012): 236–42. This article provides a great review of various hypotheses; we do not cover them all in the text but simply pick a few examples that illustrate how human interventions and actions can alter the average weight of a species.

18. David A. Kessler, "Antibiotics and the Meat We Eat," *New York Times*, March 27, 2013, accessed August 30, 2014, http://www.nytimes.com/2013/03/28/opinion/antibiotics-and-the-meat-we-eat.html?_r=0&pagewanted=print.

19. G. C. Amos, L. Zhang, P. M. Hawkey, W. H. Gaze et al., "Functional Metagenomic Analysis Reveals Rivers Are a Reservoir for Diverse Antibiotic Resistance Genes," *Veterinary Microbiology* 171, nos. 3–4 (2014): 441–47.

20. F-H. Wang, M. Qiao, J-Q. Su, Z. Chen et al., "High Throughput Profiling of Antibiotic Resistance Genes in Urban Park Soils with Reclaimed Water Irrigation," *Environmental Science & Technology* 48, no. 16 (2014): 9079–85.

21. X. Li, H. T. Pham, A. S. Janesick, and B. Blumberg, "Triflumizole Is an Obesogen in Mice That Acts Through Peroxisome Proliferator Activated Receptor Gamma (PPARy)," *Environmental Health Perspectives* 120, no. 12 (2012): 1720–26.

22. A. L. Simmons, J. J. Schlezinger, and B. E. Corkey, "What Are We Putting in Our Food That Is Making Us Fat? Food Additives, Contaminants, and Other Putative Contributors to Obesity," *Current Obesity Reports* 3, no. 2 (2014): 273–85.

23. A. Ait-Belgnaoui, A. Colon, V. Braniste, L. Ramalho et al., "Probiotic Gut Effect Prevents the Chronic Psychological Stress-Induced Brain Activity Abnormality in Mice," *Neurogastroenterology & Motility* 26, no. 4 (2014): 510–20.

24. L. Chang, S. Sundaresh, J. Elliott, P. A. Anton et al., "Dysregulation of the Hypothalamic-Pituitary-Adrenal (HPA) Axis in Irritable Bowel Syndrome," *Neurogastroenterology & Motility* 21, no. 2 (2009): 149–59.

25. X. L. Feng, Y. Wang, L. An, and C. Ronsmans, "Cesarean Section in the People's Republic of China: Current Perspectives," *International Journal of Women's Health* 6 (2014): 59–74.

An Evolutionary Anomaly: More Sex, Less Reproduction

1. Assuming, of course, that things are normal. If a big meteorite comes along or a supervolcano erupts, no amount of near-term sex will allow humans to adapt fast enough and save you or your descendants.

2. Olivia Judson, *Dr. Tatiana's Sex Advice to All Creation* (New York: Macmillan, 2003).

3. Jonathan DeHart, "Japan's Future: Less Sex, More Shoplifting," *The Diplomat*, July 22, 2013, accessed August 27, 2014, http://thediplomat.com/2013/07/japans-future-more-shoplifting.

4. "Not Enough Males Courting Fame in Japan's Porn Industry," *Malay Mail*, September 12, 2014, accessed September 8, 2014, http://www.themalaymailonline.com/features/article/not-enough-males-courting-fame-in-japans-porn-industry.

5. Gretchen Livingston and D'Vera Cohn, "Childlessness Up Among All Women; Down Among Women with Advanced Degrees," Pew Research Center, June 25, 2010, accessed August 27, 2014, http://www.pewsocialtrends.org/files/2010/11/758-childless.pdf.

6. Gretchen Livingston, "In Terms of Childlessness, U.S. Ranks Near the Top Worldwide," Pew Research Center, January 3, 2014, accessed August 27, 2014, http://www.pewresearch.org/fact-tank/2014/01/03/in-terms-of-childlessness-u-s-ranks-near-the-top-worldwide.

7. Anne Tergesen, "The Long (Long) Wait to Be a Grandparent," *Wall Street Journal*, March 30, 2014.

8. B. Chapais, "Monogamy, Strongly Bonded Groups, and the Evolution of Human Social Structure," *Evolutionary Anthropology* 22, no. 2 (2013): 52–65.

9. Nicholas Eberstadt, "Drunken Nation: Russia's Depopulation Bomb," *World Affairs*, Spring 2009, accessed August 27, 2014, http://www.worldaffairsjournal.org/article /drunken-nation-russia%E2%80%99s-depopulation-bomb.

10. Ibid.

11. Tyler Durden (pseudonym), "China Hits Key Demographic Ceiling as Working-Age Population Now Declining," *Zero Hedge* (blog), January 29, 2013, accessed August 27, 2014, http://www.zerohedge.com/news/2013-01-29/china-hits-key-demographic-ceiling -working-age-population-now-declining.

12. Dara Carr, "Is Education the Best Contraceptive?" Population Reference Bureau, accessed September 8, 2014, http://www.prb.org/pdf/IsEducat-Contracept_Eng.pdf; S. Saurabh, S. Sarkar, and D. K. Pandey, "Female Literacy Rate Is a Better Predictor of Birth Rate and Infant Mortality Rate in India," *Journal of Family Medicine and Primary Care* 2, no. 4 (2013): 349–53.

13. J. R. Goldstein, "A Secular Trend Toward Earlier Male Sexual Maturity: Evidence from Shifting Ages of Male Young Adult Mortality," *PLoS ONE* 6, no. 8 (2011), accessed August 25, 2014, http://www.plosone.org/article/info%3Adoi%2F10.1371% 2Fjournal.pone.0014826.

14. "Average Age at Menarche in Various Cultures," Museum of Menstruation and Women's Health, accessed September 14, 2014, http://www.mum.org/menarage. htm; "Growth and Puberty Secular Trends, Environmental and Genetic Factors," INSERM Collective Expertise Centre, accessed September 8, 2014, http://www .ncbi.nlm.nih.gov/books/NBK10786; "Menarche," *Wikipedia*, accessed September 8, 2014, http://en.wikipedia.org/wiki/Menarche.

15. Daniel J. DeNoon, "Earlier Puberty: Age 9 of 10 for Average U.S. Boy," WebMD, October 20, 2012, accessed September 14, 2014, http://www.webmd.com/children /news/20121020/earlier-puberty-age-9-10-average-us-boy.

16. Lily Dayton, "Trend Towards Early Puberty in Girls Continues: Researchers Ask Why," *California Health Report*, January 15, 2014, accessed September 4, 2014, http://www .healthycal.org/trend-towards-early-puberty-in-girls-continues-researchers-ask-why.

17. A. Lomniczi, A. Loche, J. M. Castellano, O. K. Ronnekleiv et al., "Epigenetic Control of Female Puberty," *Nature Neuroscience* 16, no. 3 (2013): 281–89.

18. E. Carlsen, A. Giwercman, N. Keiding, and N. E. Skakkebaek, "Evidence for Decreasing Quality of Semen During Past 50 Years," *British Medical Journal* 305, no. 6854 (1992): 609–13; S. Dindya, "The Sperm Count Has Been Decreasing Steadily for Many Years in Western Industrialised Countries: Is There an Endocrine Basis for This Decrease?" *Internet Journal of Urology* 2, no. 1 (2003), accessed October 28, 2014, http://ispub.com/IJU/2/1/7519.

19. M. Rolland, J. Le Moal, V. Wagner, D. Royère et al., "Decline in Semen Concentration and Morphology in a Sample Of 26,609 Men Close to General Population Between 1989 and 2005 in France," *Human Reproduction* 28, no. 2 (2013): 462–70.

20. Kevin B. O'Reilly, "Nobel Prize Reflects IVF's Acceptance as Medical Procedure," *American Medical News*, October 18, 2010.

21. Jen Christensen, "Record Number of Women Using IVF to Get Pregnant," CNN, February 17, 2014, accessed September 8, 2014, http://www.cnn.com/2014/02/17 /health/record-ivf-use.

22. Will McCarthy, "Thin! Tan! Hotter Than Hell!" *Wired*, June 2002, accessed September 14, 2014, http://archive.wired.com/wired/archive/10.06/melanotan_pr.html.

23. L. Illa, A. Brickman, G. Saint-Jean, M. Echenique et al., "Sexual Risk Behaviors in Late Middle Age and Older HIV Seropositive Adults," *AIDS and Behavior* 12, no. 6 (2008): 935–42; Centers for Disease Control and Prevention, "Diagnoses of HIV Infection Among Adults Aged 50 Years and Older in the United States and Dependent Areas, 2007–2010," *HIV Surveillance Supplemental Report* 18, no. 3 (2010), CDC.gov, accessed September 7, 2014, www.cdc.gov/hiv/surveillance/resources/reports/2010supp_vol18no3/index.htm.

24. O. Venn, I. Turner, I. Mathiesson, N. de Groot et al., "Strong Male Bias Drives Germline Mutation in Chimpanzees," *Science* 344, no. 6189 (2014): 1272–75.

25. G. C. McIntosh, A. F. Olshan, and P. A. Baird, "Paternal Age and the Risk of Birth Defects in Offspring," *Epidemiology* 6, no. 3 (1995): 282–88.

The Nature Versus Nurture Wars

1. One account of this history was dramatized in a play entitled *Photograph 51*. A discussion of the history and the play can be found in Robin Lloyd, "Rosalind Franklin and DNA: How Wronged Was She?" *Scientific American*, November 3, 2010, accessed July 11, 2014, http://blogs.scientificamerican.com/observations/2010/11/03/rosalind-franklin -and-dna-how-wronged-was-she.

2. Amanda Gefter, "Wilson vs. Watson: The Blessing of Great Enemies," *New Scientist*, September 10, 2009, accessed August 30, 2014, http://www.newscientist.com/article /dn17771-wilson-vs-watson-the-blessing-of-great-enemies.html.

3. "Our Tomorrow: Hopeful/Daring Talks in Session 1 of TED2016," TEDBlog, accessed April 23, 2016, http://blog.ted.com/the-hopeful-talks-in-session-1-of-ted2016/.

4. O. T. Avery, M. C. MacLeod, and M. McCarty, "Studies on the Chemical Nature of the Substance Inducing Transformation of Pneumococcal Types: Induction of Transformation by a Desoxyribonucleic Acid Fraction Isolated from Pneumococcus Type III," *Journal of Experimental Medicine* 79, no. 2 (1944): 137–58. Interestingly, Avery was never awarded the Nobel Prize, which the committee regretted: U. Deichmann, "Early Responses to Avery et al.'s Paper on DNA as Hereditary Material," *Historical Studies in the Physical and Biological Sciences* 34, no. 2 (2004): 227–31.

5. The two draft sequences were announced in 2000 at the White House and published shortly thereafter. See J. C. Venter, M. D. Adams, E. W. Myers, P. W. Li et al., "The Sequence of the Human Genome," *Science* 291, no. 5507 (2001): 1304–51. E. S. Lander, L. M. Linton, B. Birren, C. Nusbaum et al., "Initial Sequencing and Analysis of the Human Genome," *Nature* 409, no. 6822 (2001): 860–921. The first diploid genome (both chromosomes for an individual) was completed in 2007: S. Levy, G. Sutton, P. C. Ng, L. Feuk et al., "The Diploid Genome Sequence of an Individual Human," *PLoS Biology* 5, no. 10 (2007): e254.

6. I. Ezkurdia, D. Juan, J. M. Rodriguez, A. Frankish et al., "Multiple Evidence Strands Suggest That There May Be as Few as 19,000 Human Protein-Coding Genes," *Human Molecular Genetics* 23, no. 2 (2014): 5866–78.

7. And while you are mulling that one over . . . how could it possibly be that the wheat genome is five times larger than that of a human?

8. Amanda Gefter, "Wilson vs. Watson: The Blessing of Great Enemies," *New Scientist*, September 10, 2009, accessed August 30, 2014, http://www.newscientist.com/article /dn17771-wilson-vs-watson-the-blessing-of-great-enemies.html.

9. Many have accused Watson of uttering misogynistic, racist, or bullying comments, one of which got him retired as head of Cold Spring Harbor Laboratory. See

Cornelia Dean, "James Watson Retires After Racial Remarks," *New York Times*, October 25, 2007, accessed August 30, 2014, http://www.nytimes.com/2007/10/25 /science/25cnd-watson.html; Laura Blue, "The Mortification of James Watson," *Time*, October 19, 2007, accessed August 30, 2014, http://content.time.com/time/health /article/0,8599,1673952,00.html.

Missing Heredity, Mysterious Toxins

1. OMIM, or Online Mendelian Inheritance in Man, is an online catalog of human genes and genetic disorders: http://www.omim.org.

2. For a few days each summer, Victor McKusick would retreat to a small, modest cabin near the Bay of Fundy and its thirty-foot tides. Even there he would host a few of the world's great researchers and talk genetics late into the evening. In a scene out of the *Saturday Evening Post*, his smiling, wonderful wife, Anne, who married him in 1949, would come out of the kitchen with warm berry pies, place them on a red-checked cloth on the porch, and join the discussion. Only fools underestimated this soft-spoken, gentle woman: She had been a physicist on the Manhattan Project before becoming a med-school professor at Hopkins. The McKusicks' attitude toward a world in which they had seen war, disease, and want was relentlessly positive. Never complain; just face it and fix it, with good humor. When asked in 2013 about her biggest obstacle in being a female medical student and faculty member at Johns Hopkins in the 1940s and 1950s, a time when it's rumored male MDs were not particularly enlightened or supportive of female MDs, Anne's answer was, "I encountered no obstacle either before attending Johns Hopkins or here at Johns Hopkins as a faculty member." More on Anne McKusick: http://www.nlm.nih.gov/changingthefaceof medicine/physicians/biography_217.html.

3. T. A. Manolio, F. S. Collins, N. J. Cox, D. B. Goldstein et al., "Finding the Missing Heritability of Complex Diseases," *Nature* 461 (2009): 747–53.

4. J. M. Fletcher, "Why Have Tobacco Control Policies Stalled? Using Genetic Moderation to Examine Policy Impacts," *PLoS ONE* 7, no. 12 (2012): e50576.

5. Yes, there are conditions like cystic fibrosis where mutations in a single gene cause the disease, and most individuals have the exact same mutation in that gene, but these conditions tend to occur very, very rarely. And when they do crop up, they tend to cluster around specific and quite rare genetic diseases, not in the vast majority of diseases that affect most of us.

6. T. A. Manolio et al., "Finding the Missing Heritability of Complex Diseases"; E. E. Eichler, J. Flint, G. Gibson, A. Kong et al., "Missing Heritability and Strategies for Finding the Underlying Causes of Complex Disease," *Nature Reviews Genetics* 11, no. 6 (2010): 446–50.

7. A. A. Vinkhuyzen, S. van der Sluis, H. H. Maes, and D. Posthuma, "Reconsidering the Heritability of Intelligence in Adulthood: Taking Assortative Mating and Cultural Transmission into Account," *Behavioral Genetics* 42, no. 2 (2012): 187–98.

8. C. A. Rietveld, S. E. Medland, J. Derringer, J. Yang et al., "GWAS of 126,559 Individuals Identifies Genetic Variants Associated with Educational Attainment," *Science* 340, no. 6139 (2013): 1467–71.

9. Richard J. Lewis Sr., ed., *Sax's Dangerous Properties of Industrial Materials* (New York: John Wiley and Sons, 2014). The sad story of thalidomide, a drug used to treat morning sickness, which resulted in stunted limb development or death in approximately 10,000 newborns in the late 1950s and early '60s, made everyone aware of what can

happen to a fetus exposed to toxic drugs and chemicals. The FDA subsequently mandated testing of pharmaceuticals for potential embryotoxicity, and physicians today are very cautious in treating pregnant women, guided by a list of bad actors that mothers-to-be need to avoid.

10. You, too, can post the full list on your fridge door: http://www.purdue.edu/rem/ih /terat.htm.

11. American Chemistry Council, "The Business of Chemistry by the Numbers," American Chemistry Council, June 2014, accessed September 8, 2014, http://www.ameri canchemistry.com/chemistry-industry-facts.

12. Wendy Chung, "Teratogens and Their Effects," in *The New Public Health: An Introduction for the 21st Century*, chapter 23 (New York: Columbia University Press, 2004), accessed September 8, 2014, http://www.columbia.edu/itc/hs/medical/humandev /2004/Chpt23-Teratogens.pdf.

13. These are slogans from actual advertising campaigns.

14. That is why it is called a "medical practice," and it is first and foremost driven by the Hippocratic Oath, which is often phrased in Latin as *"primum non nocere"* or "First, do no harm."

15. There are many examples of conflicting data that lead to large clinical trials that show no benefit of a compound intervention, followed by the usual debates. For example, vitamins have been touted to ward off many things. Large trials and analyses of multivitamins and mineral supplements indicate no substantial health benefits on cardiovascular diseases, cognitive decline, cancer, or all-cause mortality. See E. Guallar, S. Stranges, C. Mulrow, L. J. Appel et al., "Enough Is Enough: Stop Wasting Money on Vitamin and Mineral Supplements," *Annals of Internal Medicine* 159, no. 12 (2013): 850–51. On the other hand, one of our wonderful editors, Suzanne, used to have chronic hives. The medicines her allergist provided were useless, but an 80 percent organic diet led to no hives. Whether this was her body self-healing or a change in diet, that is exactly the type of question that is maddeningly hard to answer, given so many variables and inputs.

16. J. Spoelstra, S. L. Schiff, and S. J. Brown, "Artificial Sweeteners in a Large Canadian River Reflect Human Consumption in the Watershed," *PLoS ONE* 8, no. 12 (2013): e82706.

Transgenerational Inheritance—aka "Voodoo Biology"

1. B. T. Heijmans, E. W. Tobi, A. D. Stein, H. Putter et al., "Persistent Epigenetic Differences Associated with Prenatal Exposure to Famine and Humans," *Proceedings of the National Academy of Sciences USA* 105, no. 44 (2008): 17046–49.

2. A. S. Brown, and E. S. Susser, "Prenatal Nutritional Deficiency and Risk of Adult Schizophrenia," *Schizophrenia Bulletin* 34, no. 6 (2008): 1054–63.

3. Maybe it only affects the eggs developing in the mother and daughter during famine? And the effects will go away after the third generation? We do not know enough yet. But various plant and animal studies show there really is an epigenetic effect in living things. See U. Grossniklaus, W. G. Kelly, A. C. Ferguson-Smith, M. Pembrey et al., "Transgenerational Epigenetic Inheritance: How Important Is It?" *Nature Reviews Genetics* 14, no. 3 (2013): 228–35.

4. Jayna, "AP Biology: Lamarck and Evidence for Evolution," Bench Prep, March 11, 2013, accessed August 31, 2014, https://benchprep.com/blog/ap-biology-evolution-part-2.

5. E. Jablonka and G. Raz, "Transgenerational Epigenetic Inheritance," *Quarterly Review of Biology* 84, no. 2 (2009): 131–76.

6. C. H. Waddington, "The Epigenotype," *Endeavour* 1 (1942): 18–20, reprinted in *International Journal of Epidemiology* 41, no. 1 (2012): 10–13. Waddington was an academic scientist who focused on embryology and genetics. He proposed that evolution worked via "genetic assimilation" in which certain acquired traits enabled a species to initially adapt to a new environment, and then with time these traits became hardwired into the genetic code, a state he termed "canalized." He believed Darwinian forces were at work throughout this process. A historical account of Waddington and his seminal role in originating the field of epigenetics can be found in L. Van Speybroek, "From Epigenesis to Epigenetics: The Case of C. H. Waddington," *Annals of the New York Academy of Sciences* 981 (2002): 61–81.

7. J. Kaiser, "The Epigenetics Heretic," *Science* 343, no. 6169 (2014): 361–63.

8. K. Manning, M. Tör, M. Poole, Y. Hong et al., "A Naturally Occurring Epigenetic Mutation in a Gene Encoding an SBP-Box Transcription Factor Inhibits Tomato Fruit Ripening," *Nature Genetics* 38 (2006): 348–52. See also R. M. González and N. D. Iusem, "Twenty Years of Research on Asr (ABA-stress-ripening) Genes and Proteins," *Planta* 239, no. 5 (2014): 941–49.

9. S. Zhong, Z. Fie, Y.-R. Chen, Y. Zheng, M. Huang et al., "Single-Base Resolution Methylomes of Tomato Fruit Development Reveal Epigenome Modifications Associated with Ripening," *Nature Biotechnology* 31, no. 2 (2013): 154–59.

10. Virginia Hughes wrote a great overview of what is known in the field: V. Hughes, "The Sins of the Father," *Nature* 507, no. 7490 (2014): 22–24. See also B. G. Dias and K. J. Ressler, "Parental Olfactory Experience Influences Behavior and Neural Structure in Subsequent Generations," *Nature Neuroscience* 17 (2014): 86–96.

11. M. D. Anway, A. S. Cupp, M. Uzumcu, and M. K. Skinner, "Epigenetic Transgenerational Actions of Endocrine Disruptors and Male Fertility," *Science* 308, no. 5727 (2005): 1466–69; A. C. Nottke, S. E. Beese-Sims, L. F. Pantalena, V. Reinke et al., "SPR-5 Is a Histone H3K4 Demethylase with Role in Meiotic Double-Strand Break Repair," *Proceedings of the National Academy of Sciences USA* 108, no. 31 (2011): 12805–10; E. L. Greer, S. E. Beese-Sims, E. Brookes, R. A. Spadafora et al., "A Histone Methylation Network Regulates Transgenerational Epigenetic Memory in *C. elegans*," *Cell Reports* 7, no. 1 (2014): 113–26; Jablonka and Raz, "Transgenerational Epigenetic Inheritance."

12. Transgenerational inheritance generates the same types of baffling questions faced by physicists in the past century, which means the certainty and predictability of volume after volume of the Online Mendelian Inheritance of Man is superseded by concepts akin to the uncertainty principle or a perennial favorite of quantum physics, quantum entanglement, aka spooky action at a distance.

13. J. Lim and A. Brunet. "Bridging the Transgenerational Gap with Epigenetic Memory," *Trends in Genetics* 29, no. 3 (2013): 176–86.

14. S. Saxonov, P. Berg, and D. L. Brutlag, "A Genome-Wide Analysis of CpG Dinucleotides in the Human Genome Distinguishes Two Distinct Classes of Promoters," *Proceedings of the National Academy of Sciences USA* 103, no. 5 (2006): 1412–17. H. Heyn and M. Esteller. "An Adenine Code for DNA: A Second Life for N6-Methyladenine," *Cell* 161, no. 4 (2015): 710–13. Y. Fu, G.-Z. Luo, K. Chen, X. Deng et al., "N6-Methyldeoxyadenosine Marks Active Transcription Start Sites in Chlamydomonas," *Cell* 161, no. 4 (2015): 879–92.

15. R. K. Naz and R. Sellamuthu, "Receptors in Spermatozoa: Are They Real?" *Journal of Andrology* 27, no. 5 (2006): 627–36.

16. E. L. Marczylo, A. A. Amoako, J. C. Konje, T. W. Gant et al., "Smoking Induces Differential miRNA Expression in Human Spermatozoa: A Potential Transgenerational Epigenetic Concern?" *Epigenetics* 7, no. 5 (2012): 432–39.

17. A. Sen, N. Heredia, M. C. Senut, S. Land et al., "Multigenerational Epigenetic Inheritance in Humans: DNA Methylation Changes Associated with Maternal Exposure to Lead Can Be Transmitted to the Grandchildren," *Science Report* 5, (2015): 14466. J. Feinberg, K. M. Bakulski, A. E. Jaffe, R. Tryggvadottir et al., "Paternal Sperm DNA Methylation Associated with Early Signs of Autism Risk in an Autism-Enriched Cohort," *International Journal of Epidemiology* 44, no. 4 (2015): 1199–210. P. Huypens, S. Sass, M. Wu, D. Dyckhoff et al., "Epigenetic Germline Inheritance of Diet-induced Obesity and Insulin Resistance," *Nature Genetics* 48 (2016): 497–99. I. Donkin, S. Versteyhe, L. R. Ingerslev, K. Qian et al., "Obesity and Bariatric Surgery Drive Epigenetic Variation of Spermatozoa in Humans," *Cell Metabolism* 23, no. 2 (2016): 369–78.

18. C. N. Hales and D. J. Barker, "Type 2 (Non-Insulin-Dependent) Diabetes Mellitus: The Thrifty Phenotype Hypothesis," *Diabetologia* 35 (1992): 595–601.

19. O. Viltart and C. C. A. Vanbesien-Mailliot, "Impact of Prenatal Stress on Neuroendocrine Programming," *Scientific World Journal* 7 (2007): 1493–537.

20. Hormones include insulin, leptin, IGFs, and various growth factors. For an in-depth description, see ibid.

21. Joanne Silberner, "The Khmer Rouge May Be Partly to Blame for Diabetes in Cambodia," *Public Radio International*, January 29, 2014, accessed August 31, 2014, http://www.pri.org/stories/2014-01-29/khmer-rouge-may-be-partly-blame-diabetes-cambodia.

22. C. Corcoran, M. Perrin, S. Harlap, L. Deutsch et al., "Incidence of Schizophrenia Among Second-Generation Immigrants in the Jerusalem Perinatal Cohort," *Schizophrenia Bulletin* 35, no. 3. (2009): 596–602.

23. G. Kaati, L. O. Bygren, and S. Edvinsson, "Cardiovascular and Diabetes Mortality Determined by Nutrition During Parents' and Grandparents' Slow Growth Period," *European Journal of Human Genetics* 10, no. 11 (2002): 682–88.

24. H. Heyn, N. Li, H. J. Ferreira, S. Moran et al., "Distinct DNA Methylomes of Newborns and Centenarians," *Proceedings of the National Academy of Sciences* USA 109, no. 26 (2012): 10522–27, accessed August 31, 2014, http://www.pnas.org/content/early/2012/06/05/1120658109.

25. J. Day and J. D. Sweatt, "Epigenetic Modifications in Neurons Are Essential for Formation and Storage of Behavioral Memory," *Neuropsychopharmacology* 36 (2011): 357–58.

WWIV: Nuking Our Microbes

1. There are many books on this subject; for instance, William H. McNeill, *Plagues and Peoples* (New York: Anchor, 1977).

2. V. J. Cirillo, "Two Faces of Death: Fatalities from Disease and Combat in America's Principal Wars, 1775 to Present," *Perspectives in Biology and Medicine* 51, no. 1 (2008): 121–33.

3. See, for example, Dorothy H. Crawford, *Deadly Companions: How Microbes Shaped Our History* (Oxford: Oxford University Press, 2009). Or, for a quick overview of pandemics, see "Pandemic," *Wikipedia*, accessed September 6, 2014, http://en.wikipedia.org/wiki/Pandemic.

4. Jared Diamond, "The Story of . . . Smallpox—and Other Deadly Eurasian Germs," *Guns, Germs and Steel*, PBS, accessed September 1, 2014, http://www.pbs.org/guns germssteel/variables/smallpox.html.

5. John Sweetman, *The Crimean War (Essential Histories, No 2)* (Oxford, UK: Routledge, 2001). Statistics quoted in "Statistics of Wars, Oppressions and Atrocities of the Nineteenth Century (the 1800s)," accessed September 20, 2014, http://necrometrics .com/wars19c.htm.

6. Robert K. D. Peterson, "More Feared Than All the Armies in the World: Insects, Disease, and Military History," presentation for the Michigan Mosquito Control Association, accessed September 1, 2014, http://www.mimosq.org/presentations /2014/22RobertPetersonDiseaseMilitary.pdf.

7. "Does Polio Still Exist? Is It Curable?," World Health Organization, accessed April 23, 2016, http://www.who.int/features/qa/07/en/.

8. Some worry that because there is now no herd immunity, smallpox could be a potentially deadly bioweapon if released into today's populations. See Centers for Disease Control and Prevention, "Emergency Preparedness and Response: Smallpox Fact Sheet: Vaccine Overview—The Smallpox Vaccine," CDC.gov, accessed September 1, 2014, http://www.bt.cdc.gov/agent/smallpox/vaccination/facts.asp.

9. W. W. Keen, "Before and After Lister," *Science* 41, no. 1068 (1915): 881–91.

10. J. Cohen, "Hospital Gangrene: The Scourge of Surgeons in the Past," *Infection Control & Hospital Epidemiology* 20, no. 9 (1999): 638–40.

11. "The Black Death: Bubonic Plague," TheMiddleAges.net, accessed September 1, 2014, http://www.themiddleages.net/plague.html.

12. Cirillo, "Two Faces of Death."

13. Centers for Disease Control and Prevention, "Achievements in Public Health, 1990–1999: Control of Infectious Diseases," CDC.gov, accessed September 1, 2014, http://www.cdc.gov/mmwr/preview/mmwrhtml/mm4829a1.htm.

14. The CDC health-care-associated infection (HAI) prevalence survey provides an updated national estimate of the overall problem of HAIs in U.S. hospitals. Based on a large sample of U.S. acute-care hospitals, the survey found that on any given day, about 1 in 25 hospital patients has at least one health-care-associated infection. There were an estimated 722,000 HAIs in U.S. acute-care hospitals in 2011. About 75,000 hospital patients with HAIs died during their hospitalizations. More than half of all HAIs occurred outside of the intensive care unit. Centers for Disease Control and Prevention, "Healthcare-Associated Infections (HAIs)," CDC.gov, accessed September 1, 2014, http://www.cdc.gov/HAI/surveillance/index.html.

15. Disclosure: Both Steve and Juan are investors in a young company that is working to bring new antibiotics to market. They observed for many years what antibiotic resistance can do and decided to invest in a company using an innovative technology spun out of Dr. Andy Myers's chemistry lab at Harvard to develop new antibiotics.

16. Centers for Disease Control and Prevention, "Antibiotic/Antimicrobial Resistance: Threat Report 2013," CDC.gov, accessed September 1, 2014, http://www.cdc.gov /drugresistance/threat-report-2013. See also U.S. Census, "Table 1103. Motor Vehicle Accidents—Number and Deaths: 1990 to 2009," U.S. Census, accessed September 1, 2014, http://www.census.gov/compendia/statab/2012/tables/12s1103.pdf; S. L. Murphy, J. Xu, and K. D. Kochanek, "Deaths: Final Data for 2010," *National Vital Statistics Reports* 61, no. 4 (2013), accessed September 1, 2014, http://www.cdc.gov /nchs/data/nvsr61/nvsr61_04.pdf.

17. A. M. Alicea-Serrano, M. Contreras, M. Magris, G. Hidalgo et al., "Tetracycline Resistance Genes Acquired at Birth," *Archives of Microbiology* 195, no. 6 (2013): 447–51.

18. D. M. Sievert, P. Ricks, J. R. Edwards, A. Schneider et al., for the National Healthcare Safety Network (NHSN) Team and Participating NHSN Facilities, "Antimicrobial-Resistant Pathogens Associated with Healthcare-Associated Infections: Summary of Data Reported to the National Healthcare Safety Network at the Centers for Disease Control and Prevention, 2009–2010," *Infection Control and Hospital Epidemiology* 34, no. 1 (2013): 1–14, CDC.gov, accessed September 6, 2014, http://www.jstor.org/stable/10.1086/668770.

19. Sabrina Tavernise. "FDA Restricts Antibiotics Use for Livestock," *New York Times*, December 11, 2013, accessed November 2, 2014, http://www.nytimes.com/2013/12/12 /health/fda-to-phase-out-use-of-some-antibiotics-in-animals-raised-for-meat.html.

20. Viruses abound in the world's oceans, yet researchers are only beginning to understand how they affect life and chemistry from the water's surface to the sea floor. See J. S. Weitz and S. W. Wilhelm, "An Ocean of Viruses," *The Scientist*, July 1, 2013, accessed September 1, 2014, http://www.the-scientist.com/?articles.view/articleNo /36120/title/An-Ocean-of-Viruses.

21. In fact, just to confuse you a little more, our entire microbial ecosystem is often termed the metabiome, as it includes both bacteria (the microbiome) and viruses (the virome). However, to date, most research has focused on the microbiome, and antibiotics target bacteria, not viruses, so our discussion is focused largely on the microbiome. Future research will undoubtedly include better understanding of viruses, as well as a likely greater use of virus-killing drugs as they are developed and used.

22. Here are several scientific reviews of the virome field that provide details of many of the facts and concepts mentioned in this section of the book: K. M. Wylie, G. M. Weinstock, and G. A. Storch, "Emerging View of the Human Virome," *Translational Research* 160, no. 4. (2012): 283–90; S. C. P. Williams, "The Other Microbiome," *Proceedings of the National Academy of Sciences USA* 110, no. 8 (2013): 2682–84; H. W. Virgin, "The Virome in Mammalian Physiology and Disease," *Cell* 157, no. 1 (2014): 142–50; S. R. Abeles and D. T. Pride, "Molecular Bases and Role of Viruses in the Human Microbiome," *Journal of Molecular Biology* 426, no. 23 (2014): 3892–906. For less technical discussions of the human virome, do a Google or *Wikipedia* search for terms such as "virome," "endogenous retrovirus," "horizontal gene transfer," or "viral metagenomics."

23. Scientists have studied a few nonpathogenic viruses, such as SV40, in detail. The results are fascinating. See, for instance, E. Fanning and K. Zhao, "SV40 DNA replication: From the A Gene to a Nanomachine," *Virology* 384, no. 2 (2009): 352–59.

24. D. J. Gawkrodger, M. H. Lloyd, and J. A. Hunter, "Occupational Skin Diseases in Hospital Cleaning and Kitchen Workers," *Occupational Dermatitis* 15, no. 3 (1986): 132–35.

25. S. Teo, A. T.-J. Goon, L. H. Siang, G. S. Lin et al., "Occupational Dermatoses in Restaurant, Catering and Fast-Food Outlets in Singapore," *Occupational Medicine* 59, no. 7 (2009): 466–71.

The "Yucky" Stuff Inside You

1. Until very recently, it was generally agreed that bacteria outnumbered human cells by 10 to 1, but this was recently refuted to indicate that there are 40 trillion microbes in our bodies, which is comparable to the number of human cells, of which 90 percent are blood cells. See: R. Sender, S. Fuchs, and R. Milo, "Are We Really Vastly Outnumbered? Revisiting the Ratio of Bacterial to Host Cells in Humans," *Cell* 164, no. 3 (2016): 337–40.

2. The following articles, among many others, provide overviews of the human micro-biome in either layman's or scientific language: Carl Zimmer, "How Microbes Defend and Define Us," *New York Times*, July 12, 2010, accessed August 31, 2014, http://www.nytimes.com/2010/07/13/science/13micro.html; National Institutes of Health, "Human Microbiome Project: Overview," NIH Office of Strategic Coordination Common Fund, accessed August 31, 2014, http://commonfund.nih.gov/hmp /overview; M. C. Cénit, V. Matzaraki, E. F. Tigchelaar, and A. Zhernakova, "Rapidly Expanding Knowledge on the Role of the Gut Microbiome in Health and Disease," *Biochimica et Biophysica Acta* 4439, no. 14 (2014); T. Yatsunenko, F. E. Rey, M. J. Manary, I. Trehan et al., "Human Gut Microbiome Viewed Across Age and Geography," *Nature* 486, no. 7402 (2012): 222–77.

3. Then again, even fart smells might be good for you, allowing you to ingest small amounts of hydrogen sulfide, which can prevent injury to cells: "Fart Smells Have Health Benefits, According to Exeter University Researchers," *Western Daily Press*, July 11, 2014, accessed August 31, 2014, http://www.westerndailypress.co.uk/Fart-smells -health-benefits-according-Exeter/story-21447028-detail/story.html. See also E. D'Araio, N. Shaw, A. Millward, A. Demaine et al., "Hydrogen Sulfide Induces Heme Oxygenase-1 in Human Kidney Cells," *Acta Diabetologia* 51, no. 1 (2014): 155–57.

4. Here is the rough transcript from CNN as the event was happening. It helps convey the excitement and tension of the genome announcement better than the cleaned-up versions. "President Clinton, British Prime Minister Tony Blair Deliver Remarks on Human Genome Milestone," CNN.com, June 26, 2000, accessed October 31, 2014, http://transcripts.cnn.com/TRANSCRIPTS/0006/26/bn.01.html.

5. James Shreeve, *The Genome War: How Craig Venter Tried to Capture the Code of Life and Save the World* (New York: Knopf, 2004), 48.

6. This is a nice bio that gives you a sense of Sulston's fundamental decency: The Nobel Foundation, "John E. Sulston—Biographical," NobelPrize.org, 2002, accessed August 31, 2014, http://www.nobelprize.org/nobel_prizes/medicine/laureates/2002 /sulston-bio.html.

7. Ironically, the first nearly complete genome sequence from a single human being, as opposed to all the approximations, jumbles, mixtures, and composites that had been cobbled together into "consensus genomes," was published with little fanfare in 2007. It also covered both mother's and father's contributions to each chromosome—a not-so-minor omission in other "complete" genomes. See S. Levy G. Sutton, P. C. Ng, L. Feuk et al., "The Diploid Genome Sequence of an Individual Human," *PLoS Biology* 5, no. 10 (2007): e254.

8. *Wired* did a good article on the voyage. See James Shreeve, "Craig Venter's Epic Voyage to Redefine the Origin of the Species," *Wired*, August 2004, accessed August 31, 2014, http://archive.wired.com/wired/archive/12.08/venter.html. Here are some of the *Sorcerer II* crew's adventures on Facebook: "Global Ocean Sampling Expedition," n.d., accessed August 31, 2014, https://www.facebook.com/GlobalOceanSampling. Poor Juan had to "suffer" terribly, joining Venter across the massive tides of the Bay of Fundy, retracing Darwin's visit to the Galápagos, scuba diving on pristine reefs in Hiva Oa and Fatu Hiva (the first islands across the Pacific from Ecuador), surfing in Fiji, enduring isolated and transparent bays in Australia, drinking margaritas on deserted islands in the midst of the Indian Ocean, encountering bizarre rock formations in the Seychelles . . . all for the sake of science.

9. "More than Six Million New Genes, Thousands of New Protein Families, and Incredible Degree of Microbial Diversity Discovered from First Phases of *Sorcerer II* Global Ocean Sampling Expedition," J. Craig Venter Institute, March 13, 2007, accessed August 31, 2014, http://www.jcvi.org/cms/research/projects/gos/publications; http:

//www.jcvi.org/cms/press/press-releases/full-text/article/more-than-six-million-new
-genes-thousands-of-new-protein-families-and-incredible-degree-of-microbi.

10. W. B. Whitman, D. C. Coleman, and W. J. Wiebe, "Prokaryotes: The Unseen Majority," *Proceedings of the National Academy of Sciences USA* 95, no. 12 (1998): 6578–83; Carl Zimmer, "The Microbe Factor and Its Role in Our Climate Future," *Yale Environment 360*, June 1, 2010, accessed August 31, 2014, http://e360.yale.edu/feature /the_microbe_factor_and_its_role_in_our_climate_future/2279.

11. S. R. Gill, M. Pop, R. T. Deboy, P. B. Eckburg et al., "Metagenomic Analysis of the Human Distal Gut Microbiome," *Science* 312, no. 5778 (2006): 1355–59.

12. E. M. Bik, C. D. Long, S. G. Armitage, P. Loomer et al., "Bacterial Diversity in the Oral Cavity of 10 Healthy Individuals," *ISME Journal* 4, no. 8 (2010): 962–74.

13. E. A. Grice, H. H. Kong, S. Conlan, C. B. Deming et al., "Topographical and Temporal Diversity of the Human Skin Microbiome," *Science* 324, no. 5931 (2009): 1190–92. Here is a good summary of what we know about microbial infection symbiosis in humans: J. J. Martin, J. M. Zenilman, and G. S. Lazarus, "Molecular Microbiology: New Dimensions for Cutaneous Biology and Wound Healing," *Journal of Investigative Dermatology* 130, no. 1 (2010): 38–48.

14. Carl Zimmer is a great person to follow on genomics and life sciences. He is the author of many overview articles, including the one from which this statistic came: "How Microbes Define and Defend Us," *New York Times*, July 12, 2010, accessed August 31, 2014, http://www.nytimes.com/2010/07/13/science/13micro.html.

15. M. G. Dominguez-Bell, E. K. Costello, M. Contreras, M. Magris et al., "Delivery Mode Shapes the Acquisition and Structure of the Initial Microbiota Across Multiple Body Habitats in Newborns," *Proceedings of the National Academy of Sciences USA* 107, no. 26 (2010): 11971–75; G. Biasucci, B. Benenati, L. Morelli, E. Bessi et al., "Cesarean Delivery May Affect the Early Biodiversity of Intestinal Bacteria," *Journal of Nutrition* 138, no. 9 (2008): 1796S–1800S.

16. J. Neu and J. Rushing, "Cesarean Versus Vaginal Delivery: Long-Term Infant Outcomes and the Hygiene Hypothesis," *Clinics in Perinatology* 38, no. 2 (2011): 321–31.

17. S. Salminen, G. R. Gibson, A. L. McCartney, and E. Isolauri, "Influence of Mode of Delivery on Gut Microbiota Composition in Seven Year Old Children," *Gut* 53, no. 9 (2004): 1388–89.

18. These herbs are also called *harmal* or *harmel:* Food and Agriculture Association, "Peganum Harmal (L.)," FAO.org, accessed August 31, 2014, http://www.fao.org/ag /agp/AGPC/doc/gbase/new_species/peghar.htm.

19. Joel Warner, "Mapping the Hidden Universe in Your Kitchen: An Invisible World of Microbes," *Popular Science*, April 2014, accessed August 31, 2014, http://www.popsci .com/article/science/invisible-world.

20. You can find some photos on this Web site, though the text is in Turkish: http://www .1Baba1Anne.com, accessed September 6, 2014, http://www.1baba1anne.com/bebek -bezi-yerine-holluk.

21. Perhaps the traditional mixture of dirt and herbs, intended to protect Emine from "bad airs" and evil spirits, also exposed her to allergens and microbes that today's babies don't see in their sanitized cribs. Soon after Emine's birth, her mom emigrated to the West, and her younger siblings were born in a very different environment. And at least according to Emine's mom, sleeping on the *üzerlik otu* is why Emine is so easygoing and self-confident, unlike her sisters.

22. Human Microbiome Project Consortium, "Structure, Function and Diversity of the Healthy Human Microbiome," *Nature* 486, no. 7402 (2012): 207–14.

23. "United States Life Tables, 2008," *National Vital Statistics Reports* 61, no. 3 (2012): 5.

24. United Nations, "Big Strides on Millennium Development Goals with More Targets Achievable by 2015: UN Report," *The Millennium Development Goals Report 2013,* July 1, 2013, accessed August 31, 2014, http://www.un.org/millenniumgoals/pdf/report -2013/mdg-report2013_pr_global-english.pdf.

25. A. C. Williams and R. I. Dunbar, "Big Brains, Meat, Tuberculosis and the Nicotin-amide Switches: Co-Evolutionary Relationships with Modern Repercussions on Longevity and Disease?" *Medical Hypotheses* 83, no. 1 (2014): 79–87; A. Williams and R. Dunbar, "Meaty Puzzle: Did TB Evolve to Boost Hungry Brains?," *New Scientist,* June 21, 2014, 28–29.

26. World Tourism Organization, "International Tourism Demand Exceeds Expecta-tions in the First Half of 2013," World Tourism Organization, August 25, 2013, accessed September 1, 2014, http://media.unwto.org/en/press-release/2013-08-25 /international-tourism-demand-exceeds-expectations-first-half-2013.

27. V. Tremaroli and F. Bäckhed, "Functional Interactions Between the Gut Microbiota and Host Metabolism," *Nature* 489, no. 7415 (2012): 242–49; L. A. David, C. F. Mau-rice, R. N. Carmody, D. B. Gootenberg et al., "Diet Rapidly and Reproducibly Alters the Human Gut Microbiome," *Nature* 505, no. 7484 (2014): 559–63.

28. "Túquerres," *Wikipedia*, accessed September 1, 2014, http://es.wikipedia.org/wiki /T%C3%BAquerres.

29. Washington Office on Latin America, "Little Progress in Troubled Tumaco, Colom-bia," Washington Office on Latin America, May 24, 2011, accessed September 1, 2014, http://www.wola.org/commentary/in_troubled_tumaco_little_progress.

30. The Helicobacter Foundation, "Epidemiology," The Helicobacter Foundation, accessed September 1, 2014, http://helico.com/?q=Epidemiology.

31. A. Covacci, J. L. Telford, G. Del Giudice, J. Parsonnet et al., "*Helicobacter Pylori* Vir-ulence and Genetic Geography," *Science* 284, no. 54 (1999): 1328–33.

32. N. Kodaman, A. Pazos, B. G. Schneider, M. B. Piazuelo et al., "Human and *Helico-bacter Pylori* Coevolution Shapes the Risk of Gastric Disease," *Proceedings of the National Academy of Sciences USA* 111, no. 4 (2014): 1455–60.

33. H. J. Zuo, Z. M. Xie, W. W. Zhang, W. Want et al., "Gut Bacteria Alteration in Obese People and Its Relationship with Gene Polymorphism," *World Journal of Gas-troenterology* 17, no. 8 (2011): 1076–81.

34. Tremaroli and Bäckhed, "Functional Interactions."

35. V. K. Ridaura, J. J. Faith, F. E. Rey, J. Cheng et al., "Gut Microbiota from Twins Dis-cordant for Obesity Modulate Metabolism in Mice," *Science* 341, no. 6150 (2013): 12412–14.

36. M. Sanchez, C. Darimont, V. Drapeau, S. Emady-Azar et al., "Effect of *Lactobacillus rhamnosus* CGMCC1.3724 Supplementation on Weight Loss and Maintenance in Obese Men and Women," *British Journal of Nutrition* 111, no. 8 (2013): 1507–19.

37. O. C. Aroniadis and L. J. Brandt, "Fecal Microbiota Transplantation: Past, Present and Future," *Current Opinions in Gastroenterology* 29, no. 1 (2013): 79–84; E. van Nood, A.

Vrieze, M. Nieuwdorp, S. Fuentes et al., "Duodenal Infusion of Donor Feces for Recurrent *Clostridium difficile*," *New England Journal of Medicine* 368, no. 5 (2013): 407–15.

38. S. Vinjé, E. Stroes, M. Nieuwdorp, and S. L. Hazen, "The Gut Microbiome as Novel Cardio-Metabolic Target: The Time Has Come!" *European Heart Journal* 35, no. 14 (2014): 883–87.

39. S. A. Joyce, J. MacSharry, P. G. Casey, M. Kinsella et al., "Regulation of Host Weight Gain and Lipid Metabolism by Bacterial Bile Acid Modification in the Gut," *Proceedings of the National Academy of Sciences USA* 111, no. 20 (2014): 7421–26.

40. Also, mice with colon tumors showed enrichment of certain bacterial species (*Bacteroides, Odoribacter,* and *Akkermansia*) and decreases in abundance of others (*Prevotellaceae* and *Porphyromonadaceae*). See J. P. Zackulara, N. T. Baxtera, K. D. Iversona, W. D. Sadlerb et al., "The Gut Microbiome Modulates Colon Tumorigenesis," *mBio* 4, no. 6 (2013): e00692-13.

41. L. M. Gargano and J. M. Hughes, "Microbial Origins of Chronic Diseases," *Annual Review of Public Health* 35 (2014): 65–82; Y. Wang and L. H. Kasper, "The Role of Microbiome in Central Nervous System Disorders," *Brain, Behavior, and Immunity* 38 (2014): 1–12; J. U. Scher, A. Sczesnak, R. S. Longman, N. Segata et al., "Expansion of Intestinal Prevotella Copri Correlates with Enhanced Susceptibility to Arthritis," *eLife* 2 (2013): e01202; S. M. Collins, E. Denou, E. F. Verdu, and P. Bercik, "The Putative Role of the Intestinal Microbiota in the Irritable Bowel Syndrome," *Digestive and Liver Disease* 41, no. 12 (2009): 850–53; American Society for Microbiology, "Bacterial Toxin Potential Trigger for Multiple Sclerosis," *ScienceDaily,* January 28, 2014, accessed September 1, 2014, http://www.sciencedaily.com/releases/2014/01/140128153940.htm.

42. J. Shen, M. S. Obin, and L. Zhao, "The Gut Microbiota, Obesity and Insulin Resistance," *Molecular Aspects of Medicine* 34, no. 1 (2013): 39–58.

43. Y. Wang and L. H. Kasper, "The Role of Microbiome"; J. C. Rees, "Obsessive-Compulsive Disorder and Gut Microbiota Dysregulation," *Medical Hypotheses* 82, no. 2 (2014): 163–66.

44. L. Hood. "Tackling the Microbiome," *Science* 336, no. 6086 (2012): 1209. C. A. Lopez, D. D. Kingsbury, E. M. Velazquez, and A. J. Bäumier, "Collateral Damage: Microbiota-Derived Metabolites and Immune Function in the Antibiotic Era," *Cell Host & Microbe* 16, no. 2 (2014): 156–63.

45. W. J. Lee and K. Hase, "Gut Microbiota-Generated Metabolites in Animal Health and Disease," *Nature Chemical Biology* 10, no. 6 (2014): 416–24; S. A. Joyce and C. G. Gahan, "The Gut Microbiota and the Metabolic Health of the Host," *Current Opinions in Gastroenterology* 30, no. 2 (2014): 120–27.

46. R. M. Brucker and S. R. Bordenstein, "The Hologenomic Basis of Speciation: Gut Bacteria Cause Hybrid Lethality in the Genus Nasonia," *Science* 341, no. 6146 (2013): 667–69. Another paper shows diet changes can alter the microbiome of fruit flies, which in turn changes mating preferences. G. Sharon, D. Segal, I. Zilber-Rosenberg, and E. Rosenberg, "Symbiotic Bacteria Are Responsible for Diet-Induced Mating Preference in Drosophila Melanogaster, Providing Support for the Hologenome Concept of Evolution," *Gut Microbes* 2, no. 3 (2011): 190–92.

Autism Revisited: Three Potential Drivers

1. One of Andrey Rzhetsky's first published papers addressed a simple and completely noncontroversial query: How many of my scientific colleagues' papers are B.S., or faked, and should be retracted? This was an unusual type of paper for a young academic

seeking tenure, particularly if he finds that of the 9,398,715 articles published between 1950 and 2004, a mere 596 were retracted. Being fearless, he estimated that somewhere between 10,000 and 100,000 peer-reviewed science papers should be trashed.

2. A. Rzhetsky, S. C. Bagley, K. Wang, C. S. Lyttle et al., "Environmental and State-Level Regulatory Factors Affect the Incidence of Autism and Intellectual Disability," *PLoS Computational Biology* 10, no. 3 (2014): e1003518.

3. K. S. Chul, K. S. Kyoung, and H. Y. Pyo, "Trends in the Incidence of Cryptorchidism and Hypospadias of Registry-Based Data in Korea: A Comparison Between Industrialized Areas of Petrochemical Estates and a Non-Industrialized Area," *Asian Journal of Andrology* 13, no. 5 (2011): 715–18.

4. Y. S. Kim, B. L. Leventhal, Y.-Y. Koh, E. Fombonne et al., "Prevalence of Autism Spectrum Disorders in a Total Population Sample," *American Journal of Psychiatry* 168, no. 9 (2011): 9904–12. See also "New Study Reveals Autism Prevalence in South Korea Estimated to be 2.6% or 1 in 38 Children," Autism Speaks, accessed August 30, 2014, http://www.autismspeaks.org/about-us/press-releases/new-study-reveals-autism-prevalence-south-korea-estimated-be-26-or-1-38-chil; Centers for Disease Control and Prevention, "Autism Spectrum Disorder (ASD), Data & Statistics," CDC.gov, accessed August 30, 2014, http://www.cdc.gov/ncbddd/autism/data.html.

5. L. Gaspari, D. R. Sampaio, F. Paris, F. Audran et al., "High Prevalence of Micropenis in 2710 Male Newborns from an Intensive-Use Pesticide Area of Northeastern Brazil," *International Journal of Andrology* 35, no. 3 (2012): 253–64.

6. J. J. Ryan, Z. Amirova, and G. Carrier, "Sex Ratios of Children of Russian Pesticide Producers Exposed to Dioxin," *Environmental Health Perspectives* 110, no. 11 (2002): A699–701.

7. T. Yoshimura, S. Kaneko, and H. Hayabuchi, "Sex Ratio in Offspring of Those Affected by Dioxin and Dioxin-Like Compounds: The Yusho, Seveso and Yucheng Incidents," *Occupational & Environmental Medicine* 58, no. 8 (2001): 540–41; W. J. Rogan, B. C. Gladen, Y.-L. L. Guo, and C.-C. Hsu, "Sex Ratio After Exposure to Dioxin-Like Chemicals in Taiwan," *Lancet* 353, no. 9148 (1999): 206–7.

8. C. A. Mackenzie, A. Lockridge, and M. Keith, "Declining Sex Ratio in a First Nation Community," *Environmental Health Perspectives* 113, no. 10 (2005): 1295–98.

9. "Pregnancy over Age 30," Children's Hospital of Philadelphia, accessed August 30, 2014, http://www.chop.edu/conditions-diseases/pregnancy-over-age-30.

10. L. A. Schieve, O. Devine, C. A. Boyle, J. R. Petrini et al., "Estimation of the Contribution of Non-Assisted Reproductive Technology Ovulation Stimulation Fertility Treatments to US Singleton and Multiple Births," *American Journal of Epidemiology* 170, no. 11 (2009): 1396–407; March of Dimes, "Fertility Drugs Contribute Heavily to Multiple Births," marchofdimes.org, January 15, 2010, accessed December 17, 2014, http://www.marchofdimes.org/news/fertiltity-drugs-contribute-heavily-to-multiple-births.aspx.

11. K. Lyall, D. L. Pauls, D. Spiegelman, S. L. Santangelo et al., "Fertility Therapies, Infertility and Autism Spectrum Disorders in the Nurses' Health Study II," *Paediatric Perinatal Epidemiology* 26, no. 4 (2012): 361–72; Claudia Wallis, "Studies Link Infertility Treatments to Autism," *Time*, May 20, 2010, accessed August 30, 2014, http://content.time.com/time/health/article/0,8599,1990567,00.html.

12. A. Kong, M. L. Frigge, G. Masson, S. Besenbacher et al., "Rate of De Novo Mutations and the Importance of Father's Age to Disease Risk," *Nature* 488, no. 7412 (2012): 471–75.

13. A. Reichenberg, R. Gross, M. Weiser, M. Bresnahan et al., "Advancing Paternal Age and Autism," *Archives of General Psychiatry* 63, no. 9 (2006): 1026–32; S. G. Zammit, P. Allebeck, C. Dalman, I. Lundberg et al., "Paternal Age and Risk for Schizophrenia," *British Journal of Psychiatry* 183 (2003): 405–8.

14. Emma Young wrote a great overview of the field: E. Young, "Gut Instincts: The Secrets of Your Second Brain," *New Scientist*, December 15, 2012.

15. Here is a neat interactive site where you can virtually dissect the 75 million mice brain neurons: National Institute of Mental Health (NIMH), "Mouse Brain Atlas: DBA/2J Coronal," The Mouse Brain Library, accessed August 28, 2014, http://www .mbl.org/atlas165/atlas165_start.html. (And during your next cocktail party you can also tell that pedantic gasbag, "Sir, I believe you are thinking with your enteric.")

16. Young, "Gut Instincts."

17. R. Rose'Meyer, "A Review of the Serotonin Transporter and Prenatal Cortisol in the Development of Autism Spectrum Disorders," *Molecular Autism* 4, no. 37 (2013): 1–16, accessed August 28, 2014, http://www.molecularautism.com/content/4/1/37.

18. Moises Velazquez-Manoff has written a good book on this subject: *An Epidemic of Absence* (New York: Simon & Schuster, 2012). If you want to see the arguments and counterarguments around autism see Moises Velazquez-Manoff, "Response to Autism Oped 'Smackdowns' and FAQs," Moises Velazquez-Manoff personal blog, September 9, 2012, accessed August 30, 2014, http://www.moisesvm.com/2012/09 /09/response-to-autism-oped-smackdowns-and-faqs. One can see extreme differences in the immune responses of lab rats and wild rats: William Parker, "Duke Surgical Sciences Parker Research Lab," Duke Surgical Sciences, accessed August 30, 2014, http://sciences.surgery.duke.edu/research/institutes-and-labs/parker-lab; A. Devalapalli, A. Lesher, K. Shieh, J. S. Solow et al., "Increased Levels of IgE and Autoreactive, Polyreactive IgG in Wild Rodents: Implications for the Hygiene Hypothesis," *Scandinavian Journal of Immunology* 64 (2006): 125–36.

Viruses: The Roadrunners of Evolution

1. Ironically, the debate now shifts back toward only a two-branch tree of life, with *Archaea* first, then eukaryotes. See: T. A. Williams, P. G. Foster, C. J. Cox, and T. M. Embley, "An Archaeal Origin of Eukaryotes Supports Only Two Primary Domains of Life," *Nature* 504, no. 7479 (2013): 231–36.

2. K. Vetsigian, C. Woese, and N. Goldenfeld, "Collective Evolution and the Genetic Code," *Proceedings of the National Academy of Sciences USA* 103, no. 28 (2006): 10696–701.

3. With some very recent exceptions, see Epilogue for new life codes.

4. There are a number of chemicals that closely resemble DNA and RNA and have the potential to encode information. Termed "xeno nucleic acids" (XNAs), these compounds are man-made today and are being used in the lab to explore new biology and evolutionary alternatives. For a review: M. Schmidt, "Xenobiology: A New Form of Life as the Ultimate Biosafety Tool," *BioEssays* 32, no. 4 (2010): 322–31.

5. R. A. Edwards and F. Rohwer, "Viral Metagenomics," *Nature Reviews in Microbiology* 3, no.6 (2005): 504–10.

6. Here are reviews of the virome field that provide a number of the facts mentioned in this chapter: K. M. Wylie, G. M. Weinstock, and G. A. Storch, "Emerging View of the Human Virome," *Translational Research* 160, no. 4 (2012): 283–90; S. C. P.

Williams, "The Other Microbiome," *Proceedings of the National Academy USA* 110, no. 8 (2013): 2682–84; H. W. Virgin, "The Virome in Mammalian Physiology and Disease," *Cell* 157, no. 1 (2014): 142–50; S. R. Abeles and D. T. Pride, "Molecular Bases and Role of Viruses in the Human Microbiome," *Journal of Molecular Biology* 426, no. 23 (2014): 389–906.

7. This study used the latest DNA-sequencing technology and bioinformatics to analyze viral DNA in 24 longitudinal fecal samples—i.e., deep metagenomic sequencing to evaluate 56 billion DNA bases of purified viral sequence: S. Minot, A. Bryson, C. Chehoud, G. D. Wu et al., "Rapid Evolution of the Human Gut Virome," *Proceedings of the National Academy of Sciences USA* 110, no. 30 (2013): 12450–55.

8. Jason Socrates Bardi, "The Gross Science of a Cough and a Sneeze," Live Science, June 14, 2009, accessed September 6, 2014, http://www.livescience.com/3686-gross -science-cough-sneeze.html.

9. M. Horie, T. Honda, Y. Suzuki, Y. Kobayashi et al., "Endogenous Non-Retroviral RNA Virus Elements in Mammalian Genomes," *Nature* 463, no. 7277 (2010): 84–87.

10. Note that it is called junk in part because scientists don't know exactly what it does, yet it is retained, generation after generation. Undoubtedly some of it must have a function, a point on which most scientists agree.

A Perfectly Modern Pregnancy

1. Clomid mimics LSH, the hormone that promotes ovulation. Like tens of thousands of hopeful mothers every year, Emilie did not get a regular period, so she was not regularly releasing an egg, a critical step in becoming pregnant. After two cycles of Clomid, ovulation occurred, fertilization happened, the genetic code was put into action, and the fun began with the building of a child.

2. Here is one genetic testing service for parents-to-be: Counsyl.com, accessed September 8, 2014, https://www.counsyl.com/services/family-prep-screen/carrier-screening.

3. Y. Xue, Y. Chen, Q. Ayub, N. Huang et al., "Deleterious- and Disease-Allele Prevalence in Healthy Individuals: Insights from Current Predictions, Mutation Databases, and Population-Scale Re-Sequencing," *American Journal of Human Genetics* 7, no. 6 (2012): 1022–32.

4. Centers for Disease Control and Prevention, "Births—Methods of Delivery," CDC .gov, accessed September 3, 2014, http://www.cdc.gov/nchs/fastats/delivery.htm; J. Neu and J. Rushing, "Cesarean Versus Vaginal Delivery: Long-Term Infant Outcomes and the Hygiene Hypothesis," *Clinical Perinatology* 38, no. 2 (2011): 321–31.

5. K. Aagaard, J. Ma, K. M. Antony, R. Ganu et al., "The Placenta Harbors a Unique Microbiome," *Science of Translational Medicine* 6, no. 237 (2014): 237ra65; Claire Wilson, "Baby's First Gut Bacteria May Come from Mum's Mouth," *New Scientist*, May 2014, accessed September 3, 2014, http://www.newscientist.com/article/dn25603 -babys-first-gut-bacteria-may-come-from-mums-mouth.html; Dina Fine Maron, "How Bacteria in the Placenta Could Help Shape Human Health," *Scientific American*, May 21, 2014.

6. V. Padmanabhan, H. N. Sarma, Savabieasfahani, T. L. Steckler et al., "A Developmental Reprogramming of Reproductive and Metabolic Dysfunction in Sheep: Native Steroids vs. Environmental Steroid Receptor Modulators," *International Journal of Andrology* 33, no. 2 (2010): 394–404; J. R. Rochester, "Bisphenol A and Human Health: A Review of the Literature," *Reproductive Toxicology* 42 (2013): 132–55.

7. M. D. Anway, A. S. Cupp, M. Uzumcu, and M. K. Skinner, "Epigenetic Transgener-
    ational Actions of Endocrine Disruptors and Male Fertility," *Science* 308, no. 5727
    (2005): 1466–69. Juvenile zebrafish exposed transiently to another endocrine disrup-
    tor, dioxin, transmit adverse effects to their offspring for at least two generations,
    including changes in fertility, increased proportion of females at birth, and develop-
    mental abnormalities not seen in the dioxin-exposed individuals. Dioxin's toxic
    effects are conserved across many species. See T. R. Baker, R. E. Peterson, W. Heide-
    man, "Using Zebrafish as a Model System for Studying the Transgenerational Effects
    of Dioxin," *Toxicological Sciences* 138, no. 2 (2014): 403–11.

8. National Institutes of Environmental Health Sciences, "Bisphenol A (BPA)," NIEHS,
    accessed September 9, 2014, http://www.niehs.nih.gov/health/topics/agents/sya-bpa;
    "Bisphenol A," *Wikipedia*, accessed September, 8, 2014, http://en.wikipedia.org/wiki
    /Bisphenol_A.

9. Toxins we know we should avoid, but often choose not to: Alcohol. Tobacco. eCigs.
    Caffeine. Drugs. Your fail-safe poison defenses tried to warn you long ago—remember
    that first sip of Dad's scotch, or that cough of your first cigarette? As you would expect,
    there is a strong literature tying these vices to bad outcomes. These toxins change an
    infant's epigenome. Caveat emptor.

10. Human cells from virtually any organ can be grown in a laboratory using microfluid-
    ics and other engineering techniques to mimic the environment of the human body.
    These "organs on a chip" reproduce the physiology, biochemistry, and microanatomy
    of the true organ. There is a wonderful TED talk by Dr. Geraldine Hamilton
    describing this technology at the Wyss Institute: Geraldine Hamilton, "Body Parts
    on a Chip," TEDx Boston 2013, accessed September 3, 2014, https://www.ted.com
    /talks/geraldine_hamilton_body_parts_on_a_chip.

11. D. S. Feig, J. Hwee, B. R. Shah, G. L. Booth et al., "Trends in Incidence of Diabetes
    in Pregnancy and Serious Perinatal Outcomes: A Large, Population-Based Study in
    Ontario, Canada, 1996–2010," *Diabetes Care* 37, no. 6 (2014): 1590–96.

12. M. F. Rolland-Cachera, M. Deheeger, M. Maillot, and F. Bellisle, "Early Adiposity
    Rebound: Causes and Consequences for Obesity in Children and Adults," *Interna-
    tional Journal of Obesity* (London) 30, Supplement 4 (2006): S11–17.

13. P. Dominguez-Salas, S. E. Moore, M. S. Baker, A. W. Bergen et al., "Maternal Nutri-
    tion at Conception Modulates DNA Methylation of Human Metastable Epialleles,"
    Nature Communications 5, no. 3746 (2014): doi:10.1038/ncomms4746; Helen Briggs,
    "Pre-pregnancy Diet 'Permanently Influences Baby's DNA," BBC News, April 29,
    2014, accessed September 14, 2014, http://www.bbc.com/news/health-27211153.

14. D. Martínez, T. Pentinat, S. Ribó, C. Daviaud et al., "In Utero Undernutrition in
    Male Mice Programs Liver Lipid Metabolism in the Second-Generation Off-
    spring Involving Altered LXRA DNA Methylation," *Cell Metabolism* 19, no. 6 (2014):
    941–51.

15. "Epigenetic Nutrients Table," Learn.Genetics, accessed September 4, 2014, http:
    //learn.genetics.utah.edu/content/epigenetics/nutrition/table.html.

16. Paul A Offit and Sarah Erush, "Skip the Supplements," *New York Times*, December
    14, 2013; S. G. Newmaster, M. Grguric, D. Shanmughanandhan, S. Ramalingam et al.,
    "DNA Barcoding Detects Contamination and Substitution in North American
    Herbal Products," *BioMed Central Medicine* 11 (2013): 222.

17. J. L. Jacobson, S. W. Jacobson, G. Muckle, M. Kaplan-Estrin et al., "Beneficial
    Effects of a Polyunsaturated Fatty Acid on Infant Development: Evidence from the
    Inuit of Arctic Quebec," *Journal of Pediatrics* 152, no. 3 (2008): 356–64.

18. A. Ren, L. Zhang, L. Hao, A. Li et al., "Comparison of Blood Folate Levels Among Pregnant Chinese Women in Areas with High and Low Prevalence of Neural Tube Defects," *Public Health Nutrition* 10, no. 8 (2007): 762–68.

19. J. Selhub, P. F. Jacques, A. G. Bostom, P. W. Wilson et al., "Relationship Between Plasma Homocysteine and Vitamin Status in the Framingham Study Population. Impact of Folic Acid Fortification," *Public Health Reviews* 28, nos. 1–4 (2000): 117–45.

20. One can certainly overthink and over-worry. At one point, Steve grew anxious as he watched the mother-to-be (and baby-to-be) eat a big helping of tofu. Rich in protein, but also high in genistein, which has been linked to epigenetic regulation. Fortunately, a quick Google search showed that genistein is considered beneficial.

21. H. Beydoun and A. F. Saftlas, "Physical and Mental Health Outcomes of Prenatal Maternal Stress in Human and Animal Studies: A Review of Recent Evidence," *Paediatric Perinatal Epidemiology* 22, no. 5 (2008): 438–66; N. M. Talge, C. Neal, V. Glover, and the Early Stress, Translational Research and Prevention Science Network: Fetal and Neonatal Experience on Child and Adolescent Mental Health, "Antenatal Maternal Stress and Long-Term Effects on Child Neurodevelopment: How and Why?" *Journal of Child Psychology and Psychiatry* 48, nos. 3–4 (2007): 245–61; K. O'Donnell, T. G. O'Connor, and V. Glover, "Prenatal Stress and Neurodevelopment of the Child: Focus on the HPA Axis and Role of the Placenta," *Developmental Neuroscience* 31, no. 4 (2009): 285–92; B. Billack, R. Serio, I. Silva, and C. H. Kinsley, "Epigenetic Changes Brought About by Perinatal Stressors: A Brief Review of the Literature," *Journal of Pharmacology and Toxicology Methods* 66, no. 3 (2012): 221–31; F. Veru, D. P. Laplante, G. Luheshi, and S. King, "Prenatal Maternal Stress Exposure and Immune Function in the Offspring," *Stress* 17, no. 2 (2014): 133–48; L. C. von Hertzen, "Maternal Stress and T-Cell Differentiation of the Developing Immune System: Possible Implications for the Development of Asthma and Atopy," *Journal of Allergy and Clinical Immunology* 109, no. 6 (2002): 923–28.

22. R. C. Painter, T. J. Roseboom, and S. R. de Rooij, "Long-Term Effects of Prenatal Stress and Glucocorticoid Exposure," *Birth Defects Research Part C: Embryo Today* 96, no. 4 (2012): 315–24.

23. M. K. Bhasin, J. A. Dusek, B.-H. Chang, M. G. Joseph et al., "Relaxation Response Induces Temporal Transcriptome Changes in Energy Metabolism, Insulin Secretion and Inflammatory Pathways," *PLoS ONE* 8, no. 5 (2013): e62817.

24. A. M. Jukic, D. A. Lawlor, M. Juhl, and K. M. Owe, "Physical Activity During Pregnancy and Language Development in the Offspring," *Paediatric Perinatal Epidemiology* 27, no. 3 (2013): 283–93.

25. Different bodies are coded differently in terms of how they react to stress. For instance, four specific gene variants statistically predict how prone someone is to post-traumatic stress disorder (PTSD). The more copies you have of certain gene variants of FKBP5, COMT, CHRNA5, or CRHR1, the likelier you are to have lifetime issues post-trauma. Having none of these variants makes you almost immune to the long-term effects of severe combat stress. Two or three copies lead to a 13 percent lifetime probability. Four or more, and it is over 20 percent. But it is not the core gene code alone that determines the outcome. Having these specific DNA variants indicates a higher *probability* of PTSD, not a certainty of PTSD. Even among individuals who carry four, five, or even six of the predisposing gene variants, 80 percent do not come down with PTSD following times of adverse stress. Some genes may predispose, trigger, or prevent, but there is likely also an epigenetic/environmental driver of PTSD. See J. A. Boscarino, P. M. Erlich, S. N. Hoffman, and X. Zhang, "Higher FKBP5, COMT, CHRNA5, and CRHR1 Allele Burdens Are Associated with PTSD

and Interact with Trauma Exposure: Implications for Neuropsychiatric Research and Treatment," *Neuropsychiatric Disease Treatment* 8 (2012): 131–39.

26. J. E. Koenig, A. Spor, N. Scalfone, A. D. Fricker et al., "Succession of Microbial Consortia in the Developing Infant Gut Microbiome," *Proceedings of the National Academy of Sciences USA* 108 (Suppl. 1) (2011): 4578–85; C. Palmer, E. M. Bik, D. B. DiGiulio, D. A. Relman et al., "Development of The Human Infant Intestinal Microbiota," *PLoS Biology* 5, no. 7 (2007): e177.

27. M. B. Azad, T. Konya, H. Maughan, D. S. Guttman et al., "Gut Microbiota of Healthy Canadian Infants: Profiles by Mode of Delivery and Infant Diet at 4 Months," *Canadian Medical Association Journal* 185, no. 5 (2013): 385–94.

28. Yet the United States still spends billions to buy mother's-milk substitutes, marketed almost entirely by three companies. See V. Oliveira, E. Frazao, and D. Smallwood, "The Infant Formula Market: Consequences of a Change in the WIC Contract Brand," *United States Department of Agriculture Economic Research Report* 124 (2011): 5, accessed September 4, 2014, http://www.ers.usda.gov/media/121286/err124.pdf. But as awareness has grown on the benefits of breastfeeding, sales of baby formula have trended down. See I. Le Huërou-Luron, S. Blat, and G. Boudry, "Breast- v. Formula-Feeding: Impacts on the Digestive Tract and Immediate and Long-Term Health Effects," *Nutrition Research Reviews* 23, no. 1 (2010): 23–26. So now the U.S. Dairy Council is focused on exports, especially to China, which spends $12 billion and holds "the fate of the global baby milk formula in its hands." See Mark Astley, "China Will Continue to Drive Global Infant Formula Sales—Euromonitor," Dairy Reporter.com, December 4, 2012, accessed September 4, 2014, http://www.dairyre porter.com/Markets/China-will-continue-to-drive-global-infant-formula-sales -Euromonitor. Given concerns about using dairy-based formula, scientists are busily inventing ways to synthesize human milk, without the mother—unnaturally natural? See L. P. Won-Heong, P. Pathanibul, J. Quarterman, J.-H. Jo et al., "Whole Cell Biosynthesis of a Functional Oligosaccharide, 2'-Fucosyllactose, Using Engineered *Escherichia coli*," *Microbial Cell Factories* 11, no. 1 (2011): 48.

29. Sam Miller and Alexandra Waldhorn, "New York City Health Department Launches 'Latch on NYC' Initiative to Support Breastfeeding Mothers," New York City Department of Health and Mental Hygiene, May 9, 2012, accessed September 4, 2014, http://www.nyc.gov/html/doh/html/pr2012/pr013-12.shtml.

30. Lauran Neergaard, "Animal Moms Customize Milk Depending on Baby's Sex," Associated Press, February 14, 2014, accessed September 3, 2014, http://bigstory .ap.org/article/animal-moms-customize-milk-depending-babys-sex; "Exploring the Many Effects of Mother's Milk," AAAS 2014 Annual Meeting, accessed September 8, 2014, http://membercentral.aaas.org/announcements/2014-annual-meeting-exploring -many-effects-mother-s-milk; K. Hinde, A. J. Carpenter, J. S. Clay, B. J. Bradford, "Holsteins Favor Heifers, Not Bulls: Biased Milk Production Programmed During Pregnancy as a Function of Fetal Sex," *PLoS ONE* 9, no. 2 (2014): e86169.

31. C. G. Perrine, A. J. Sharma, M. E. Jefferds, M. K. Serdula et al., "Adherence to Vitamin D Recommendations Among US Infants," *Pediatrics* 125, no. 4 (2010): 627–32.

32. M. Al-Assmakh, F. Anuar, F. Zadjali, J. Rafter, and S. Pettersson, "Gut Microbial Communities Modulating Brain Development and Function," *Gut Microbes* 3, no. 4 (2012): 366–73.

33. Robert Preidt, "Scientists Probe Dark Chocolate's Health Secrets: Heart Benefits May Stem from Reaction in Stomach Bacteria, Research Suggests," WebMD News from HealthDay, March 18, 2014, accessed August 29, 2014 http://www.webmd.com/food -recipes/news/20140318/scientists-probe-dark-chocolates-health-secrets.

34. K. Raikkonen, A. V. Pesonen, A. L. Jarvenpaa, and T. E. Strandberg, "Sweet Babies: Chocolate Consumption During Pregnancy and Infant Temperament at Six Months," *Early Human Development* 76, no. 2 (2004): 139–45.

35. Moises Velasquez-Manoff, "A Cure for the Allergy Epidemic?" *New York Times*, November 9, 2013.

36. Centers for Disease Control and Prevention, "Healthy Pets, Healthy People," CDC .gov, January 6, 2012, accessed September 4, 2014, http://www.cdc.gov/healthypets /health_benefits.htm.

37. Y. Wang and L. H. Kasper, "The Role of Microbiome in Central Nervous System Disorders," *Brain, Behavior, and Immunity* 38 (2012): 1–12.

38. R. D. Heijtz, S. Wang, F. Anuar, Y. Qian et al., "Normal Gut Microbiota Modulates Brain Development and Behavior," *Proceedings of the National Academy of Sciences USA* 108, no. 7 (2011): 3047–52.

39. P. Bercik, A. J. Park, D. Sinclair, A. Khoshdel et al., "The Anxiolytic Effect of *Bifidobacterium Longum* NCC2001 Involves Vagal Pathways for Gut-Brain Communication," *Neurogastroenterology and Motility* 23, no. 12 (2011): 1132–39.

40. F. Indrio, A. Di Mauro, G. Riezzo, E. Civardi et al., "Prophylactic Use of a Probiotic in the Prevention of Colic, Regurgitation, and Functional Constipation," *Journal of the American Medical Association Pediatrics* 168, no. 8 (2014): 778.

41. Cord blood preservation is not available in every hospital, such as where Esme was delivered. The storage can cost $5,000, plus an ongoing annual fee.

42. But what if Esme develops a food or pollen allergy? A Stanford study showed that children who ate a minuscule amount of peanuts, and then increased the amount over many months, could cure their debilitating allergy. See G. P. Yu, B. Weldon, S. Neale-May, and K. C. Nadeau, "The Safety of Peanut Oral Immunotherapy in Peanut-Allergic Subjects in a Single-Center Trial," *International Archives of Allergy and Immunology* 159, no. 2 (2012): 179–82; A. Syed, M. A. Garcia, S.-C. Lyu, R. Bucayu et al., "Peanut Oral Immunotherapy Results in Increased Antigen-Induced Regulatory T-Cell Function and Hypomethylation of Forkhead Box Protein 3 (FOXP3)," *Journal of Allergy and Clinical Immunology* 133, no. 2 (2014): 500–510.

43. K. Hill, D. You, M. Inoue, and M. Z. Oestergaard, "Child Mortality Estimation: Accelerated Progress in Reducing Global Child Mortality, 1990–2010," *PLoS Medicine* 9, no. 8 (2012): e1001303; Moises Velasquez-Manoff, "A Cure for the Allergy Epidemic?"

44. The Flynn effect, while hotly debated, describes the gradual rise in IQ scores observed in many cohorts since IQ testing began in the 1930s—too fast for genetic selection to have occurred. See J. R. Flynn, "Massive IQ Gains in 14 Nations: What IQ Tests Really Measure," *Psychological Bulletin* 101, no. 2 (1987): 171–91.

Bringing It All Together—DESTINY Is Propelling Evolution

1. The hologenome concept was proposed in 2007 by Eugene Rosenberg and Ilana Zilber-Rosenberg to describe the symbiotic coevolution of a coral/bacterial community. It described the interplay of microbes and corals, and was subsequently extended more widely to include the genes of all the microbiota and the host organism. We are respectfully expanding on this concept to include the epigenome and the complex interactions that exist among the four coevolving genomes of an organism within an

environment. E. Rosenberg, O. Koren, L. Reshef, R. Efrony et al., "The Role of Microorganisms in Coral Health, Disease and Evolution," *Nature Reviews Microbiology* 5, no. 5 (2007): 355–62; E. Rosenberg, G. Sharon, I. Zilber-Rosenberg, "The Hologenome of Evolution Contains Lamarckian Aspects Within a Darwinian Framework," *Environmental Microbiology* 11, no. 12 (2009): 2959–62.

2. M. Lynch, "Rate, Molecular Spectrum, and Consequences of Human Mutation," *Proceedings of the National Academy of Sciences USA* 107, no. 3 (2010): 961–68; P. D. Keightley, "Rates and Fitness Consequences of New Mutations in Humans," *Genetics* 190, no. 2 (2012): 295–304.

3. A. Scally and R. Durbin, "Revising the Human Mutation Rate: Implications for Understanding Human Evolution," *Nature Reviews Genetics* 13, no. 10 (2012): 745–53.

4. Cavemen weren't so concerned about inbreeding: K. Prüfer, F. Racimo, N. Patterson, F. Jay et al., "The Complete Genome Sequence of a Neanderthal from the Altai Mountains," *Nature* 505, no. 7481 (2014): 43–49.

5. Darwin was agnostic (even declaring ignorance) regarding mechanism, e.g., mutation, so he spoke in terms of variation and slow time frames. He recognized that things were more rapid in domesticated animals, but did not consider them to be different species from their feral progenitors. In fact, he suggested that domesticated animals showed more individual variation when living under more stable conditions.

6. Chimpanzee Sequencing and Analysis Consortium, "Initial Sequence of the Chimpanzee Genome and Comparison with the Human Genome," *Nature* 43, no. 7055 (2005): 69–87; D.-D. Wu, D. M. Irwin, and Y.-P. Zhang, "De Novo Origin of Human Protein-Coding Genes," *PLoS Genetics* 7, no. 11 (2011): e1002379.

7. J. Zeng, G. Konopka, B. G. Hunt, T. M. Preuss et al., "Divergent Whole-Genome Methylation Maps of Human and Chimpanzee Brains Reveal Epigenetic Basis of Human Regulatory Evolution," *American Journal of Human Genetics* 91, no. 3 (2012): 455–65.

8. R. Haygooda, C. C. Babbitt, O. Fedrigo, and G. A. Wray, "Contrasts Between Adaptive Coding and Noncoding Changes During Human Evolution," *Proceedings of the National Academy of Sciences USA* 107, no. 17 (2010): 7853–57.

9. O. Viltart and C. C. A. Vanbesien-Mailliot, "Impact of Prenatal Stress on Neuroendocrine Programming," *Scientific World Journal* 7 (2007): 1493–537.

10. For historical context, environment, the epigenome, and the microbiome really matter. The same SIM1 gene variants that predispose the Pima to obesity don't seem to have a similar effect in France. The R64 allele of the ADRB3 gene leads to weight gain in Asians but not Europeans. H. Choquet and D. Meyre, "Genetics of Obesity: What Have We Learned?" *Current Genomics* 12, no. 3 (2011): 169–79.

Playing with the Building Blocks of Life

1. One way to think about gene therapy is through a parable sometimes told by one of the great deans of Harvard Business School, John McArthur. A grizzled, tough businessman and a gentle teacher and mentor, McArthur initially comes across as a slow-talking western Canadian who wears short-sleeve dress shirts in almost any weather. Careful; he masks a spectacular intelligence and aggressiveness behind a soft-spoken "Eh?" When McArthur finds something really odd, unnatural, or out of place, he sometimes slowly unfurls a parable about a farmer walking the fences and coming upon a turtle atop a fencepost. One thing the farmer can be sure of: Someone put that turtle there. Turtles don't just do that; it requires human intervention.

2. D. A. Jackson, R. H. Symons, and P. Berg, "Biochemical Method for Inserting New Genetic Information into DNA of Simian Virus 40: Circular SV40 DNA Molecules Containing Lambda Phage Genes and the Galactose Operon of *Escherichia Coli*," *Proceedings of the National Academy of Sciences USA* 69, no. 10 (1972): 2904–9; P. Berg, D. Baltimore, H. W. Boyer, S. N. Cohen et al., "Potential Biohazards of Recombinant DNA Molecules," *Science* 185, no. 148 (1974): 303. This critical meeting took place at a retreat named Refuge from the Sea—in Spanish: Asilomar—which was founded by the Young Women's Christian Association.

3. David Arnold, "Roots of a Quarrel: Vellucci, Lampoon Wage Feud over a Tree," *Boston Globe*, April 6, 1991. Mayor Vellucci also tried to get the entire headquarters building of the bothersome and mocking *Harvard Lampoon* to be designated a public urinal.

4. The seven drugs: Humira, Remicade, Rituxan, Enbrel, Lantus, Avastin, and Herceptin. See Cell Culture Dish, "Biologics Still on Top in Best Selling Drugs of 2013," The Dish, accessed September 2, 2014, http://cellculturedish.com/2014/03/top-ten-biologics-2013-us-pharmaceutical-sales-2, which cites "The Top 25 Best-Selling Drugs of 2013," *Genetic Engineering & Biotechnology News*, March 3, 2014, accessed September 2, 2014, http://www.genengnews.com/insight-and-intelligenceand153/the-top-25-best-selling-drugs-of-2013/77900053/.

5. Announcement by Cambridge city manager Robert W. Healy, "Cambridge Launches New Biotech Web Site." March 30, 2007, accessed November 1, 2014, http://www2.cambridgema.gov/deptann.cfm?story_id=1318; here is the link to the actual Cambridge Biotech Web site: http://www.cambridgema.gov/CDD/EconDev/LifeSciencesAndTechnologyBusinesses/.

6. iGEM: Synthetic Biology Based on Standard Biological Parts, accessed November 1, 2014, http://igem.org/About.

7. "Letter from W. G. Katzenmeyer to George McDonald Church, January 16, 1976," Harvard Molecular Technologies, accessed September 2, 2014, http://arep.med.harvard.edu/gmc/F.jpg. The dean who signed the termination letter has no publications in PubMed and very few relevant publications. Church, however, is one of the most cited scientists in the world.

Humans Hijacking Viruses

1. M. Horie, T. Honda, Y. Suzuki, Y. Kobayashi et al., "Endogenous Non-Retroviral RNA Virus Elements in Mammalian Genomes," *Nature* 463, no. 7277 (2010): 84–87.

2. Sheryl Gay Stolberg, "The Biotech Death of Jesse Gelsinger," *New York Times*, November 28, 1999, accessed September 18, 2014, http://www.nytimes.com/1999/11/28/magazine/the-biotech-death-of-jesse-gelsinger.html.

3. Someday we may dispense with disease vectors altogether. For instance, ingenious methods devised by David Kaplan, a big-mustached, twinkly-eyed Tufts professor, proposes using engineered spider's silk that floats through your body and attaches to specific diseased cells. See K. Numata, M. R. Reagan, R. H. Goldstein, M. Rosenblatt et al., "Spider Silk-Based Gene Carriers for Tumor Cell-Specific Delivery," *Bioconjugate Chemistry* 22, no. 8 (2011): 1605–10.

4. John Wiley and Sons Ltd., "Gene Therapy Clinical Trials Worldwide: Vectors Used in Gene Therapy Clinical Trials," *Journal of Gene Medicine*, accessed September 20, 2014, http://www.abedia.com/wiley/vectors.php.

5. Not that any of this is cheap and easy. Time and safety constraints slightly increased the cost of Glybera therapy per patient to a mere $1.6 million, making it the most expensive treatment in the world—but it is likely a one-time treatment that cures the condition.

6. John Wiley and Sons Ltd., "Gene Therapy Clinical Trials Worldwide."

7. A. Biffi, E. Montini, L. Lorioli, M. Cesani et al., "Lentiviral Hematopoietic Stem Cell Gene Therapy Benefits Metachromatic Leukodystrophy," *Science* 341, no. 6148 (2013): 1233158, accessed October 28, 2014, http://www.sciencemag.org/content/341 /6148/1233158.short; A. Aiuti, L. Biasco, S. Scaramuzza, F. Ferrua et al., "Lentiviral Hematopoietic Stem Cell Gene Therapy in Patients with Wiskott-Aldrich Syndrome," *Science* 341, no. 6148 (2013): 1233151, accessed October 28, 2014, http://www .sciencemag.org/content/341/6148/1233151.

8. R. E. MacLaren, M. Groppe, A. R. Barnard, C. L. Cottriall et al., "Retinal Gene Therapy in Patients with Choroideremia: Initial Findings from a Phase 1/2 Clinical Trial," *Lancet* 383, no. 9923 (2014): 1129–37. Also, the *Journal of Gene Medicine* is a comprehensive and up-to-date source regarding current trials.

9. K. Mancuso, W. W. Hauswirth, Q. Li, T. B. Connor et al., "Gene Therapy for Red-Green Colour Blindness in Adult Primates," *Nature* 461, no. 7265 (2009): 784–87.

10. Amy Oliver, "Birds, Bees and Claude Monet: How the Famed Artist Could See in Ultraviolet Just Like Some Animals," *Daily Mail*, April 18, 2012, accessed September 20, 2014, http://www.dailymail.co.uk/sciencetech/article-2131608/Claude-Monet -How-famed-artist-ultraviolet-just-like-animals.html.

11. C. Williams, "All-Seeing Eyes," *New Scientist* 221, no. 2961 (2014): 40.

12. "Louise Brown," *Wikipedia*, accessed September 1, 2014, http://en.wikipedia.org /wiki/Louise_Brown.

Editing Life on a Grand Scale

1. P. Mali, L. Yang, K. M. Esvelt, J. Aach et al., "RNA-Guided Human Genome Engineering Via Cas9," *Science* 339, no. 6121 (2013): 823–26. CRISPR is short for CRISPR-Cas system and related technologies.

2. J. D. Sander and J. K. Joung, "CRISPR-Cas Systems for Editing, Regulating and Targeting Genomes," *Nature Biotechnology* 32, no. 4 (2014): 347–55.

3. "Nature: Prehistoric Life: History of Life on Earth," BBC, accessed September 1, 2014, http://www.bbc.co.uk/nature/history_of_the_earth.

4. E. V. Koonin and Y. Wolf, "Is Evolution Darwinian or/and Lamarckian?" *Biology Direct* 4 (2009): 42.

5. For example: L. Ye, J. Wang, A. I. Beyer, F. Teque et al., "Seamless Modification of Wild-Type Induced Pluripotent Stem Cells to the Natural CCR5Δ32 Mutation Confers Resistance to HIV Infection," *Proceedings of the National Academy of Sciences USA* 111, no. 26 (2014): 9591–96; Q. Zheng, X. Cai, M. H. Tan, S. Schaffert et al., Precise Gene Deletion and Replacement Using the CRISPR/Cas9 System in Human Cells," *Biotechniques* 57, no. 3 (2014): 115–24; M. Jinek, A. East, A. Cheng, S. Lin et al., "RNA-Programmed Genome Editing in Human Cells," *eLife* 2 (2013): e00417.

6. Conversation with Steve Gullans, July 11, 2014, in Boston.

7. Y. Azoulay, S. Druyan, L. Yadgary, Y. Hadad et al., "The Viability and Performance Under Hot Conditions of Featherless Broilers Versus Fully Feathered Broilers," *Poultry Science* 90, no. 1 (2011): 19–29.

8. Emma Young, "Featherless Chicken Creates a Flap," *New Scientist*, May 21, 2002, accessed September 1, 2014, http://www.newscientist.com/article/dn2307-featherless -chicken-creates-a-flap.html.

9. H. Kim and J. S. Kim, "A Guide to Genome Engineering with Programmable Nucleases," *Nature Reviews Genetics* 15, no. 5 (2014): 321–34; P. Mali, K. M. Esvelt, and G. M. Church, "Cas9 as a Versatile Tool for Engineering Biology," *Nature Methods* 10, no. 10 (2013): 957–63.

10. Glowing Plant, "Natural Lighting Without Electricity," Glowing Plant, accessed September 1, 2014, http://www.glowingplant.com.

11. Full disclosure: Genovia Bio is a subsidiary of Synthetic Genomics, a company that both Steve and Juan have invested in. For a list of companies the authors are involved in, see the Excel Venture Management Web site: www.excelvm.com.

12. When you undertake massive, parallel recombination, the extremely unlikely becomes common. . . . See Del Harvey, "The Strangeness of Scale at Twitter," TED, accessed September 1, 2014, http://www.ted.com/talks/del_harvey_the_strangeness _of_scale_at_twitter.

13. Gantz and Bier and their team show that a gene drive can be greater than 98 percent effective in mosquitoes using CRISPR in a single generation to address malaria. V. M. Gantza, N. Jasinskieneb, O. Tatarenkovab, A. Fazekasb et al., "Highly Efficient Cas9-Mediated Gene Drive for Population Modification of the Malaria Vector Mosquito *Anopheles stephensi*," *Proceedings of the National Academy of Sciences USA*, 112, no. 49 (2015): E6736–43.

14. A. Hammond, R. Galizi, K. Kyrou, A. Simoni et al., "A CRISPR-Cas9 Gene Drive System Targeting Female Reproduction in the Malaria Mosquito Vector *Anopheles gambiae*," *Nature Biotechnology* 34 (2016): 78–83; doi:10.1038/nbt.3439.

15. Bill Gates, "The Deadliest Animal in the World," *GatesNotes*, April 25, 2014, https: //www.gatesnotes.com/Health/Most-Lethal-Animal-Mosquito-Week.

16. Twenty-six scientists wrote principles for gene drives in the laboratory setting: O. S. Akbari, H. J. Bellen, E. Bier, S. L. Bullock et al., "Safeguarding Gene Drive Experiments in the Laboratory," *Science* 349, no. 6251 (2016): 927–29. Includes a table for how to practice safe science.

17. Rob Stein, "How Could Releasing More Mosquitoes Help Fight Zika?," NPR, March 25, 2016, http://www.npr.org/sections/goatsandsoda/2016/03/25/471304974/how -could-releasing-more-mosquitoes-help-fight-zika.

18. Oxitec, "FDA Publishes Preliminary Finding of No Significant Impact on Oxitec's Self-Limiting Mosquito," March 11, 2016, http://www.oxitec.com/fda-preliminary -finding-no-significant-impact-oxitecs-self-limiting-mosquito.

19. J. Champer, A. Buchman, O. Akbari, "Cheating Evolution: Engineering Gene Drives to Manipulate the Fate of Wild Populations," *Nature Reviews Genetics* 17, no. 3 (2016): 146–59; doi:10.1038/nrg.2015.34.

20. P. Liang, Y. Xu, X. Zhang, C. Ding et al., "CRISPR/Cas9-Mediated Gene Editing in Human Tripronuclear Zygotes," *Protein & Cell* 6, no. 5 (2015): 363–72.

21. E. Lanphier, F. Urnov, S. E. Haecker, M. Werner et al., "Don't Edit the Human Germ Line," *Nature* 519, no. 7544 (2015): 410–11.

22. "Statement on Genome Editing Technologies and Human Germline Genetic Modification," The Hinxton Group, http://www.hinxtongroup.org/Hinxton2015_Statement.pdf.

23. R. Isasi, E. Kleiderman, B. M. Knoppers, "Editing Policy to Fit the Genome?" *Science* 351, no. 6271 (2016): 337–39.

24. X. Kang, W. He, Y. Huang, Q. Yu, et al., "Introducing Precise Genetic Modifications into Human 3PN Embryos by CRISPR/Cas-Mediated Genome Editing," *Journal of Assisted Reproduction and Genetics* 33, no. 5 (2016): 581–88. P. Liang, Y. Xu, X. Zhang, C. Ding et al., "CRISPR/Cas9-Mediated Gene Editing in Human Tripronuclear Zygotes," *Protein & Cell* 6, no. 5 (2015): 363–72. Here is perspective on the ethical situation regarding human embryo editing: E. Lanphier, F. Urnov, S. E. Haecker, M. Werner et al., "Don't Edit the Human Germ Line," *Nature* 519, no. 7544 (2015): 410–11. See also: National Human Genome Research Institute, "Germline Gene Transfer," National Human Genome Research Institute, accessed September 20, 2014, https://www.genome.gov/10004764; Charles Coutelle, "Prospects for Prenatal Gene Therapy," Imperial College London, London, UK, published online February 2014, doi:10.1002/9780470015902.a0025275; K. R. Smith, "Gene Therapy: The Potential Applicability of Gene Transfer Technology to the Human Germline," *Internal Journal of Medical Sciences* 1, no. 2 (2004): 76–91. Finally, the UK recently approved basic research on human embryos using CRISPR, which, parenthetically, will provide greater insights on how to make it safe and effective for future use in humans, if society agrees. E. Callaway, "UK Scientists Gain Licence to Edit Genes in Human Embryos," *Nature* 530, no. 7588 (2016): 18.

Unnatural Acts, Designer Babies, and Sex 2.0

1. M. Brännström, L. Johannesson, P. Dahm-Kähler, A. Enskog et al., "The First Clinical Uterus Transplantation Trial: A Six-Month Report," *Fertility and Sterility* 101, no. 5 (2014): 1228–36. M. Brännström, L. Johannesson, H. Bokström, N. Kvarnström, et al., "Livebirth After Uterus Transplantation," *Lancet* (2014), ePub ahead of print, doi:10.1016/S0140-6736(14)61728-1.

2. Steve Connor, "UK Becomes First Country in World to Approve IVF Using Genes of Three Parents," *Independent*, June 28, 2013, accessed September 9, 2014, http://www.independent.co.uk/news/science/uk-becomes-first-country-in-world-to-approve-ivf-using-genes-of-three-parents-8677595.html. Two examples of ethical perspectives: Michael Cook, "UK to Allow Research into Three-Parent Embryos," *BioEdge*, June 29, 2013, accessed September 20, 2014, http://www.bioedge.org/index.php/bioethics/bioethics_article/10578; Jessica Cussins, "Eight Misconceptions About 'Three-Parent Babies,'" *Biopolitical Times*, July 9, 2013, accessed September 20, 2014, http://www.biopoliticaltimes.org/article.php?id=7003.

3. T. Rito, M. B. Richards, V. Fernandes, F. Alshamali et al., "The First Modern Human Dispersals Across Africa," *PLoS ONE* 8, no. 11 (2013): e80031.

4. You can get a sense of the devastating effects that mutation in genes like DAD, LHON, MELAS, MERRF, MNGIE, NARP, and WPW have by reviewing the symptoms of mitochondrial diseases at United Mitochondrial Disease Foundation, "Types of Mitochondrial Disease," http://www.umdf.org/site/c.8qKOJ0MvF7LUG/b.7934629/k.4C9B/Types_of_Mitochondrial_Disease.htm.

5. National Institutes of Health, "Mitochondrial DNA," accessed September 20, 2014, http://ghr.nlm.nih.gov/mitochondrial-dna.

6. L. Craven, H. A. Tuppen, G. D. Greggains, S. J. Harbottle et al., "Pronuclear Transfer in Human Embryos to Prevent Transmission of Mitochondrial DNA Disease," *Nature* 465, no. 7294 (2010): 82–85. In unusual cases where a child needs a bone-marrow transplant and has no living donor match, parents have opted to have a new baby that is pretested with IVF techniques to ensure it has optimal immune-system genes so when he/she is born, stem cells can be transplanted to the waiting child. We also can conduct genetic testing to select embryos for IVF prior to implantation to ensure that the new baby does not carry a parent's known mutant gene. Thus today we are one step away from selecting embryos without these diseases before birth, and then just one step further away from genetically modifying for disease resistance.

7. Wellcome Trust, "UK Researchers Successfully Transfer Genetic Material Between Two Fertilised Eggs," April 14, 2010, accessed September 2, 2014, http://www.wellcome.ac.uk/News/Media-office/Press-releases/2010/WTX059136.htm.

8. Department of Health and Human Fertilisation and Embryology Authority, "Innovative Genetic Treatment to Prevent Mitochondrial Disease," HFEA, June 28, 2013, accessed September 2, 2014, https://www.gov.uk/government/news/innovative-genetic-treatment-to-prevent-mitochondrial-disease.

9. Jewish Genetic Disease Consortium, "Genetics and Carrier Screening," accessed September 9, 2014, http://www.jewishgeneticdiseases.org/genetics-and-carrier-screening.

10. "Conditions Screened by State: Massachusetts," Baby's First Test, accessed September 2, 2014, http://www.babysfirsttest.org/newborn-screening/states/massachusetts.

11. A. E. Sanders, C. Wang, M. Katz, C. A. Derby et al., "Association of a Functional Polymorphism in the Cholesteryl Ester Transfer Protein (CETP) Gene with Memory Decline and Incidence of Dementia," *Journal of the American Medical Association* 303, no. 2 (2010): 150–58.

12. Y. He, C. R. Jones, N. Fujiki, Y. Xu et al., "The Transcriptional Repressor DEC2 Regulates Sleep Length in Mammals," *Science* 325, no. 5942 (2009): 866–70.

13. T. I. Pollin, C. M. Damcott, H. Shen, S. H. Ott et al., "A Null Mutation in Human APOC3 Confers a Favorable Plasma Lipid Profile and Apparent Cardioprotection," *Science* 322, no. 5908 (2008): 1702–5; G. Atzmon, M. Rincon, C. B. Schechter, A. R. Shuldiner et al., "Lipoprotein Genotype and Conserved Pathway for Exceptional Longevity in Humans," *PLoS Biology* 4, no. 4 (2006): e113.

14. B. J. Willcox, T. A. Donlon, Q. He, R. Chen et al., "FOXO3A Genotype Is Strongly Associated with Human Longevity," *Proceedings of the National Academy of Sciences USA* 105, no. 37 (2008): 13987–92.

15. S. Fishel, "Evidence-Based Medicine and the Role of the National Health Service in Assisted Reproduction," *Reproductive BioMedicine Online* 27, no. 5 (2013): 568–69.

16. But the United States may lag in this effort. A 2016 National Institute of Medicine report issued an extraordinary finding: We only allow mitochondrial replacement therapies under the direst cases, despite the fact that not acting can lead to seizures, weakness, blindness, and death. (And if you do a transplant, only do so on male babies.) National Academies of Sciences, Engineering, and Medicine, *Mitochondrial Replacement Techniques: Ethical, Social, and Policy Considerations* (Washington, DC: National Academies Press, 2016); doi:10.17226/21871.

Boyden Brains

1. M. Karayiorgou, J. Flint, J. A. Gogos, and R. C. Malenka, "The Best of Times, the Worst of Times for Psychiatric Disease," *Nature Neuroscience* 15, no. 6 (2112): 811–12.

2. With this seminal paper, the field of optogenomics transformed neuroscience research leading to Boyden, Karl Deisseroth, and their colleagues being awarded The Brain Prize of 2013: X. Han, J. G. Bernstein, et al, "Millisecond-Timescale Optical Control of Neural Dynamics in the Nonhuman Primate Brain," *Neuron* 62, no. 2 (2009): 191–98; doi:10.1016/j.neuron.2009.03.011.

3. G. Nagel, T. Szellas, W. Huhn, S. Kateriya et al., "Channelrhodopsin-2: A Directly Light-Gated Cation-Selective Membrane Channel," *Proceedings of the National Academy of Sciences USA* 100, no. 24 (2003): 13940–45.

4. Light, both internal and external, matters; it affects us in the most profound ways. Nature, through our inborn visual system, has taught and conditioned us in such a way that light controls our brains, moods, and circadian rhythms. Now, by generating our own internal light, reshaping how it enters our brains and what it signals, we can modulate the natural rhythms and signals common to all animals, including hominids.

5. A. Gerits, R. Farivar, B. R. Rosen, L. L. Wald et al., "Optogenetically Induced Behavioral and Functional Network Changes in Primates," *Current Biology* 22, no. 18 (2012): 17220–28; R. A. Berman and R. H. Wurtz, "Signals Conveyed in the Pulvinar Pathway from Superior Colliculus to Cortical Area MT," *Journal of Neuroscience* 31, no. 2 (2011): 373–84; "Activation of Brain Regions Can Change a Monkey's Choice," Phys.org, May 29, 2014, accessed September 3, 2014, http://phys.org/news/2014-05 -brain-region-monkey-choice.html.

6. David H. Freedman, "Brain Control. Ed Boyden Is Learning How to Alter Behavior by Using Light to Turn Neurons On and Off," *MIT Technology Review*, October 27, 2010, accessed September 20, 2014, http://www.technologyreview.com/featured story/421400/brain-control; Ed Boyden, "A Light Switch for Neurons," TED, March 2011, accessed September 20, 2014, https://www.ted.com/talks/ed_boyden.

7. IEDs—improvised explosive devices, often encountered by soldiers during conflict in Afghanistan and Iraq.

8. Ed Boyden's work in PTSD with humans is not yet published, but his animal studies are described in his TED talk. This article describes use of electrical stimulation for PTSD: Patrick Tucker, "The Military Is Building Brain Chips to Treat PTSD," *Atlantic*, May 2014, accessed September 21, 2014, http://www.theatlantic.com/tech nology/archive/2014/05/the-military-is-building-brain-chips-to-treat-ptsd/371855.

9. J. Guintivano, T. Brown, A. Newcomer, M. Jones et al., "Identification and Replication of a Combined Epigenetic and Genetic Biomarker Predicting Suicide and Suicidal Behaviors," *American Journal of Psychiatry* (2014), ePub ahead of print, doi:10. 1176/appi.ajp.2014.14010008; Robert Glatter, "Genetic Biomarker Identified That May Predict Suicide Risk," *Forbes*, August 9, 2014, accessed September 21, 2014, http://www.forbes.com/sites/robertglatter/2014/08/09/genetic-biomarker-identified -that-may-predict-suicide-risk.

10. Sara Reardon, "Suicidal Behaviour Is a Disease, Psychiatrists Argue," *New Scientist*, May 17, 2013, accessed September 3, 2014, http://www.newscientist.com/article /dn23566-suicidal-behaviour-is-a-disease-psychiatrists-argue.html#.VAcwFyh8sr4.

11. Because it penetrates our tissues so effectively, eventually red light may become the key color for treating some brain disorders. Ever placed a flashlight against your palm or behind your fingers? Remember the light coming through? Red light penetrates far into soft tissues, making it easier to use for various therapies.

12. M. M. Doroudchi, K. P. Greenberg, J. Liu, K. A. Silka et al., "Virally Delivered Channelrhodopsin-2 Safely and Effectively Restores Visual Function in Multiple Mouse Models of Blindness," *Molecular Therapy* 19, no. 7 (2011): 1220–29.

13. Meanwhile, back in the realm of traditional couch psychiatry, little has changed, which is why in 2014 you see articles with titles like "What Is Going On in Psychiatry When Nothing Seems to Happen?" A. Sfera, *Front Psychiatry* 4 (2013): 178. The publication of the DSM-5, the bible for mental disorders, highlighted once again just how big a divide exists between quantitative brain researchers and many mental health practitioners—so much so that the head of the National Institute of Mental Health argued, "Unlike our definitions of ischemic heart disease, lymphoma, or AIDS, our DSM diagnoses are based on a consensus about a cluster of clinical symptoms, not any objective laboratory measure. . . . In the rest of medicine this would be equivalent to creating diagnostic symptoms based on the nature of chest pains or the quality of a fever." See Thomas Insel, "Director's Blog: Transforming Diagnosis," National Institute of Mental Health, April 29, 2013, accessed September 3, 2014, http://www.nimh.nih.gov/about/director/2013/transforming-diagnosis.shtml.

14. S. Ramirez, X. Liu, P.-A. Lin, J. Suh et al., "Creating a False Memory in the Hippocampus," *Science* 341, no. 6144 (2013): 387–91.

15. S. Nabavi, R. Fox, R., C. D. Proulx, J. Y. Lin et al., "Engineering a Memory with LTD and LTP," *Nature* 511, no. 7509 (2014): 348–52.

16. British Neuroscience Association, "How 'Free Will' Is Implemented in the Brain and Is It Possible to Intervene in the Process?" AlphaGalileo, April 8, 2013, accessed September 3, 2014, http://www.alphagalileo.org/ViewItem.aspx?ItemId=130016&CultureCode=en.

Better Living Through Chemistry?

1. Mary Lou Jepsen, "Screens to Change Our Visions," Mary Lou Jepsen's personal Web site, accessed September 3, 2014, maryloujepsen.com; Mary Lou Jepsen, "Bringing Back My Real Self with Hormones," *New York Times*, November 23, 2013, accessed September 3, 2014, http://www.nytimes.com/2013/11/24/opinion/sunday/bringing-back-my-real-self-with-hormones.html?pagewanted=all&_r=0.

2. "Monitoring the Future Study: Trends in Prevalence of Various Drugs," National Institute on Drug Abuse, accessed September 3, 2014, http://www.drugabuse.gov/trends-statistics/monitoring-future/monitoring-future-study-trends-in-prevalence-various-drugs.

3. "Ecstasy Effects," The Good Drugs Guide, accessed September 3, 2014, http://www.thegooddrugsguide.com/ecstasy/effects.htm. Along with the pleasure, Ecstasy also has some significant and serious side effects, although there's quite a bit of debate on their severity and prevalence because of how few clinical trials have been allowed. One friend of ours, who is prone to unapproved chemical experimentation late at night, calls Ecstasy his "worship-my-wife" drug.

4. Ronald D. Miller, ed., "Mood Altering Drugs," *Anesthesia* (Fifth Edition), Section 3, Anesthesia Management, accessed September 3, 2014, http://web.squ.edu.om/med-Lib/MED_CD/E_CDs/anesthesia/site/content/v03/030135r00.htm.

5. "Grapefruit-Medication Interactions: Forbidden Fruit or Avoidable Consequences?" Canadian Medical Association, November 26, 2012, accessed September 3, 2014, http://www.cmaj.ca/content/early/2012/11/26/cmaj.120951.

6. Miller, ed., "Mood Altering Drugs."

7. W. Zhong, H. Maradit-Kremers, J. L. St. Sauver, B. P. Yawn et al., "Age and Sex Patterns of Drug Prescribing in a Defined American Population," *Mayo Clinic Proceedings* 88, no. 7 (2013): 697–707.

8. S. Billioti de Gage, Y. Moride, T. Ducruet, T. Kurth et al., "Benzodiazepine Use and Risk of Alzheimer's Disease: Case-Control Study," *British Medical Journal* 349 (2014): g5205.

9. National Institute on Drug Abuse, "DrugFacts: Prescription and Over-the-Counter Medications," National Institute on Drug Abuse, rev. May 2013, accessed September 3, 2014, http://www.drugabuse.gov/publications/drugfacts/prescription-over-counter -medications.

10. "Vital Signs: Overdoses of Prescription Opioid Pain Relievers—United States, 1999– 2008," *Morbidity and Mortality Weekly Report (MMWR)*, November 4, 2011, accessed September 21, 2014, http://www.cdc.gov/mmwr/preview/mmwrhtml/mm6043a4 .htm#fig2.

11. K. Hasegawa, D. F. M. Brown, Y. Tsugawa, and C. A. Camargo, "Epidemiology of Emergency Department Visits for Opioid Overdose: A Population-Based Study," *Mayo Clinic Proceedings* 89, no. 4 (2014): 462–71.

12. National Center for Health Statistics, "10 Leading Causes of Injury Deaths by Age Group Highlighting Unintentional Injury Deaths, United States—2010," CDC.gov, National Center for Health Statistics, National Vital Statistics System, accessed September 21, 2014, http://www.cdc.gov/injury/wisqars/pdf/10LCID_Unintentional _Deaths_2010-a.pdf.

13. National Institute on Drug Abuse, "DrugFacts."

Forever Young, Beautiful, and Fearless?

1. P. Kapahi, S. N. Murphy, A. B. Goldfine, R. W. Grant et al., "Germline Signaling Mediates the Synergistically Prolonged Longevity by Double Mutations in daf-2 and rsks-1 in *C. elegans*," *Cell Reports* 5, no. 6 (2013): 1600–1610.

2. "Serge Voronoff," *Wikipedia*, accessed September 9, 2014, http://en.wikipedia.org /wiki/Serge_Voronoff.

3. M. Sinha, Y. C. Jang, J. Oh, D. Khong et al., "Restoring Systemic GDF11 Levels Reverses Age-Related Dysfunction in Mouse Skeletal Muscle," *Science* 344, no. 6184 (2014): 649–52.

4. G. Zhang, J. Li, S. Purkayastha, Y. Tang et al., "Hypothalamic Programming of Systemic Ageing Involving IKK-β, NF-κB and GnRH," *Nature* 497, no. 7448 (2013): 211–16.

5. Douglas Heaven, "Master Key Opens Door to Longer Life," *New Scientist*, May 4, 2013, 8–9.

6. K.-J. Min, C.-K. Lee, and H.-N. Park, "The Lifespan of Korean Eunuchs," *Current Biology* 22, no. 18 (2012): R792–93.

7. A. P. Gomes, N. L. Price, A. J. Y. Ling, J. J. Moslehi et al., "Declining NAD+ Induces a Pseudohypoxic State Disrupting Nuclear-Mitochondrial Communication During Aging," *Cell* 155, no. 7 (2013): 1624–38; Linda Geddes, "Reverse Ageing by Boosting

Cells' Energy Factories," *New Scientist*, September 27, 2013, accessed September 21, 2014, http://www.newscientist.com/article/mg21929364.200-reverse-ageing-by-boosting -cells-energy-factories.html#.VBBsOSh8sr4; NIH Research Matters, "How Resveratrol May Fight Aging," NIH.gov, accessed September 10, 2014, http://www.nih.gov/research matters/march2013/03252013resveratrol.htm.

8. G. Hannum, J. Guinney, L. Zhao, L. Zhang et al., "Genome-Wide Methylation Profiles Reveal Quantitative Views of Human Aging Rates," *Molecular Cell* 49, no. 2 (2013): 359–67.

9. Ibid.

10. K. Christensen, M. Thinggaard, A. Oksuzyan, T. Steenstrup et al., "Physical and Cognitive Functioning of People Older Than 90 Years: A Comparison of Two Danish Cohorts Born 10 Years Apart," *Lancet* 382, no. 9903 (2013): 1507–13.

11. A. Radix and M. Silva, "Beyond the Guidelines: Challenges, Controversies, and Unanswered Questions," *Pediatric Annals* 43, no. 6 (2014): e145–50.

12. G. Jenkins, L. J. Wainwright, R. Holland, K. E. Barrett et al., "Wrinkle Reduction in Post-Menopausal Women Consuming a Novel Oral Supplement: A Double-Blind Placebo-Controlled Randomized Study," *International Journal of Cosmetic Science* 36, no. 1 (2014): 22–31.

13. P. Claes, D. K. Liberton, K. Daniels, K. M. Rosana et al., "Modeling 3D Facial Shape from DNA," *PLoS Genetics* 10, no. 3 (2014): e1004224.

14. P. Aldhous, "DNA Mug Shot Gives Cops Another Lead," *New Scientist* 221, no. 2961 (2014): 14.

15. One of the best uses of DNA has been to free the innocent. Through the end of 2013, more than 314 people who had been rotting away in jail for an average of 13.5 years were exonerated and freed after DNA tests proved they were innocent. Given the cumulative 4,202 unjust years served by these people, one would hope some kind of voluntary testing accelerates and spreads. See "DNA Exonerations Nationwide," Innocence Project, accessed September 10, 2014, http://www.innocenceproject.org /Content/DNA_Exonerations_Nationwide.php.

16. L. Armengol, M. Gratacòs, M. A. Pujana, M. Ribasés et al., "5' UTR-Region SNP in the NTRK3 Gene Is Associated with Panic Disorder," *Molecular Psychiatry* 7, no. 9 (2002): 928–30; M. H. Pollack, J. Tiller, F. Xie, and M. H. Trivedi, "Tiagabine in Adult Patients with Generalized Anxiety Disorder: Results from 3 Randomized, Double-Blind, Placebo-Controlled, Parallel-Group Studies," *Journal of Clinical Psychopharmacology* 28, no. 3 (2008): 308–16.

Unnatural Attraction

1. J. Zilhão, D. E. Angelucci, E. Badai-Garcia, F. d'Errico et al., "Symbolic Use of Marine Shells and Mineral Pigments by Iberian Neandertals," *Proceedings of the National Academy of Sciences USA* 107, no. 3 (2010): 1023–28.

2. "Global Cosmetics Manufacturing Industry Market Research Report from IBIS-World Has Been Updated," PRWeb, March 25, 2013, accessed August 27, 2014, http://www.prweb.com/releases/2013/3/prweb10563036.htm.

3. Richard Corson, *Fashions in Makeup, from Ancient to Modern Times* (London: P. Owen, 1972). Quoted in N. L. Etcoff, S. Stock, L. Haley, S. Vickery et al., "Cosmetics as a Feature of the Extended Human Phenotype: Modulation of the Perception of Biologically Important Facial Signals." *PLoS ONE* 6, no. 10 (2011): e25656.

4. K. Scott, "Cheating Darwin: The Genetic and Ethical Implications of Vanity and Cosmetic Plastic Surgery," *Journal of Evolution & Technology* 20, no. 2 (2009): 1–8.

5. Tehran Bureau, *Guardian*, "The Beauty Obsession Feeding Iran's Voracious Cosmetic Surgery Industry," March 1, 2013, accessed August 27, 2014, http://www .guardian.co.uk/world/iran-blog/2013/mar/01/beauty-obsession-iran-cosmetic -surgery.

6. N. L. Etcoff, S. Stock, L. E. Haley, S. A. Vickery et al., "Cosmetics as a Feature of the Extended Human Phenotype: Modulation of the Perception of Biologically Important Facial Signals," *PLoS ONE* 6, no. 10 (2011), accessed August 24, 2014, http: //www.plosone.org/article/info%3Adoi%2F10.1371%2Fjournal.pone.0025656.

7. "Global Cosmeceuticals Market Outlook 2016," PRNewsire, February 19, 2013, accessed August 28, 2014, http://www.prnewswire.com/news-releases/global-cosmeceuticals -market-outlook-2016-191889571.html.

8. Inas Rashad, "Height, Health and Income in the U.S., 1984–2006," Georgia State University, working paper, March 2008, accessed September 7, 2014, http://uwrg.gsu .edu/files/2014/01/08-3-1_Paper_Height_Rashad.pdf.

9. C. Bushdid, M. O. Magnasco, L. B. Vosshall, and A. Keller, "Humans Can Discriminate More Than 1 Trillion Olfactory Stimuli," *Science* 343, no. 6177 (2014): 1370–72.

10. M. Kadohisa, "Effects of Odor on Emotion, with Implications," *Frontiers in Systems Neuroscience* 7 (2013): 66, accessed August 27, 2014, http://www.ncbi.nlm.nih.gov /pmc/articles/PMC3794443.

11. C. M. Gendron, T. Kuo, Z. M. Harvanek, B. Y. Chung et al., "Drosophila Life Span and Physiology Are Modulated by Sexual Perception and Reward," *Science* 342, no. 6170 (2013): 544–48.

12. Not that this ever stopped hucksters. Look in the back of magazines specifically targeted at men or women. Often you will see an ad for a magical substance, involving pheromones, which will supposedly make you irresistible to the opposite sex.

13. At this stage it's fair to ask, "How can we know so little about gross anatomy?" The answer is we are still quite arrogant and ignorant about the human body. Not a good combination. In 2013, two Belgian doctors discovered a new ligament in the knee that turned out to be crucial for ACL surgeries. Never mind that this body part had been studied and operated on for decades and that the "new" anatomical discovery was in plain sight. See S. Claes, E. Vereecke, M. Maes, J. Victor et al., "Anatomy of the Anterolateral Ligament of the Knee," *Journal of Anatomy* 223, no. 4 (2013): 321–28.

14. C. J. Wysocki and G. Preti, "Facts, Fallacies, Fears, and Frustrations with Human Pheromones," *The Anatomical Record. Part A, Discoveries in Molecular, Cellular, and Evolutionary Biology* 281, no. 1 (2004): 1201–11.

15. R. L. Doty "Human Pheromones: Do They Exist?" In C. Mucignat-Caretta, ed., *Neurobiology of Chemical Communication* (Boca Raton, FL: CRC Press, 2014), Chapter 19.

16. M. McClintock, "Menstrual Synchrony and Suppression," *Nature* 229 (1971): 244–45.

17. C. Wedekind and S. Furi, "Body Odour Preferences in Men and Women: Do They Aim for Specific MHC Combinations or Simply Heterozygosity?" *Proceedings of the Royal Society B: Biological Sciences* 264, no. 1387 (1997): 1471–79.

18. C. Ober, L. R. Weitkamp, N. Cox, H. Dytch et al., "HLA and Mate Choice in Humans," *American Journal of Human Genetics* 61, no. 3 (1997): 497–504.

19. C. Wedekind, T. Seebeck, F. Bettens, and A. J. Paepke, "MHC-Dependent Mate Preferences in Humans," *Proceedings of the Royal Society B: Biological Sciences* 260, no. 1359 (1995): 245–49.

20. Ibid.

21. C. Opie, Q. D. Atkinson, R. I. M. Dunbarc, and S. Shultz, "Male Infanticide Leads to Social Monogamy in Primates," *Proceedings of the National Academy of Sciences USA* 110, no. 33 (2013): 13328–32.

22. Zoe Cormier, "Gene Switches Make Prairie Voles Fall in Love," June 2, 2013, *Nature Online*, http://www.nature.com/news/gene-switches-make-prairie-voles-fall-in-love-1 .13112; H. Wang, F. Duclot, Y. Liu, Z. Wang et al., "Histone Deacetylase Inhibitors Facilitate Partner Preference Formation in Female Prairie Voles," *Nature Neuroscience* 16, no. 7 (2013): 919–24.

23. H. Walum, P. Lichtenstein, J. M. Neiderhiser, D. Reiss et al., "Variation in the Oxytocin Receptor Gene (OXTR) Is Associated with Pair-Bonding and Social Behavior," *Biological Psychiatry* 71, no. 5 (2012): 419–28.

24. Larry Young and Brian Alexander, *The Chemistry Between Us: Love, Sex, and the Science of Attraction* (New York: Current, 2014); Larry Young, "Why Do Voles Fall in Love?" YouTube, accessed August 28, 2014, http://www.youtube.com/watch?v=Oh8x9 KDkYTc; H. E. Ross, C. D. Cole, Y. Smith, I.D. Neumann et al., "Characterization of the Oxytocin System Regulating Affiliative Behavior in Female Prairie Voles," *Neuroscience* 162, no. 4 (2009): 892–903.

25. H. Walum, L. Westberg, S. Henningsson, J. M. Neiderhiser et al., "Genetic Variation in the Vasopressin Receptor 1a Gene (AVPR1A) Associates with Pair-Bonding Behavior in Humans," *Proceedings of the National Academy of Sciences USA* 105, no. 37 (2008): 14153–56.

26. Helen Fisher, "10 Facts About Infidelity," *TED Blog*, January 23, 2014, accessed October 28, 2014, http://blog.ted.com/2014/01/23/10-facts-about-infidelity-helen -fisher, cites the Walum et al. study.

27. S. E. Taylor, S. Sapire-Bernstein, and T. E. Seeman, "Are Plasma Oxytocin in Women and Plasma Vasopressin in Men Biomarkers of Distressed Pair-Bond Relationships?" *Psychological Science* 21, no. 1 (2010): 3–7.

28. "Oxytocin Factor—30 ml Nasal Spray," Amazon.com, accessed September 7, 2014, http://www.amazon.com/Oxytocin-Factor-30-Nasal-Spray/dp/B007K8LTD2.

29. K. Wišniewski, S. Alagarsamy, R. Galyean, H. Tariga et al., "New, Potent, and Selective Peptidic Oxytocin Receptor Agonists," *Journal of Medicinal Chemistry* 57, no. 12 (2014): 5306–17.

30. PubMed search for "oxytocin human sexuality," accessed September 21, 2014, http: //www.ncbi.nlm.nih.gov/pubmed/?term=oxytocin+human+sexuality.

31. For instance, he recently used a Doppler signal, emitted from a sonogram machine, to prove that women have two types of orgasms. O. Buisson, P. Foldes, E. Jannini, and S. Mimoun, "Coitus as Revealed by Ultrasound in One Volunteer Couple," *Journal of Sexual Medicine* 7, no. 8 (2010): 2750–54; E. A. Jannini, A. Rubio-Casillas, B. Whipple, O. Buisson et al., "Female Orgasm(s): One, Two, Several," *Journal of Sexual Medicine* 9, no. 4 (2012): 956–65.

32. E. A. Jannini, R. Blanchard, A. Camperio-Ciani, and J. Bancroft. "Male Homosexuality: Nature or Culture?" *Journal of Sexual Medicine* 7 (2010): 3245–53. This article provides opinions and overviews from three experts in the field.

33. A. Camperio-Ciani, and E. Pellizzari, "Fecundity of Paternal and Maternal Non-Paternal Female Relatives of Homosexual and Heterosexual Men," *PLoS ONE* 7, no. 12 (2012), accessed August 25, 2014, http://www.plosone.org/article/info%3Adoi%2F10.1371%2Fjournal.pone.0051088.

34. A. F. Bogaert, "Number of Older Brothers and Sexual Orientation: New Tests and the Attraction/Behavior Distinction in Two National Probability Samples," *Journal of Personal and Social Psychology* 84, no. 3 (2003): 644–52; R. Blanchard and A. F. Bogaert, "Homosexuality in Men and the Number of Older Brothers," *American Journal of Psychiatry* 153, no. 1 (1996): 27–31; D. P. Vanderlaan, R. Blanchard, H. Wood, and K. J. Zucker, "Birth Order and Sibling Sex Ratio of Children and Adolescents Referred to a Gender Identity Service," *PLoS ONE* 9, no. 32014 (2014): e90257. (This does not mean the third brother will be gay. It means if the eldest has a 2 percent overall chance of being born gay, the next has 2 percent x 1.38, and the next 2 percent x 1.38 x 1.38 . . .)

35. D. F. Swaab and M. A. Hofman, "An Enlarged Superchiasmatic Nucleus in Homosexual Men," *Brain Research* 537, nos. 1–2 (1990): 141–48.

36. D. F. Swaab, "Sexual Orientation and Its Basis in Brain Structure and Function," *Proceedings of the National Academy of Sciences USA* 105, no. 30 (2008): 10273–74.

37. I. Savic, A. Garcia-Falgueras, and D. F. Swaab, "Sexual Differentiation of the Human Brain in Relation to Gender Identity and Sexual Orientation," *Progress in Brain Research* 186 (2010): 41–62.

Sports Quandaries and Beyond . . .

1. "IOC Awards US Broadcast Rights for 2014, 2016, 2018 and 2020 Olympic Games to NBCUniversal," International Olympic Committee, July 6, 2011, accessed September 8, 2014, http://www.olympic.org/news/ioc-awards-us-broadcast-rights-for-2014-2016-2018-and-2020-olympic-games-to-nbcuniversal/130827.

2. Total wages estimated at up to $380 billion per year: U.S. House of Representatives, 106th Congress, "House Report 106-903, Student Athlete Protection Act," Government Printing Office, September 27, 2000, accessed September 8, 2014, http://www.gpo.gov/fdsys/pkg/CRPT-106hrpt903/html/CRPT-106hrpt903.htm; compared to GDP of various countries: World Bank, "Gross Domestic Product 2013," World Data-Bank, accessed September 8, 2014, http://databank.worldbank.org/data/download/GDP.pdf.

3. Here is just one of the ways that ambiguity hurts everyone: Samantha Shapiro, "Caught in the Middle: A Failed Gender Test Crushed Santhi Soundarajan's Olympic Dreams," ESPN, August 1, 2012, accessed September 8, 2014, http://espn.go.com/olympics/story/_/id/8192977/failed-gender-test-forces-olympian-redefine-athletic-career-espn-magazine.

4. P. Fénichel, F. Paris, P. Phillbert, S. Hiéronimus et al., "Molecular Diagnosis of 5α-Reductase Deficiency in 4 Elite Young Female Athletes Through Hormonal Screening for Hyperandrogenism," *Journal of Clinical Endocrinology and Metabolism* 98, no. 6 (2013): E1055–059.

5. Search for "hyperandrogenism sports," PubMed, accessed September 8, 2014, http://www.ncbi.nlm.nih.gov/pubmed/?term=hyperandrogenism+sports.

6. "Caster Semenya: Male or Female?" *The Science of Sport* (blog), August 19, 2009, accessed September 8, 2014, http://www.sportsscientists.com/2009/08/caster-semenya-male-or-female. Here is how the Olympic Committee dealt with the issue:

"IOC Addresses Eligibility of Female Athletes with Hyperandrogenism," International Olympic Committee, April 5, 2011, accessed September 8, 2014, http://www.olympic.org/content/press-release/ioc-addresses-eligibility-of-female-athletes-with-hyperandrogenism/.

7. Myers-Briggs tests are personality tests based on the work of Carl Jung: "MBTI Basics," MyersBriggs.org, accessed September 8, 2014, http://www.myersbriggs.org/my-mbti-personality-type/mbti-basics.

8. T. B. Hayesa, V. Khourya, A. Narayana, M. Nazira et al., "Atrazine Induces Complete Feminization and Chemical Castration in Male African Clawed Frogs (*Xenopus laevis*)," *Proceedings of the National Academy of Sciences USA* 107, no. 10 (2010): 4612–17.

9. W. G. Foster, S. Maharaj-Briceño, and D. G. Cyr, "Dioxin-Induced Changes in Epididymal Sperm Count and Spermatogenesis," *Environmental Health Perspectives*, 118, no. 4 (2010): 458–64; W. Julliard, J. H. Fechner, and J. D. Mezrich, "The Aryl Hydrocarbon Receptor Meets Immunology: Friend or Foe? A Little of Both," *Frontiers in Immunology* 5, no. 458 (2014): 1–6; L. E. Gray Jr. and J. Ostby, "Effects of Pesticides and Toxic Substances on Behavioral and Morphological Reproductive Development: Endocrine Versus Nonendocrine Mechanisms," *Toxicology & Industrial Health* 14, no. 1–2 (1998): 159–84.

10. Claude J. Migeon and Amy B. Wisniewski, "Congenital Adrenal Hyperplasia Due to 21-Hydroxylase Deficiency: A Guide for Patients and Their Families," Johns Hopkins Children's Center, accessed September 8, 2014, https://www.hopkinschildrens.org/cah/printable.html.

11. J. Kota, C. R. Handy, A. M. Haidet, C. L. Montgomery et al., "Follistatin Gene Delivery Enhances Muscle Growth and Strength in Nonhuman Primates," *Science Translational Medicine* 1, no. 6 (2009): 6ra15.

12. T. van der Gronde, O. de Hon, H. J. Haisma, and T. Pieters, "Gene Doping: An Overview and Current Implications for Athletes," *British Journal of Sports Medicine* 47 (2013): 670–78. A partial list of peptide/protein enhancers and genes that are known to work in animals or humans includes: erythropoietin, insulinlike growth factor, growth hormone, myostatin, vascular endothelial growth factor, fibroblast growth factor, endorphin and enkephalin, α actinin 3, peroxisome proliferator-activated receptor-delta (PPARδ), and cytosolic phosphoenolpyruvate carboxykinase (PEPCK-C).

13. R. W. Hanson and P. Hakimi, "Born to Run: The Story of the PEPCK-Cmus Mouse," *Biochimie* 90, no. 6 (2008): 838–42.

14. Australian Crime Commission, "New Generation Performance and Image Enhancing Drugs and Organised Criminal Involvement in Their Use in Professional Sport," Australian Crime Commission, accessed September 8, 2014, http://www.crimecommission.gov.au/sites/default/files/files/organised-crime-and-drugsin-sports-feb2013.pdf.

15. CJC-1295, GHRP-2, GHRP-6, and Hexarelin.

16. P. Wang, S. Dong, J. H. Shieh, E. Peguero et al., "Erythropoietin Derived by Chemical Synthesis," *Science* 342, no. 6164 (2013): 1357–60.

17. L. C. Hsieh-Wilson and M. E. Griffin, "Improving Biologic Drugs Via Total Chemical Synthesis," *Science* 342, no. 6164 (2013): 1332–33.

18. "Eero Mantyranta, Ski Champ Who Later Failed Doping Test," *Boston Globe*, December 31, 2013, accessed September 8, 2014, http://www.bostonglobe.com/metro/obituaries/2013/12/31/eero-mantyranta-failed-drug-test-after-winning-gold/hscil2uSL47V37o2M8iuPM/story.html.

19. A. de la Chapelle, A. L. Traskelin, and E. Juvonen, "Truncated Erythropoietin Receptor Causes Dominantly Inherited Benign Human Erythrocytosis," *Proceedings of the National Academy of Sciences USA* 90, no. 10 (1993): 4495–99.

20. M. H. De Moor, T. D. Spector, L. F. Cherkas, M. Faichi et al., "Genome-Wide Linkage Scan for Athlete Status in 700 British Female DZ Twin Pairs," *Twin Research and Human Genetics* 10, no. 6 (2007): 812–20.

21. Want more details? Among fifteen climbers who had ascended to 8,000 meters without oxygen, none were D/D homozygotes, and the very top performers were all I/I homozygotes.

22. Katrina Karkazis and Rebecca Jordan-Young, "The Trouble with Too Much T," *New York Times*, April 10, 2014, accessed September 8, 2014, http://www.nytimes.com /2014/04/11/opinion/the-trouble-with-too-much-t.html?_r=0.

23. J. Enriquez and S. Gullans, "Olympics: Genetically Enhanced Olympics Are Coming," *Nature* 487, no. 7407 (2012): 297.

24. S. Li and I. Laher, "Exercise Pills: At the Starting Line," *Trends in Pharmacological Sciences* 36, no. 12 (2015): 906–17.

Designer Organs and Cloned Humans

1. "Wake Forest Physician Reports First Human Recipients of Laboratory-Grown Organs," Wake Forest Baptist Health, May 2006, accessed September 5, 2014, http: //www.wakehealth.edu/News-Releases/2006/Wake_Forest_Physician_Reports _First_Human_Recipients_of_Laboratory-Grown_Organs.htm; Anthony Atala, "Growing New Organs," TED, October 2009, accessed September 15, 2014, https://www.ted .com/talks/anthony_atala_growing_organs_engineering_tissue.

2. K. Takahashi, and S. Yamanaka, "Induction of Pluripotent Stem Cells from Mouse Embryonic and Adult Fibroblast Cultures by Defined Factors," *Cell* 126, no. 4 (2006): 663–76.

3. To grossly oversimplify, it is like erasing every program on your computer and then reloading every factory-preloaded program anew. (But not including any of your personal files, documents, photographs, music . . .) See the scientific background note to the 2012 Nobel Prize ceremony, "Mature Cells Can Be Reprogramed to Become Pluripotent," Nobel Media, accessed September 25, 2014, http://www.nobelprize.org /nobel_prizes/medicine/laureates/2012/advanced-medicineprize2012.pdf.

4. Although we have no clue how this procedure changes/transmits/alters an epigenome.

5. "List of animals that have been cloned," *Wikipedia*, accessed September 5, 2014, http://en.wikipedia.org/wiki/List_of_animals_that_have_been_cloned.

6. S. Wakayama, T. Khoda, H. Obokata, M. Tokoro et al., "Successful Serial Recloning in the Mouse over Multiple Generations," *Cell Stem Cell* 12, no. 3 (2013): 293–97.

Evolving Brains Revisited

1. Sulston is an extraordinarily laudable human being. Here is one of many deservedly admiring profiles of his work: Andrew Brown, "One Man and His Worm," *Guardian*, October 9, 2002, accessed September 5, 2014, http://www.theguardian.com/science /2002/oct/09/genetics.science.

2. A TEDx talk on new mathematics: "A New Type of Mathematics: David Dalrymple at TEDx Montreal," TEDx, June 21, 2012, YouTube, accessed September 5, 2014, https://www.youtube.com/watch?v=vh-FSX8jm90&feature=player_embedded. David got seduced away from brain research by Twitter. His project was completed by Robert Prevedel at the University of Vienna and MIT grad student Young Gyu Yoon. You can follow the progress on modeling an entire worm brain here: James Pearn, "OpenWorm," Artificial Brains, accessed September 5, 2014, http://www.arti ficialbrains.com/openworm.

3. C. R. Madan, "Augmented Memory: A Survey of the Approaches to Remembering More," *Frontiers in System Neuroscience* 8 (2014): 30.

4. S. Canavero, "HEAVEN: The Head Anastomosis Venture Project Outline for the First Human Head Transplantation with Spinal Linkage (GEMINI)," *Surgical Neurology International* 4, suppl. 1 (2013): S335–42.

5. I. Feiz-Erfan, L. F. Gonzalez, and C. A. Dickman, "Atlantooccipital Transarticular Screw Fixation for the Treatment of Traumatic Occipitoatlantal Dislocation. Technical Note," *Journal of Neurosurgery: Spine* 3 (2005): 381–85; "Doctors Reattach Teen's Head After Car Wreck," ABC News, accessed September 5, 2014, http://abc news.go.com/GMA/DrJohnson/story?id=125410.

6. Chuck Shepherd, "This Teenage Boy Is One Lucky Person," *Reading Eagle*, February 15, 2003, page C4.

7. Y. S. Lee, C. Y. Lin, H. H. Jiang, M. Depaul et al., "Nerve Regeneration Restores Supraspinal Control of Bladder Function After Complete Spinal Cord Injury," *Journal of Neuroscience* 33, no. 26 (2013): 10591–606.

8. B. R. Ksander, P. E. Kolovou, B. J. Wilson, K. R. Saab et al., "ABCB5 Is a Limbal Stem Cell Gene Required for Corneal Development and Repair," *Nature* 511, no. 7509 (2014): 353–57.

9. K. Lee, D.-N. Kwon, T. Ezashi, Y.-J. Choi et al., "Engraftment of Human iPS Cells and Allogeneic Porcine Cells into Pigs with Inactivated RAG2 and Accompanying Severe Combined Immunodeficiency," *Proceedings of the National Academy of Sciences USA* 111, no. 20 (2014): 7260–65.

10. W. Zhang, P. J. Wang, H. Y. Sha, J. Ni et al., "Neural Stem Cell Transplants Improve Cognitive Function Without Altering Amyloid Pathology in an APP/PS1 Double Transgenic Model of Alzheimer's Disease," *Molecular Neurobiology* 50, no. 2 (2014): 423–37; J. Rekha, L. R. Veena, N. Prem, P. Kalaivani et al., "NIH-3T3 Fibroblast Transplants Enhance Host Regeneration and Improve Spatial Learning in Ventral Subicular Lesioned Rats," *Behavioural Brain Research* 218, no. 2 (2011): 315–24.

11. X. Han, M. Chen, F. Wang, M. Windrem et al., "Forebrain Engraftment by Human Glial Progenitor Cells Enhances Synaptic Plasticity and Learning in Adult Mice," *Cell Stem Cell* 12, no. 3 (2013): 342–53; A. Benraiss, M. J. Toner, Q. Xu, E. Bruel-Jungerman et al., "Sustained Mobilization of Endogenous Neural Progenitors Delays Disease Progression in a Transgenic Model of Huntington's Disease," *Cell Stem Cell* 12, no. 6 (2013): 787–99; United Therapeutics, "Synthetic Genomics Inc. Signs Collaborative Research and Development Agreement with Lung Biotechnology Inc., a Subsidiary of United Therapeutics Corporation, to Develop Humanized Pig Organs to Revolutionize Transplantation Field United Therapeutics," Press Release, May 6, 2014, accessed September 7, 2014, http://ir.unither.com/releasedetail.cfm?releaseid=845454.

12. P. J. Hallett, O. Cooper, D. Sadi, H. Robertson et al., "Long-Term Health of Dopaminergic Neuron Transplants in Parkinson's Disease Patients," *Cell Reports* 7, no. 6 (2014): 1755–61.

13. R. E. Hampson, D. Song, I. Opris, L. M. Santos et al., "Facilitation of Memory Encoding in Primate Hippocampus by a Neuroprosthesis That Promotes Task-Specific Neural Firing," *Journal of Neural Engineering* 10, no. 6 (2013): 066013. In the typically wild and florid scientific description that often accompanies science discoveries, the Wake Forest team stated, "These findings provide the first successful application of a neuroprosthesis designed to enhance and/or repair memory encoding in primate brain."

14. N. Suthana, Z. Haneef, J. Stern, R. Mukamel et al., "Memory Enhancement and Deep-Brain Stimulation of the Entorhinal Area," *New England Journal of Medicine* 366 (2012): 502–10.

15. MIT Synthetic Neurobiology Group Protocols, accessed November 2, 2014, http://syntheticneurobiology.org/protocols/protocoldetail/35/9.

16. S. Kohli, S. G. Fisher, Y. Tra, M. J. Adams et al., "The Effect of Modafinil on Cognitive Function in Breast Cancer Survivors," *Cancer* 115 (2009): 2605–16.

17. S. A. Deadwyler, T. W. Berger, A. J. Sweatt, D. Song et al., "Donor/Recipient Enhancement of Memory in Rat Hippocampus," *Frontiers in Systems Neuroscience* 7 (2013): 120.

18. Mary Meeker and Liang Wu, "2013 Internet Trends," Kleiner, Perkins, Caufield and Byers, May 29, 2013, accessed September 12, 2014, http://www.kpcb.com/insights/2013-internet-trends.

19. W. Glannon, "Prostheses for the Will," *Frontiers in Systems Neuroscience* 8 (2014): 79.

The Robot-Computer-Human Interface

1. The Arc Fusion dinners were private events hosted in several U.S. cities in 2014 by David Ewing Duncan and Stephen Petranek. Here is a link to the Arc Fusion Summit in April 2015, accessed November 3, 2014, http://www.arcprograms.net.

2. One thing that made the 2014 contest challenging was that each team was not just competing with its robot alone; in each round of the competition, each robot was paired at random with two other robots. So not only did kids have to build a great and robust robot, but every round they also had to coordinate and compete along with a two other teams that they had just met against another team of three robots and their teenage overlords. The "frenemies" aspect of the competition meant each team might want to help the other two teams they were paired with to improve their design and tactics, while also realizing that those same teams could be a competitor the next day. One also had to be careful with one's opponents; foul somebody's robot now and the hurt machine might be part of your team for the next bracket. So each team adapted in real time, cooperating, passing, blocking, and scoring with many designs and styles of play.

3. Here is a video that provides a small taste of what the Finals of the 2014 FIRST Robotics Championship were like: "FIRST, #OMGROBOTS!!!!" Vimeo.com, accessed September 8, 2014, http://vimeo.com/93021844.

4. Along with Dean Kamen, cofounder Woody Flowers developed FIRST's extraordinary culture of respect, generosity, sharing, and caring. Understated to the extreme, Woody hangs out on the sidelines looking like a slightly out-of-place 1960s flower child: ponytail, round glasses, gray hair, gentle demeanor. Not evident to outsiders is the fact that he's one of the top-rated and most revered MIT professors, teaching the most appropriately named 2.007 class, "Introduction to Manufacturing," which produces peaceful prototypes of machines James Bond would admire. Until you get close and look at his denim jacket, you would have no idea how much the kids love

and respect him; then you see their signatures: Every kid there has taken a pen and come by to autograph Flowers's jacket, to thank him, to show how much they care for and respect the living soul of FIRST. Neither Flowers nor Kamen ever scream or rant as do many "coaches." They just encourage, provide guidance and occasionally some tough love and help when a chip doesn't work or an arm doesn't launch. The motto and driver of the competition: "Gracious Professionalism."

5. Dean's entire wardrobe, including for White House visits, consists of blue jeans and blue-jean shirts. On the other hand, while many of us enjoy two-car garages, his is a two-homemade-helicopter garage.

6. By that point, entire categories of human jobs, like "chauffeur," "truck driver," "delivery person," and "stockroom clerk," will be well on their way to extinction. See Dylan Love, "Children Born Today Will Never Have to Drive a Car, Says Robotics Expert," *Business Insider*, April 10, 2014, accessed September 8, 2014, http://www.business insider.com/driverless-cars-2014-4.

7. This U.S. Navy video is a historic breakpoint in piloted vs. nonpiloted combat aircraft: "X047B Completes First Carrier-Based Arrested Landing," YouTube, July 10, 2013, accessed September 8, 2014, http://www.youtube.com/watch?v=Rc2k6G8LuqY. Best bet on the drones going forward. More and more flying is done by machines, not humans. Likely no human could safely and consistently perform the myriad minute adjustments now required by today's commercial airliners, flying in the thin air of 35,000 feet, close to the speed of sound. No human could safely land in CAT IIIb conditions, with only 150 feet of forward visibility from the cockpit (during a landing, a 737 covers this distance in less than half a second). But planes self-land in these conditions all the time, without the pilot touching any of the controls. The only limitation, after landing, is whether the pilot can see the central yellow line and signs that guide the plane off the runway and to the terminal. P.S.: If you ever need to convert one unit of measure into another weird unit, here is a great site: "Speed Units Converter," Convert-to.com, accessed September 8, 2014, http://convert-to.com/202 /speed-units.html.

8. "Statistics," PlaneCrashInfo.com, accessed September 8, 2014, http://www.plane crashinfo.com/cause.htm.

9. MIT statistics professor Arnold Barnett quoted here: Jad Mouawad and Christopher Drew, "Airline Industry at Its Safest Since the Dawn of the Jet Age," *New York Times*, February 11, 2013, accessed September 8, 2014, http://www.nytimes.com/2013/02 /12/business/2012-was-the-safest-year-for-airlines-globally-since-1945.html?_r=0. Also interesting: Peter Jacobs, "12 Reasons Why Flying Is the Safest Way to Travel," *Business Insider*, July 9, 2013, accessed September 8, 2014, http://www.businessinsider .com/flying-is-still-the-safest-way-to-travel-2013-7?op=1.

10. O. Svenson, "Are We All Less Risky and More Skillful Than Our Fellow Drivers?" *Acta Psychologica* 47, no. 2 (1981): 143–48.

11. J. Kruger and D. Dunning, "Unskilled and Unaware of It: How Difficulties in Recognizing One's Own Incompetence Lead to Inflated Self-Assessments," *Journal of Personality and Social Psychology* 77, no. 6 (1999): 1121–34.

12. By the time most of a society realizes that driving has been displaced by robots, the game has been over for a long time. In 2013, driverless cars were only legal in a few pioneering states: Nevada, Florida, California, and Michigan. Yet robots have been chauffeuring consistently better than humans for a while. Many new cars are tending toward autonomy already; one professional car reviewer took out a new Infiniti and could sit back and "simply watch, even on mildly curving highways, for three or more miles at a stretch," Daniel P. Howley, "The Race to Build Self-Driving Cars," *New York Times*, August 23, 2012.

13. Peter Jacobs, "12 Reasons Why Flying Is the Safest Way to Travel"; Mouawad and Drew, "Airline Industry at Its Safest."

14. Here is an interesting robotics road map: "A Roadmap for U.S. Robotics: From Internet to Robotics, 2013 Edition," Robotics in the United States of America, March 20, 2013, accessed September 8, 2014, http://robotics-vo.us/sites/default/files/2013%20Robotics%20Roadmap-rs.pdf.

15. A. Radas, M. Mackey, A. Leaver, A. L. Bouvier et al., "Evaluation of Ergonomic and Education Interventions to Reduce Occupational Sitting in Office-Based University Workers: Study Protocol for a Randomized Controlled Trial," *Trials* 14 (2013): 330.

16. A. J. Gaskins, J. Mendiola, M. Afeiche, N. Jørgensen et al., "Physical Activity and Television Watching in Relation to Semen Quality in Young Men," *British Journal of Sports Medicine* (2013): 10.1136/bjsports-2012-091644.

17. Hugh Herr, "The New Bionics That Let Us Run, Climb and Dance," TED, March 28, 2014, accessed September 8, 2014, https://www.ted.com/talks/hugh_herr_the _new_bionics_that_let_us_run_climb_and_dance.

18. Adam Liebendorfer, "Sound Science: Li Xu Examines How Cochlear Implants Can Be Fine-Tuned for Speakers of Tonal Languages," *Ohio University Perspectives,* Spring/Summer 2012, accessed September 20, 2014, http://www.ohio.edu/research /communications/soundscience.cfm.

19. "Federation Deems Prosthesis Unfair," Associated Press, July 30, 2014, accessed September 20, 2014, http://espn.go.com/olympics/trackandfield/story/_/id/11285062 /german-federation-drops-amputee-long-jumper-markus-rehm-team.

20. K. K. Whitcome, L. J. Shapiro, and D. E. Lieberman, "Fetal Load and the Evolution of Lumbar Lordosis in Bipedal Hominins," *Nature* 450 (2007): 1075–78. Daniel Lieberman has written extensively about the evolution of the human body over millions of years and how it is not always well suited to aspects of modern life. See his seminal book: Daniel E. Lieberman, *The Story of the Human Body* (New York: Pantheon, 2013).

21. "Humanity+," HumanityPlus, accessed September 8, 2014, http://humanityplus.org/.

Perhaps an Ethical Question or Two?

1. Eugenics has been practiced since ancient times, often as infanticide. But it was Galton who began the process of making eugenics a science.

2. Steve Selden, "Eugenics Popularization," Image Archive on the American Eugenics Movement, accessed September 9, 2014, http://www.eugenicsarchive.org/html /eugenics/essay6text.html. (Article refers to Harvard, Columbia, and Cornell.) Here is Victor McKusick's take on this: "There are some who may view the genome in a determinist way, believing that the human condition will ultimately be seen entirely as the manifestation of sequence information and computation. We do not subscribe to such a view." F. S. Collins and V. McKusick, "Implications of the Human Genome Project for Medical Science," *Journal of the American Medical Association* 285, no. 5 (2001): 540–44.

3. Edwin Black wrote a great book on this most nasty of subjects: Edwin Black, *The War Against the Weak: Eugenics and America's Campaign to Create a Master Race* (New York: Four Walls Eight Windows, 2003).

4. Paul Lombardo, "Eugenic Sterilization Laws," Image Archive on the American Eugenics Movement, accessed September 9, 2014, http://www.eugenicsarchive.org/html/eugenics/essay8text.html.

5. "Cold Spring Harbor Laboratory's Image Archive on the American Eugenics Movement," Cold Spring Harbor Laboratory, accessed September 9, 2014, http://www.eugenicsarchive.org/eugenics/.

6. J. A. Clayton and F. S. Collins, "Policy: NIH to Balance Sex in Cell and Animal Studies," *Nature* 509, no. 7500 (2014): 282–83.

7. A. K. Beery and I. Zucker, "Sex Bias in Neuroscience and Biomedical Research," *Neuroscience & Biobehavioral Reviews* 35, no. 3 (2011): 565–72.

8. L. Du, H. Bayir, Y. Lai, X. Zhang et al., "Innate Gender-Based Proclivity in Response to Cytotoxicity and Programmed Cell Death Pathway," *Journal of Biological Chemistry* 279, no. 37 (2004): 38563–70.

9. As one of our smart editors, Suzie LaFleur, comments: "I worked in a classroom where the teacher would say, 'I treat everyone fairly, but not equally.' She used it to mean that if a kid had, say, an attention issue, he might have different systems of punishments and rewards. How do we use the word 'equal'? It should mean human life is all equally precious—but true 'fair' treatment would mean making sure that doctors find the right drug to treat women, and maybe a different one for men. Only then will both genders have been treated fairly and given true equality." This makes it even more concerning that some scientists flee from addressing controversial topics.

10. R. Bowden, T. S. MacFie, S. Myers, G. Hellenthal et al., "Genomic Tools for Evolution and Conservation in the Chimpanzee: *Pan Troglodytes Ellioti* Is a Genetically Distinct Population," *PLoS Genetics* 8, no. 3 (2012): e1002504.

11. M. C. Marchetto, I. Narvaiza, A. M. Denli, C. Benner et al., "Differential L1 Regulation in Pluripotent Stem Cells of Humans and Apes," *Nature* 503, no. 7477 (2013): 525–29.

12. Erika Check Hayden, "Ethics: Taboo Genetics: Probing the Biological Basis of Certain Traits Ignites Controversy. But Some Scientists Choose to Cross the Red Line Anyway," *Nature* 502, no. 7469 (2013): 26–28.

13. The authors accessed ongoing online polls on January 3, 2014. The number of people voting ran from more than 3,685 on the violence question to 4,783, who mostly opposed delving into intelligence.

14. Ironically, one of the communities that dare advocate research on these controversial topics is one of the most discriminated-against communities historically: gays. The logic is that many know from birth that "I was just born that way," and want to truly understand what leads to differences between groups.

15. M. U. Yood, B. D. McCarthy, J. Kempf, G. P. Kucera et al., "Racial Differences in Reaching Target Low-Density Lipoprotein Goal Among Individuals Treated with Prescription Statin Therapy," *American Heart Journal* 152, no. 4 (2006): 777–84.

16. L. C. Edelstein, L. M. Simn, R. T. Montoya, M. Holinstat et al., "Racial Differences in Human Platelet PAR4 Reactivity Reflect Expression of PCTP And miR-376c," *Nature Medicine* 19, no. 12 (2013): 1609–16.

17. K. Ng, J. B. Scott, B. F. Drake, A. T. Chan et al., "Dose Response to Vitamin D Supplementation in African Americans: Results of A 4-Arm, Randomized, Placebo-Controlled Trial," *American Journal of Clinical Nutrition* 99, no. 3 (2014): 587–98;

M. F. Holick, "Bioavailability of Vitamin D and Its Metabolites in Black and White Adults," *New England Journal of Medicine* 369, no. 21 (2013): 2047–48.

18. L. Wade. "Genomics. Initiative Aims to Minister to Mexico's Unique Genetic Heritage," *Science* 342, no. 6160 (2013): 788.

19. Pam Chwedyk, "BiDil Controversy Continues as FDA Approves First 'Race-Specific' Drug," *Minority Nurse*, Fall 2005, accessed September 9, 2014, http://www.minoritynurse.com/article/bidil-controversy-continues-fda-approves-first-%E2%80%9Crace-specific%E2%80%9D-drug#sthash.cuogSEl1.dpuf.

20. Tara Bannow, "Race-Related Controversy Causes Drug Flop, BiDil Was Approved by the FDA in 2005 to Treat Heart Failure," *Minnesota Daily*, March 9, 2010, accessed September 9, 2014, http://www.mndaily.com/2010/03/09/race-related-controversy-causes-drug-flop.

21. B. M. Rusert and C. D. Royal, "Grassroots Marketing in a Global Era: More Lessons from BiDil," *Journal of Law, Medicine & Ethics: A Journal of the American Society of Law, Medicine & Ethics* 39, no. 1 (2011): 79–90.

22. There are warnings in some of the top science magazines about research involving ethnicity, gender, and related topics; for example, "Dangerous Work: Behavioural Geneticists Must Tread Carefully to Prevent Their Research Being Misinterpreted," Nature.com, October 2, 2013, accessed September 9, 2014, http://www.nature.com/news/dangerous-work-1.13861.

23. P. D. Evans, S. L. Gilbert, N. Mekel-Bobrov, E. J. Vallender et al., "Microcephalin, a Gene Regulating Brain Size, Continues to Evolve Adaptively in Humans," *Science* 309, no. 5741 (2005): 1717–20; N. Mekel-Bobrov, S. L. Gilbert, P. D. Evans, E. J. Vallender et al., "Ongoing Adaptive Evolution of ASPM, a Brain Size Determinant in *Homo sapiens*," *Science* 309, no. 5741 (2005): 1720–22.

24. Here is an odd one: D. Dediu and D. R. Ladd, "Linguistic Tone Is Related to the Population Frequency of the Adaptive Haplogroups of Two Brain Size Genes, ASPM and Microcephalin," *Proceedings of the National Academy of Sciences of the USA* 104, no. 26 (2007): 10944–49. The argument is that the distribution of tonal languages may be tied to these mutations. For one blogger's take on this, read James Winters, "ASPM, Microcephalin and Tone," *A Replicated Typo* (blog), January 24, 2009, accessed September 9, 2014, http://replicatedtypo.wordpress.com/2009/01/24/aspm-microcephalin-tone.

25. Antonio Regalado, "Scientist's Study of Brain Genes Sparks a Backlash," *Wall Street Journal*, June 16, 2006, accessed September 21, 2014, http://online.wsj.com/news/articles/SB115040765329081636?mod=blogs. This again reinforces the need for an ethical backbone and rules for this type of research and discovery. And a need to further educate.

26. Karen Kaplan, "Jewish Legacy Inscribed on Genes?," *Los Angeles Times*, April 18, 2009, accessed April 16, 2016, http://www.latimes.com/science/la-sci-jewish-iq18-2009apr18-story.html. More recently, *New York Times* science reporter Nick Wade relearned this painful lesson after publishing his book about race that included considerable speculation without enough caveats, qualifiers, and rebuttals. Nicholas Wade, *A Troublesome Inheritance: Genes, Race and Human History* (New York: Penguin Press, 2014). Here is the open letter from 143 scientists published in the *New York Times*, which concludes, "We are in full agreement that there is no support from the field of population genetics for Wade's conjectures," See "Letters: 'A Troublesome Inheritance,'" *New York Times*, August 10, 2014, accessed September 12, 2014, http://www.nytimes.com/2014/08/10/books/review/letters-a-troublesome-inheritance.html. An updated foreword in Wade's book addressed this concern.

27. X. Yi, Y. Liang, E. Huerta-Sanchez, X. Jin et al., "Sequencing of 50 Human Exomes Reveals Adaptation to High Altitude," *Science* 329, no. 5987 (2010): 75–78; T. S. Simonson, Y. Yang, C. D. Buff, H. Yun et al., "Genetic Evidence for High-Altitude Adaptation in Tibet," *Science* 329, no. 5987 (2010): 72–75; C. M. Beall, G. L. Cavalleri, L. Deng, R. C. Elston et al., "Natural Selection On EPAS1 (HIF2α) Associated with Low Hemoglobin Concentration in Tibetan Highlanders," *Proceedings of the National Academy of Sciences USA* 107, no. 25 (2010): 11459–64.

28. S. Tzur, S. Rosset, R. Shemer, G. Yudkovsky et al., "Missense Mutations in the APOL1 Gene Are Highly Associated with End Stage Kidney Disease Risk Previously Attributed to the MYH9 Gene," *Human Genetics* 128, no. 3 (2010): 345–50. G. Genovese, D. J. Friedman, M. D. Ross, L. Lecordier et al., "Association of Trypanolytic Apol1 Variants with Kidney Disease in African Americans," *Science* 329, no. 5993 (2010): 841–45.

29. B. T. Lahn and L. Ebenstein, "Let's Celebrate Human Genetic Diversity," *Nature* 461, no. 7265 (2009): 726–68.

30. Luigi Luca Cavalli-Sforza, "Genes, Peoples, and Languages," *Scientific American* 265, no. 5 (1991): 104–10. This article is described and put in the context of its criticism in a paper by B. M. Rusert and C. D. Royal, "Grassroots Marketing in a Global Era: More Lessons from BiDil," *Journal of Law Medicine & Ethics* 39, no. 1 (2011): 79–90, in which the authors note, "Cavalli-Sforza suggests that geneticists should collect indigenous DNA samples now and ask questions later, since aboriginal populations are rapidly disappearing under the forces of modernization. He concludes, 'Priceless evidence is slipping through our fingers as aboriginal populations lose their identity. Growing interest in the Human Genome Project may, however, stimulate workers to gather evidence of human genetic diversity before it disappears.'" Rusert and Royal noted that "social scientists like Troy Duster insisted that the new genetics was in danger of ushering in insidious practices of eugenics."

Technically Life, Technically Death

1. J. A. Martin, S. Kirmeyer, M. Osterman, and R. A. Shepherd, "Born a Bit Too Early: Recent Trends in Late Preterm Births," *National Center for Health Statistics Data Brief* 24 (2009): 1–8.

2. S. Rahman, W. E. Ansari, N. Nimeri, S. Tinay et al., "Achieving Excellence in Maternal, Neonatal and Perinatal Survival: Executive Synopsis of PEARL Study Annual Report 2011," *Journal of Postgraduate Medical Institute* 27, no. 1 (2012): 4–7.

3. H. Blencowe, S. Cousens, D. Chou, M. Oestergaard et al., "Born Too Soon: The Global Epidemiology of 15 Million Preterm Births," *Reproductive Health* 10, Supplement 1 (2013): S2.

4. G. Rocha and H. Guimarães, "On the Limit of Viability Extremely Low Gestational Age at Birth," *Acta Médica Portuguesa* 24, Supplement 2 (2011): 181–88; A. Kugelman, D. Bader, L. Lerner-Geva, V. Boyko et al., "Poor Outcomes at Discharge Among Extremely Premature Infants: A National Population-Based Study," *Archives of Pediatrics & Adolescent Medicine* 166, no. 6 (2012): 543–50.

5. N. Marlow, D. Wolke, M. A. Bracewell, and M. Samara, for the EPICure Study Group, "Neurologic and Developmental Disability at Six Years of Age After Extremely Preterm Birth," *New England Journal of Medicine* 352, no. 1 (2005): 9–19.

6. E. Korvenranta, L. Lehtonen, L. Rautava, U. Häkkinen et al., "Impact of Very Preterm Birth on Health Care Costs at Five Years of Age," *Pediatrics* 125, no. 5 (2010): e1109–14.

7. R. E. Behrman and A. S. Butler, eds., *Committee on Understanding Premature Birth and Assuring Healthy Outcomes, Preterm Birth: Causes, Consequences, and Prevention* (Washington, DC: National Academies Press, 2007), 389–93; K. M. Stjernqvist, "Extremely Low Birth Weight Infants Less Than 901 g. Impact on the Family During the First Year," *Scandinavian Journal of the Society of Medicine* 20, no. 4 (1992): 226–33.

8. Katy Butler, "The Ultimate End-of-Life Plan: How One Woman Fought the Medical Establishment and Avoided What Most Americans Fear: Prolonged, Plugged-In Suffering," *Wall Street Journal*, September 5, 2013.

9. Joseph Carroll, "Public Continues to Support Right-to-Die for Terminally Ill Patients," Gallup, June 19, 2006, accessed September 20, 2014, http://www.gallup .com/poll/23356/public-continues-support-righttodie-terminally-ill-patients.aspx.

10. Megan Greene and Leslie R. Wolfe, "Pregnancy Exclusions in State Living Will and Medical Proxy Statutes," Center for Women Policy Studies, August 2012, accessed September 4, 2014, http://www.centerwomenpolicy.org/programs/health/statepolicy /documents/REPRO_PregnancyExclusionsinStateLivingWillandMedicalProxyStat utesMeganGreeneandLeslieR.Wolfe.pdf. You can wish or sign whatever you want, but a key sentence at the end of Section 166.001 of the Advance Directives Act reads, "Under Texas law this directive has no effect if I have been diagnosed as pregnant." "Health and Safety Code, Title 2. Health, Subtitle H. Public Health Provisions, Chapter 166. Advance Directives, Subchapter A. General Provisions," Statutes.Legis.State .TX.us, accessed September 5, 2014, http://www.statutes.legis.state.tx.us/Docs/HS /htm/HS.166.htm. And it's not just Texas; legislatures in twelve states have mandated to hospitals that it's not your choice, it's not your body; too bad if you are brain dead, terminal, or in extreme suffering, you are hereby legislated to remain a living incubator.

Trust Whom?

1. Pew Research, Religion & Public Life Project, "The Global Religious Landscape," Pew Forum, December 18, 2012, accessed September 9, 2014, http://www.pewforum .org/2012/12/18/global-religious-landscape-exec/.

2. Jerry Coyne, "Biblical Morality Part 2: Killing Non-Virgin Brides and Rebellious Kids," *Why Evolution Is True* (book blog), June 26, 2012, accessed September 9, 2014, http://whyevolutionistrue.wordpress.com/2012/06/26/biblical-morality -part-2-killing-non-virgin-brides-and-rebellious-kids; Steve Wells, "What the Bible Says About Stoning," Skeptics Annotated Bible, accessed September 9, 2014, http: //skepticsannotatedbible.com/says_about/stoning.html.

3. Virginia Nicholson, "Singled Out: How Two Million Women Survived Without Men After the First World War," *Daily Mail*, September 15, 2007, accessed September 9, 2014, http://www.dailymail.co.uk/femail/article-481882/Condemned-virgins -The-million-women-robbed-war.html#ixzz2eK9v1UvS.

4. "Why Did the Prophet Have So Many Wives?" *OnIslam*, accessed September 9, 2014, http://www.onislam.net/english/ask-about-islam/ethics-and-values/muslim -character/166258-why-did-the-prophet-have-so-many-wives.html.

5. G. Prachi, "Neil deGrasse Tyson: Science Is True 'Whether or Not You Believe in It,'" *Salon*, March 11, 2014, accessed September 9, 2014, http://www.salon.com/2014 /03/11/neil_degrasse_tyson_science_is_true_whether_or_not_you_believe_in_it.

6. Ara Norenzayan, *Big Gods: How Religion Transformed Cooperation and Conflict* (Princeton: Princeton University Press, 2013).

7. Google Spain SL, *Google Inc. v. Agencia Española de Protección de Datos*, Mario Costeja González Case number C-131/12 ECLI ECLI:EU:C:2014:317, Judgment of the Court, May 13, 2013, accessed November 3, 2014 http://curia.europa.eu/juris /document/document_print.jsf?doclang=EN&docid=152065.

8. "HIPAA Violations and Enforcement," American Medical Association, accessed September 9, 2014, http://www.ama-assn.org/ama/pub/physician-resources/solu tions-managing-your-practice/coding-billing-insurance/hipaahealth-insurance -portability-accountability-act/hipaa-violations-enforcement.page.

9. Jamie Heywood and Ben Heywood, "Patients Like Me," PatientsLikeMe.com, accessed September 9, 2014, http://www.patientslikeme.com/.

10. P. W. Wilson, R. D. Abbott, and W. P. Castelli, "High Density Lipoprotein Cholesterol and Mortality. The Framingham Heart Study," *Arteriosclerosis* 8, no. 6 (1988): 737–41.

11. Ibid.

12. J. S. Brownstein, S. N. Murphy, A. B. Goldfine, R. W. Grant et al., "Rapid Identification of Myocardial Infarction Risk Associated with Diabetes Medications Using Electronic Medical Records," *Diabetes Care* 33, no. 3 (2010): 526–31.

The Future of Life

1. A hotshot MIT economist friend recalls that his daughter once asked, "Where is Mom?" He answered, "I don't know, can I help you?" The ensuing dialogue went something like this: "No, thanks." "Are you sure? What do you need?" "Help with my math homework." Big pause. Dad, being one of the top finance professors in the United States, was a little miffed and surprised, so he said, "Are you sure I can't help you?" "No, Dad, I don't want to know too much. You just complicate it." (Recounted to the authors by a professor of economics, MIT, April 10, 2014.)

I Don't Remember You . . . De-Extinction

1. "Species Extinction—The Facts," IUCN Red List, International Union for Conservation of Nature, May 2007, accessed August 29, 2014, http://cmsdata.iucn.org/down loads/species_extinction_05_2007.pdf.

2. Tom Wolfe, *The Electric Kool-Aid Acid Test* (New York: Farrar, Straus and Giroux, 1968).

3. Given this history, one might be tempted to stereotype what positions Brand might take on various issues, and one could be very wrong. Depending on evidence and technology, Brand continually surprises; he was a core founder of the environmental movement, but is also a strong advocate of nuclear power, even post Fukushima. He looked at data, trends, and overall effects, and he concluded that yes, uranium's dangerous, but in the long term less so than the costs of continuing to spew carbon into the air and warming the planet.

4. If you go to Tyrol, you can study the world's oldest tattoo, on a 5,300-year-old hipster mummy. (His citizenship is unclear; frozen in the Ötztal Alps, he lived most of his life in what is now Italy but was excavated from a glacier in Austria.)

5. V. Nyström, J. Humphrey, P. Skoglund, N. J. McKeown et al., "Microsatellite Genotyping Reveals End-Pleistocene Decline in Mammoth Autosomal Genetic Variation," *Molecular Ecology* 21, no. 14 (2012): 3391–402.

6. L. Orlando, A. Ginolhac, G. Zhang, D. Froese et al., "Recalibrating Equus Evolution Using the Genome Sequence of an Early Middle Pleistocene Horse," *Nature* 499, no. 7456 (2013): 74–78; Jane J. Lee, "World's Oldest Genome Sequenced from 700,000-Year-Old Horse DNA," *National Geographic*, June 2013, accessed September 20, 2014, http://news.nationalgeographic.com/news/2013/06/130626-ancient-dna-oldest-sequenced-horse-paleontology-science; Neil Clarkson, "Why Did Horses Die Out in North America? *Horse Talk*, November 29, 2012, accessed September 20, 2014, http://horsetalk.co.nz/2012/11/29/why-did-horses-die-out-in-north-america/#axzz3 Dtud8CJM.

7. R. E. Green, J. Krause, A. W. Briggs, T. Maricic et al, "A Draft Sequence of the Neandertal Genome," *Science* 328, no. 5979 (2010): 710–22; Ker Than, "Neanderthals, Humans Interbred—First Solid DNA Evidence," *National Geographic*, May 2010, accessed September 20, 2014, http://news.nationalgeographic.com/news/2010/05/100506-science-neanderthals-humans-mated-interbred-dna-gene.

8. D. Gokhman, E. Lavi, K. Prüfer, M. F. Fraga et al., "Reconstructing the DNA Methylation Maps of the Neandertal and the Denisovan," *Science* 344, no. 6183 (2014): 523–27.

9. "Interview with George Church: Can Neanderthals Be Brought Back from the Dead?" *Der Spiegel*, January 18, 2013, accessed September 10, 2014, http://www.spiegel.de/international/zeitgeist/george-church-explains-how-dna-will-be-construction-material-of-the-future-a-877634.html.

10. R. Nudel and D. F. Newbury, "FOXP2," *Wiley Interdisciplinary Reviews: Cognitive Science* 4, no. 5 (2013): 547–60.

11. J. Zhang, D. M. Webb, and O. Podlaha, "Accelerated Protein Evolution and Origins of Human-Specific Features: FOXP2 As an Example," *Genetics* 162, no. 4 (2002): 1825–35; W. Enard, M. Przeworski, S. E. Fisher, C. S. Lai et al., "Molecular Evolution of FOXP2, a Gene Involved in Speech and Language," *Nature* 418, no. 6900 (2002): 869–72.

12. G. Konopka, J. M. Bomar, K. Winden, G. Coppola et al., "Human-Specific Transcriptional Regulation of CNS Development Genes by FOXP2," *Nature* 462, no. 7270 (2009): 213–17.

13. S. Niermeyer, P. Yang, Shanmina, Drolkar et al., "Arterial Oxygen Saturation in Tibetan and Han Infants Born in Lhasa, Tibet," *New England Journal of Medicine* 333, no. 19 (1995): 1248–52.

14. L. G. Moore, D. Young, R. E. McCullough, T. Droma et al., "Tibetan Protection from Intrauterine Growth Restriction (IUGR) and Reproductive Loss at High Altitude," *American Journal of Human Biology* 13, no. 5 (2001): 635–44.

15. E. Huerta-Sánchez, X. Jin, Asan, Z. Bianba et al., "Altitude Adaptation in Tibetans Caused by Introgression of Denisovan-like DNA," *Nature* 512, no. 7513 (2014): 194–97.

Humanity's Really Short Story

1. Despite the work of Sir Isaac Newton, Albert Einstein, and other brilliant physicists, we still do not know exactly what gravity is. We have never seen gravity waves or gravity particles (gravitons).

2. L. Gao and T. Theuns, "Lighting the Universe with Filaments," *Science* 317, no. 5844: 1527–30. For readers seeking a more technical explanation, here is how the authors of the article actually describe the early birth of stars: "The first stars in the universe formed

when chemically pristine gas heated as it fell into dark-matter potential wells, cooled radiatively because of the formation of molecular hydrogen, and became self-gravitating."

3. There are many Web sites that provide terrific information about the universe: "Hubble, ESA," Hubble News, SpaceTelescope.org, accessed September 7, 2014; "Seyfert's Sextet," *Wikipedia*, accessed September 7, 2014; "Astronomy Picture of the Day," NASA, APOD.NASA.gov, December 10, 2013, accessed September 7, 2014; "Hubble Watches Galaxies Engage in Dance of Destruction," Hubble Site, Hubblesite.org, December 12, 2002, accessed September 7, 2014, http://hubblesite.org/newscenter/archive/releases/2002/22. How much is a trillion? It would take you about fifty years to count from one to a billion if you worked at it ten hours per day, hence fifty thousand years to count to one trillion. "Activity: Count to a Billion," Math Is Fun, accessed September 7, 2014, http://www.mathsisfun.com/activity/count-billion.html. But actually you can get really nerdy about this topic. Look at Dr. Math: "Really Counting to One Billion," The Math Forum at Drexel, accessed September 7, 2014, http://mathforum.org/library/drmath/view/59179.html.

4. "Spitzer Space Telescope," JPL/Caltech, accessed September 7, 2014, http://www.spitzer.caltech.edu/search/image_set/20?search=milky+way. Here is how NASA estimates the age of the universe as of April 16, 2010: "How Old Is the Universe?" NASA, accessed September 7, 2014, http://map.gsfc.nasa.gov/universe/uni_age.html.

5. J. Valley, W., A. J. Cavosie, T. Ushikubo, D. A. Reinhard et al., "Hadean Age for a Post-Magma-Ocean Zircon Confirmed by Atom-Probe Tomography," *Nature Geoscience* 7 (2014): 219–23.

6. Michael J. Benton, *When Life Nearly Died: The Greatest Mass Extinction of All Time* (New York: Thames and Hudson, 2003). Japan's sacred gingko tree is one of the few plants that survived that extinction.

7. T. D. White, B. Asfaw, Y. Beyene, Y. Haile-Selassie et al., "Ardipithecus Ramidus and the Paleobiology of Early Hominids," *Science* 326, no. 5949 (2009): 75–86.

8. There was a bottleneck in the size of the human population 1 million years ago, when it is estimated that fewer than 55,500 individuals were alive at the same time. This is smaller than the current population of chimpanzees or western lowland gorillas in the world. Interestingly, the ancient human population had more genetic diversity than modern humans across the globe. For a very readable summary of this research, see Carina Storrs, "Endangered Species: Humans Might Have Faced Extinction 1 Million Years Ago," *Scientific American*, January 20, 2010, accessed August 29, 2014, http://www.scientificamerican.com/article/early-human-population-size-genetic-diversity. Primary research article: D. Huff, J. Xing, A. R. Rogers, D. Witherspoon et al., "Mobile Elements Reveal Small Population Size in the Ancient Ancestors of Homo Sapiens," *Proceedings of the National Academy of Sciences USA* 107, no. 5 (2010): 2147–52.

Evolving Hominins . . .

1. Ernest Mayr, *What Evolution Is* (New York: Basic Books, 2002), 261.

2. G. J. Sawyer, Viktor Deak, Esteban Sarmiento, Richard Milner et al., *The Last Human: A Guide to Twenty-Two Species of Extinct Humans* (New Haven: Yale University Press, 2007); L. R. Berger, D. J. de Ruiter, S. E. Churchill, P. Schmid et al., *"Australopithecus Sediba*: A New Species of Homo-Like Australopith from South Africa," *Science* 328, no. 5975 (2010): 195–204. Here is the list: *Sahelanthropus tchadensis, Orrorin tugenensis, Ardipithecus ramidus/kadabba, Australopithecus anamensis, Kenyanthropus platyops, Australopithecus afarensis, Paranthropus aethiopicus, Australopithecus garhi, Australopithecus africanus, Australopithecus sediba, Paranthropus robustus/crassidens, Homo rudolfensis, Homo habilis, Paranthropus boisei, Homo ergaster, Homo georgicus, Homo erectus, Homo pekinensis, Homo floresiensis, Homo*

antecesor, Homo rhodesiensis, Homo heidelbergensis, Homo neandethalensis . . . (And that's just one list, one version. Often other protohumans are added, disputed, reincluded, excluded . . .)

3. One clue in the puzzle is that the mitochondrial DNA of a Denisovan compared with a human varies by 385 DNA bases, whereas Neanderthal to human varies by 202, meaning divergence from Denisovans may have taken place about twice as long ago as it did from Neanderthal. See J. Krause, Q. Fu, J. M. Good, B. Viola et al., "The Complete Mitochondrial DNA Genome of an Unknown Hominin from Southern Siberia," *Nature* 464, no. 7290 (2010): 894–97.

4. John Wenz, "The Other Neanderthal," *Atlantic*, August 27, 2014, accessed September 7, 2014, http://www.theatlantic.com/technology/archive/2014/08/the-other-neanderthal/375916/2/.

5. D. Reich, N. Patterson, M. Kircher, F. Delfin et al., "Denisova Admixture and the First Modern Human Dispersals into Southeast Asia and Oceania," *American Journal of Human Genetics* 89, no. 4 (2011): 516–28.

6. Krause et al., "The Complete Mitochondrial DNA Genome of an Unknown Hominin." Debates within the paleontology/anthropology community are the academic equivalent of the WWE. All who dare claim that they have found a new hominin species will most assuredly be denigrated, attacked, and pummeled with no-holds-barred brutality. Folks are still arguing whether the small-bodied humans from Micronesia—the "Hobbits"—are a mixture of *sapiens* and *floresiensis*, or one, or the other, or a new species. L. R. Berger, S. E. Churchill, B. De Klerk, and R. L. Quinn, "Small Bodied Humans from, Palau, Micronesia," *PLoS ONE* 3, no. 3 (2008): e1780. Even Ardi, *Science* magazine's discovery of the year, was repeatedly mugged by various detractors in a fight that continues today. See T. E. Cerling, N. E. Levin, J. Quade, J. G. Wynn et al., "Comment on the Paleoenvironment of *Ardipithecus Ramidus*," *Science* 328, no. 5982 (2010): 1105. The recent discovery of Denisovans really pushed the edge of technology. Unlike other fossils that were skulls, Denisovans were identified as a separate species based on a single finger bone fragment from a Russian cave. We don't know what this species' skull looks like. But we do know, using DNA sequencing on the fossilized bone marrow, that it's a different species of humanoid. But the point is not whether there were exactly twenty-four or twenty-seven separate hominins, or whether exactly five or seven coexisted with us. The point is there have been many hominins, and many overlapped with us.

7. David Whitehouse, "When Humans Faced Extinction," BBC News, June 9, 2003, accessed September 7, 2014, http://news.bbc.co.uk/2/hi/science/nature/2975862.stm.

8. Not that you are wondering, but the name "Neanderthal" comes from the discovery of non-*sapiens* bones in the Neander Thal (Valley) of Germany in 1901.

9. Thirteen virtually identical regions on chromosomes: 1, 4, 5, 6, 9, 10, 15, 17, 20, 22, whose genesis traces to Neanderthals.

10. Q. Fu, H. Li, P. Moorjani, F. Jay, et al., "Genome Sequence of a 45,000-Year-Old Modern Human from Western Siberia," *Nature* 514 (2014): 445–49; Here is a lay report of the discovery: Dan Vergano, "45,000-Year-Old Bone Pinpoints Era of Human-Neanderthal Sex," *National Geographic*, October 22, 2014, accessed November 2, 2014, http://news.nationalgeographic.com/news/2014/10/141022-siberian-genome-ancient-science-discovery. Here is the original study of interbreeding between humans and Neanderthals: S. Sankararaman, N. Patterson, H. Li, S. Pääbo et al., "The Date of Interbreeding Between Neandertals and Modern Humans," *PLoS Genetics* 8, no. 10 (2012): e1002947.

11. M. Meyer, Q. Fu, A. Aximu-Petri, I. Glocke, "A Mitochondrial Genome Sequence of a Hominin from Sima De Los Huesos," *Nature* 505, no. 7483 (2014): 403–6.

12. Svante Pääbo is one of the key figures in evolutionary genetics pioneering techniques for isolating DNA from the inside (marrow) of ancient hominin bones and then analyzing the DNA sequence for comparison to modern humans. The biggest problem, aside from finding some very old bones to begin with, is isolating pristine DNA without contaminating it with modern human DNA and then discarding those DNA sequences that are from ancient microbes that invaded the decaying bones millennia ago.

13. "When Populations Collide," Howard Hughes Medical Institute, January 29, 2014, accessed September 7, 2014, http://www.hhmi.org/news/when-populations-collide.

14. The draft of the entire genome of the Neanderthal was provided in: R. E. Green, J. Krause, A. W. Briggs, T. Maricic et al., "A Draft Sequence of the Neandertal Genome," *Science* 328, no. 5979 (2010): 710–22. A second paper published simultaneously provided a more directed analysis of mutations in genes of three separate Neanderthals from different regions and found eighty-eight gene differences between Neanderthals and humans. See H. A. Burbano, E. Hodges, R. E. Green, A. W. Briggs et al., "Targeted Investigation of the Neandertal Genome by Array-Based Sequence Capture," *Science* 328, no. 5979 (2010): 723–25.

15. B. Vernot and J. M. Akey, "Resurrecting Surviving Neandertal Lineages from Modern Human Genomes," *Science* 343, no. 6174 (2014): 1017–21.

16. D. Reich, N. Patterson, D. Campbell, A. Tandon et al., "Reconstructing Native American Population History," *Nature* 488, no. 7411 (2012): 370–74.

17. DNA from humans whose ancestors migrated out of Africa to Europe, Asia, and beyond all contain 1 to 3 percent Neanderthal and/or Denisova DNA. As a result, only a few Africans are 100 percent *Homo sapiens* at the DNA level.

18. Karl Landsteiner won the Nobel for this discovery. See "Blood Groups, Blood Typing and Blood Transfusions," Nobel Media, accessed September 7, 2014, http://nobelprize.org/educational/medicine/landsteiner/readmore.html.

19. "Blood Type," *Wikipedia*, accessed September 7, 2014, http://en.wikipedia.org/wiki/Blood_type.

20. "Rh Blood Types," Palomar Community College, accessed September 7, 2014, http://anthro.palomar.edu/blood/Rh_system.htm.

21. She might develop antigens to the father's blood and then, in turn, fight the fetus. Want a tutorial on blood type from the folks who award the Nobels?: See "The Blood-Typing Game," Nobel Media, http://www.nobelprize.org/educational/medicine/bloodtyping game/1.html.

22. The great evolutionary biologist and avid baseball fan Stephen J. Gould championed the idea of punctuated equilibrium, in which evolution has long periods of relative stability that are interrupted by periods of radical change in morphology or other traits. See Warren D. Allmon, Patricia H. Kelley, and Robert H. Ross, eds., *Stephen Jay Gould: Reflections on His View of Life* (New York: Oxford University Press, 2009).

23. I. Olalde, M. E. Allentoft, F. Sánchez-Quinto, G. Santpere et al., "Derived Immune and Ancestral Pigmentation Alleles in a 7,000-Year-Old Mesolithic European," *Nature* 507, no. 7491 (2014): 225–28.

24. H. Eiberg, J. Troelsen, M. Nielsen, A. Mikkelsen et al., "Blue Eye Color in Humans May Be Caused by a Perfectly Associated Founder Mutation in a Regulatory Element Located Within the HERC2 Gene Inhibiting OCA2 Expression," *Human Genetics* 123, no. 2 (2008): 177–87.

25. Ibid.; Nicholas Wade, "East Asian Physical Traits Linked to 35,000-Year-Old Mutation," *New York Times*, February 14, 2013; P. A. Gerber, P. Hevezi, B. A. Buhren, C. Martinez et al., "Systematic Identification and Characterization of Novel Human Skin-Associated Genes Encoding Membrane and Secreted Proteins," *PLoS ONE* 8, no. 6 (2013): e63949; Blackwell Publishing Ltd., "Disease Resistance May Be Genetic," *Science Daily*, August 31, 2007, accessed September 7, 2014, http://www.sciencedaily.com/releases/2007/08/070830150014.htm; R. W. Michelmore, I. Paran, and R. V. Kesseli, "Identification of Markers Linked to Disease-Resistance Genes by Bulked Segregant Analysis: A Rapid Method to Detect Markers in Specific Genomic Regions by Using Segregating Populations," *Proceedings of the National Academy of Sciences USA* 88 (1991): 9828–32.

26. Every year *Nature*, Tim O'Reilly, and Google host a three-day gathering: Science Foo. Three hundred of the smartest, weirdest, and quirkiest scientists and inventors come together. On day one they all introduce themselves. One title. Three things that interest them. No sentences. All in ten to fifteen seconds apiece. (A gong sounds loudly and mercilessly on those who attempt verbiage.) Here is a small sample of the topics people were interested in and working on: consciousness, limits of technology, planets, evolution, nanotech, children's toys, bioinformatics, amusement parks for the brain, puzzles, ubiquitous public programming, economics of climate change, deep uncertainty, democracy, dark matter, imaging, prime numbers, skepticism, transforming criminal science, preemptive problem solving, anything 5 percent chocolate, summer of code, sailing, synthetic immune systems, meteorites, computing molecules, protein origami, self-driving cars, squid, bridging science and humanity, microbes, quantum spacetime, everything, combinatorial robotics, film, future identity, silk, risk taking, collective identity, global threats, warm beer experiments, genocide/racism and reality television, contagion, proteomics, monkeys and mistakes, 10,000-year clocks, accountable predictions, multiverses, very-long-distance physics, radical moderation, oxymorons . . . (And that was only the first few minutes of intros.) Over the next five years we will double the amount of data generated by our species, across all time. *Digital Science*, "Science Foo Camp," accessed October 21, 2014, http://www.digital-science.com/sciencefoo.

27. J. Hawks, E. T. Wang, G. M. Cochran, H. C. Harpending et al., "Recent Acceleration of Human Adaptive Evolution," *Proceedings of the National Academy of Sciences USA* 104 (2007): 20753–58; K. McAuliffe, "They Don't Make *Homo sapiens* Like They Used To," *Discover*, March 2009.

28. C. D. Huff, J. Xing, A. R. Rogers, D. Witherspoon et al., "Mobile Elements Reveal Small Population Size in the Ancient Ancestors of *Homo Sapiens*," *Proceedings of the National Academy of Sciences USA* 107, no. 5 (2010): 2147–52.

29. Del Harvey, "The Strangeness of Scale at Twitter," TED, March 2014, accessed September 7, 2014, http://www.ted.com/talks/del_harvey_the_strangeness_of_scale_at_twitter#t-129464.

30. For examples, search PubMed under "Rapid Speciation," accessed September 7, 2014, http://www.ncbi.nlm.nih.gov/pubmed/?term=%22rapid+speciation%22.

31. Wouldn't it be sad if nothing ever improved or changed? You can read more of Jones's non-argument in the Royal Society of Edinburgh 2002 debate "Is Evolution Over?" Dr. Jones's position in the debate is cited by a number of reports, such as Peter Ward, "What Will Become of Homo Sapiens?" *Scientific American*, accessed September 7, 2014, http://www.mukto-mona.com/Special_Event_/Darwin_day/2009/english/SA_human_future.pdf.

Synthetic Life

1. The ten fastest-growing U.S. industries from 2003 to 2013 in rank order were: social-network game development, e-book publishing, social networking sites, online fashion sample sales, online payment-processing software developers, online greeting-card sales, online photo printing, online shoe sales, online household-furniture sales, fantasy sports services. "Top 10 Fastest-Growing US Industries: The Internet Makes Its Mark," *Marketing Charts*, April 24, 2013, accessed September 21, 2014, http://www.marketingcharts.com/online/top-10-fastest-growing-us-industries-the-internet-makes-its-mark-28968.

2. D. G. Gibson, J. I. Glass, C. Lartigue, V. N. Noskov et al., "Creation of a Bacterial Cell Controlled by a Chemically Synthesized Genome," *Science* 329, no. 5987 (2010): 52–56; Ewen Callaway, "Immaculate Creation: Birth of the First Synthetic Cell," *New Scientist*, May 20, 2010, accessed September 8, 2014, http://www.newscientist.com/article/dn18942-immaculate-creation-birth-of-the-first-synthetic-cell.html#.VA28HCh8sr4.

3. Disclosure: Juan Enriquez is a cofounder of SGI and equity holder. Steve holds equity indirectly as well. There are also a host of other young synthetic biology companies, such as Amyris, Solazyme, Codexis, and LS9, in which neither Juan nor Steve have an interest. Chris de Morsella, "12 Synthetic Biology Biofuel & Biochemical Companies to Watch," *Green Economy Post* (blog), accessed September 8, 2014, http://greeneconomypost.com/synthetic-biology-biofuel-biochemical-company-17244.htm.

Humans and Hubris: Does Nature Win in the End?

1. Our only real historical rivals arose in the "boring billion" years 1.8 to 0.8 billion years ago, when single-cell marine organisms dominated and there was little evolution that can be seen in the fossil record. Then a few oxygen-excreting bugs multiplied and multiplied. Slowly they oxidized iron, ran out of places to store loose oxygen, and began bleeding their gases into the atmosphere. Gradually a planet that was covered by 1 to 2 percent oxygen became a planet with 21 percent oxygen; most previous life went extinct as the new inhabitants changed the atmosphere. The previously dominant extremophiles, *Archaea*, that did survive were exiled to particular niches.

2. Dan Charles, "In the Making of Megafarms, a Mixture of Pride and Pain," NPR, June 16 2014, accessed September 7, 2014, http://www.npr.org/blogs/thesalt/2014/06/16/321705130/in-the-making-of-megafarms-a-few-winners-and-many-losers.

Leaving Earth?

1. Dwight Garner, "Into the Nothing, After Something: 'Why Does the World Exist?' by Jim Holt," *New York Times*, August 2, 2012, accessed September 7, 2014, http://www.nytimes.com/2012/08/03/books/why-does-the-world-exist-by-jim-holt.html?pagewanted=all&_r=0. Actual quote from the article: "'My own position,' he writes, seeking middle ground between the beliefs of Christians and Gnostics, is 'that the universe was created by a being that is 100 percent malevolent but only 80 percent effective.'"

2. D. H. Rothman, G. P. Fournier, K. L. French, E. J. Alm et al., "Methanogenic Burst in the End-Permian Carbon Cycle," *Proceedings of the National Academy of Sciences USA* 111, no. 15 (2013).

3. D. Schulze-Makuch, S. Haque, M. R. de Sousa Antonio, D. Ali et al., "Microbial Life in a Liquid Asphalt Desert," *Astrobiology* 11, no. 3 (2011): 241–58.

4. Joe Palca, "Crazy Smart: When a Rocker Designs a Mars Lander," NPR, August 3, 2012, accessed September 7, 2014, http://www.npr.org/2012/08/03/157597270/crazy-smart-when-a-rocker-designs-a-mars-lander. Watch this video: "The Landing of Curiosity," CoconutScienceLab channel, YouTube, August 7, 2012, http://www.youtube.com/watch?v=nSGbmtdg5y0, which explains the technical parts of the landing. There are multiple videos on YouTube showing the emotions and reactions of the team as each stage of the landing took place. It is just as emotional and exciting as any sports final you have ever seen. Search for "mars landing JPL crew reaction," on Bing, Bing.com, accessed September 7, 2014, http://www.bing.com/videos/search?q=mars+landing+JPL+crew+reaction&FORM=VIRE2#view=detail&mid=814102BF688E53252166814102BF688E53252166.

5. M. Basner, D. F. Dinges, D. Mollicone, A. Ecker et al., "Mars 520-d Mission Simulation Reveals Protracted Crew Hypokinesis and Alterations of Sleep Duration and Timing," *Proceedings of the National Academy of the Sciences USA* 110, no. 7 (2013): 2635–40.

6. D. J. Dijk, D. F. Neri, J. K. Wyatt, J. M. Ronda et al., "Sleep, Performance, Circadian Rhythms, and Light-Dark Cycles During Two Space Shuttle Flights," *American Journal of Physiology: Regulatory and Integrative Comparative Physiology* 281, no. 5 (2001): R1647–64.

7. Raising a child, as just a few parents know, is a slightly time- and resource-intensive occupation. Space travel, with its very limited space and resource constraints, brings up interesting questions as to how many children a mission might have, and when. . . .

8. A. Caspi, K. Sugden, T. E. Moffitt, A. Taylor et al., "Influence of Life Stress on Depression: Moderation by a Polymorphism in the 5-HTT Gene," *Science* 301, no. 5631 (2003): 386–89.

9. J. E. De Neve, "Functional Polymorphism (5-HTTLPR) in the Serotonin Transporter Gene Is Associated with Subjective Well-Being: Evidence from a US Nationally Representative Sample," *Journal of Human Genetics* 56, no. 6 (2011): 456–59.

10. T. Powell, R. P. McGuffin, U. M. D'Souza, S. Cohen-Woods et al., "Putative Transcriptomic Biomarkers in the Inflammatory Cytokine Pathway Differentiate Major Depressive Disorder Patients from Control Subjects and Bipolar Disorder Patients," *PLoS ONE* 9, no. 3 (2014): e91076.

11. "Radiation Dose Chart," American Nuclear Society, accessed September 7, 2014, http://www.ans.org/pi/resources/dosechart/.

12. "Atomic Radiation Is More Harmful to Women," Nuclear Information and Resource Service, accessed November 2, 2014, http://www.nirs.org/radiation/radhealth/radiationwomen.pdf.

13. There are an estimated 40 billion Earthlike habitable planets within our Milky Way Galaxy. Today, the closest known planets that are potentially habitable are twelve to twenty light-years away. See "List of Nearest Terrestrial Exoplanet Candidates," *Wikipedia*, accessed October 21, 2014, http://en.wikipedia.org/wiki/List_of_nearest_terrestrial_exoplanet_candidates. The fastest man-made spacecraft today is Juno, which travels 25 miles per second or 789 million miles per year. At this speed, a spacecraft could reach the nearest known habitable planet in 90,000 to 150,000 years, or 4,500 to 7,500 generations of humans. If we can reach much greater travel speeds, which many scientists believe is possible, the time will be much shorter. (Thanks to Dr. David Sinclair for alerting us to these calculations and prospects for reaching other habitable planets.)

14. A. Gurnett, W. S. Kurth, L. F. Burlaga, and N. F. Ness, "In Situ Observations of Interstellar Plasma with Voyager 1D," *Science* 341, no. 6153 (2013): 1489–92.

15. N. F. Ness, M. H. Acuña, R. P. Lepping, J. E. Connerney et al., "Magnetic Field Studies by Voyager 1: Preliminary Results at Saturn," *Science* 212, no. 4491 (1981): 211–17.

16. R. T. Byrne, A. J. Klingele, E. L. Cabot, W. S. Schackwitz et al., "Evolution of Extreme Resistance to Ionizing Radiation via Genetic Adaptation of DNA Repair," *eLife* 3 (2014): e01322.

17. Ibid.; T. L. Park, K. Anderson, V. Lailai-Tasmania et al., "Smoking During Pregnancy Causes Double-Strand DNA Break Damage to the Placenta," *Human Pathology* 45, no. 1 (2014): 17–26; R. Nowarski and M. Kotler, "APOBEC3 Cytidine Deaminases In Double-Strand DNA Break Repair and Cancer Promotion," *Cancer Research* 73, no. 12 (2013): 3494–98; B. L. Mahaney, M. Hammel, K. Meek, J. A. Tainer et al., "XRCC4 And XLF from Long Helical Protein Filaments Suitable for DNA End Protection and Alignment to Facilitate DNA Double Strand Break Repair," *Biochemistry and Cell Biology* 91, no. 1 (2013): 31–41.

18. J. N. Kheir, L. A. Scharp, M. A. Borden, E. J. Swanson et al., "Oxygen Gas-Filled Microparticles Provide Intravenous Oxygen Delivery," *Science Translational Medicine* 4, no. 140 (2012): 140ra88.

19. "Ecuadorean Dwarfs May Unlock Cancer Clues," ABC News, February 17, 2011, accessed September 7, 2014, http://abcnews.go.com/Health/OnCall/ecuadorean -dwarfs-unlock-cancer-clues/story?id=12940816&singlePage=true.

20. A. Kaneda, C. J. Wang, R. Cheong, W. Timp et al., "Enhanced Sensitivity to IGF-II Signaling Links Loss of Imprinting of IGF2 to Increased Cell Proliferation and Tumor Risk," *Proceedings of the National Academy of Sciences USA* 104, no. 52 (207): 20926–31.

21. Hernan Lorenzi, J. Craig Venter Institute, January 2014, personal communication with author (Juan Enriquez); Madison Dunitz, "The Astronaut Microbiome," microBE net, January 23, 2014, accessed September 22, 2014, http://microbe.net/2014/01 /23/the-astronaut-microbiome; "Astronaut Microbiome," J. Craig Venter Institute, accessed September 22, 2014, http://www.jcvi.org/cms/research/projects/astronaut -microbiome/overview/.

22. H. Y. Li, H. Zhang, G. Y. Miao, Y. Xie et al., "Simulated Microgravity Conditions and Carbon Ion Irradiation Induce Spermatogenic Cell Apoptosis and Sperm DNA Damage," *Biomedical and Environmental Sciences* 26, no. 9 (2013): 726–34.

23. Neuroskeptic, "Gene-Guided Antidepressants?" *Neuroskeptic* (blog), *Discover*, October 25, 2012, accessed September 7, 2014, http://blogs.discovermagazine.com/neuro skeptic/category/5htt/.

24. Dr. Ting Wu of Harvard was likely the first to propose sending a DNA printer to other planets during her Renaissance Weekend talk on December 31, 2004. This story was related to the authors in an e-mail from Dr. George Church, September 28, 2014. She and Dr. Susan Dymecki were also the brains behind the space symposium.

25. Assuming life appeared 3.8 billion years ago and civilization arose 10,000 years ago.

New Evolutionary Trees

1. D. A. Malyshev, K. Dhami, T. Lavergne, T. Chen et al., "A Semi-Synthetic Organism with an Expanded Genetic Alphabet," *Nature* 509, no. 7500 (2014): 385–88.

2. It is actually five base pairs, because uracil (U) substitutes for thymine when transcribing RNA.

3. Bradley J. Fikes, "Life Engineered with Expanded DNA Code," *San Diego Union Tribune*, May 7, 2014, accessed September 8, 2014, http://www.utsandiego.com/news/2014/May/07/romesberg-dna-scripps-d5SICSTP/2/.

4. K. Sefah, Z. Yang, K. M. Bradley, S. Hoshika et al., "In Vitro Selection with Artificial Expanded Genetic Information Systems," *Proceedings of the National Academy of Sciences USA* 111, no. 4 (2014): 1449–54.

5. P. Marlière, J. Patrouix, V. Döring, P. Herdewijn et al., "Chemical Evolution of a Bacterium's Genome," *Angewandte Chemie* (International Edition in English) 50, no. 31 (2011): 7109–14.

6. "Interview: Genetic alphabets," Royal Society of Chemistry, *Highlights in Chemical Biology*, August 10, 2009, accessed September 8, 2014, http://www.rsc.org/Publishing/Journals/cb/Volume/2009/9/genetic_alphabets.asp; M. Kimoto, R. Yamashige, K. Matsunaga, S. Yokoyama et al., "Generation of High-Affinity DNA Aptamers Using an Expanded Genetic Alphabet," *Nature Biotechnology* 31 (2013): 453–57.

7. Roberta Kwok, "Chemical Biology: DNA's New Alphabet," *Nature* 491, no. 7425 (2012): 516–18.

8. Ibid.

9. Steven Benner, Westheimer Institute of Science and Technology, University of Gainesville, quoted in Colin Barras, "Home and Dry," *New Scientist* 222, no. 2965 (2014): 36–39.

10. S. A. Benner, H. J. Kim, and M. A. Carrigan, "Asphalt, Water, and the Prebiotic Synthesis of Ribose, Ribonucleosides, and RNA," *Accounts of Chemical Research* 45, no. 12 (2012): 2025–34.

11. M. J. Russell, L. M. Barge, R. Bhartia, D. Bocanegra et al, "The Drive to Life on Wet and Icy Worlds," *Astrobiology* 14, no. 4 (2014): 308–43.

12. Barras, "Home and Dry."

13. H. J. Kim and S. A. Benner, "Comment on 'The Silicate-Mediated Formose Reaction: Bottom-Up Synthesis of Sugar Silicates,'" *Science* 329, no. 5994 (2010): 902.

14. J. D. Stephenson, L. J. Hallis, K. Nagashima, S. J. Freeland et al., "Boron Enrichment in Martian Clay," *PLoS ONE* 8, no. 6 (2013): e64624.

15. Peter Ward, "What Will Become of Homo Sapiens?" *Scientific American*, January 2009, 68–73.

16. J. Hawks, E. T. Wang, G. M. Cochran, H. C. Harpending et al., "Recent Acceleration of Human Adaptive Evolution," *Proceedings of the National Academy of Sciences USA* 104, no. 52 (2007): 20753–58.

17. N. Annaluru, H. Muller, L. A. Mitchell, S. Ramalingam et al., "Total Synthesis of a Functional Designer Eukaryotic Chromosome," *Science* 344, no. 6179 (2014): 55–58.

18. P. Genovese, G. Schiroli, G. Escobar, T. Di Tomaso et al., "Targeted Genome Editing in Human Repopulating Haematopoietic Stem Cells," *Nature* 510, no. 7504 (2014): 235–40.

19. F. González, Z. Zhu, Z. D. Shi, K. Lelli et al., "An iCRISPR Platform for Rapid, Multiplexable, and Inducible Genome Editing in Human Pluripotent Stem Cells," *Cell Stem Cell* 15, no. 2 (2014): 215–26.

20. "Synthetic Genomics Inc. Signs Collaborative Research and Development Agreement with Lung Biotechnology Inc., a Subsidiary of United Therapeutics Corporation, to Develop Humanized Pig Organs to Revolutionize Transplantation Field," United Therapeutics, May 6, 2014, accessed September 8, 2014, http://ir.unither.com/releasedetail.cfm?ReleaseID=845454.

21. "The Resilience Project: A Search for Unexpected Heroes," Resilience Project, accessed September 8, 2014, http://www.resilienceproject.me.

22. L. Ye, J. Wang, A. I. Beyer, F. Teque et al., "Seamless Modification of Wild-Type Induced Pluripotent Stem Cells to the Natural CCR5Δ32 Mutation Confers Resistance to HIV Infection," *Proceedings of the National Academy of Sciences USA* 111, no. 26 (2014): 9591–96.

23. M. J. LaJoie, A. J. Rovner, D. B. Goodman, H.-R. Aerni et al., "Genomically Recoded Organisms Expand Biological Functions," *Science* 342, no. 6156 (2013): 357–60.

24. J. Couzin-Frankel, "New Company Pushes the Envelope on Pre-Conception Testing," *Science* 338, no. 6105 (2012): 315–16.

25. "Method and System for Generating a Virtual Progeny Genome, US 8620594 B2," Google Patents, accessed September 8, 2014, http://www.google.com/patents/US8620594.

26. Some people don't know that the Nobel Prize in Economics was not originally conceived by Alfred Nobel. Hence it is not officially a Nobel Prize. It was established in 1968 by the Sveriges Riksbank (the central bank of Sweden) and is officially known as the Prize in Economic Sciences in Memory of Alfred Nobel. Similar to the Nobel Prizes in physics, chemistry, physiology and medicine, literature, and peace, this prize is awarded by the Royal Swedish Academy of Sciences, reflecting its preeminent significance in the field. NobelPrize.org, accessed November 11, 2014, http://www.nobelprize.org/nobel_prizes/economic-sciences/.

Epilogue: *Eppur Si Muove*

1. Juan Enriquez interview with Floyd Romesberg, Scripps Institute, March 30, 2016.

2. M. M. Georgiadis, I. Singh, W. F. Kellett, S. Hoshika et al., "Structural Basis for a Six Nucleotide Genetic Alphabet," *Journal of the American Chemical Society* 137, no. 21 (2015): 6947–55.

3. K. Byrne and R. A. Nichols, "*Culex pipiens* in London Underground Tunnels: Differentiation Between Surface and Subterranean Populations," *Heredity* 82, no. 1 (1999): 7–15.

4. "Florida Lizards Evolve Rapidly, Within 15 Years and 20 Generations," *UT Austin News*, October 23, 2014, accessed May 1, 2016, http://news.utexas.edu/2014/10/23/anole-lizards-evolution-florida.

5. You can spend a fun afternoon perusing twenty-five years of research at "The Guppy Project" Web site: http://cnas.ucr.edu/guppy/.

6. S. P. Egan, G. J. Ragland, L. Assour, T. H. Q. Powell et al., "Experimental Evidence of Genome-Wide Impact of Ecological Selection During Early Stages of Speciation-with-Gene-Flow," *Ecology Letters* 18, no. 8 (2015): 817–25.

7. B. S. Bhullar, Z. S. Morris, E. M. Sefton, A. Tok et al., "A Molecular Mechanism for the Origin of a Key Evolutionary Innovation, the Bird Beak and Palate, Revealed by

an Integrative Approach to Major Transitions in Vertebrate History," *Evolution* 69, no. 7 (2015): 1665–77.

8. Amy Harmon, "Open Season Is Seen in Gene Editing of Animals," *New York Times*, November 26, 2015; V. M. Gantz, N. Jasinskiene, O. Tatarenkova, A. Fazekas et al., "Highly Efficient Cas9-Mediated Gene Drive for Population Modification of the Malaria Vector Mosquito *Anopheles stephensi*," *Proceedings of the National Academy of Sciences USA* 112, no. 49 (2015): E6736–43.

9. Danny Hillis, "The Enlightenment Is Dead, Long Live Entanglement," accessed May 9, 2016, http://jods.mitpress.mit.edu/pub/enlightenment-to-entanglement. This is a wonderfully nuanced and smart essay, not just about time but about the promise and peril of future technology. Do yourself a favor and read it.

10. This is what true courage looks like: William J. Cole, "Bombing Survivor Adrianne Haslet-Davis Ran a Grueling Odyssey," *Boston Globe*, April 20, 2016, accessed May 9, 2016, https://www.bostonglobe.com/metro/2016/04/20/bombing-survivor-adrianne -haslet-davis-ran-grueling-odyssey/wPUXx6e2jn81SpBOfrUeWL/story.html.

11. Amy Golod, "3-D Printing Comes to Boston's Museum of Fine Arts," *U.S. News and World Report*, March 15, 2016, accessed May 9, 2016, http://www.usnews.com/news /articles/2016-03-15/3-d-printing-comes-to-bostons-museum-of-fine-arts.

12. Doubt that you would ever be cowed/awed by the ability of the so-called disabled? Watch this video and then we can chat: "Viktoria Modesta—Prototype," YouTube, accessed May 9, 2016, https://www.youtube.com/watch?v=jA8inmHhx8c.

13. Juan Enriquez interview with Hugh Herr, MIT Media Lab, March 10, 2016.

14. "What If There Was No Such Thing as Human Disability?" MIT Center for Extreme Bionics proposal document, April 2016.

15. A. M. Oliveira, T. J. Hemstedt, H. E. Freitag, and H. Bading, "Dnmt3a2: A Hub for Enhancing Cognitive Functions," *Molecular Psychiatry* (2015): 1–7.

16. Anthony Cuthbertson, "Rabbit Brain Returns Successfully from Cryopreservation," *Newsweek*, February 10, 2016, accessed May 9, 2016, http://www.newsweek.com /rabbit-brain-first-mammal-brain-return-successfully-cryopreservation-424913.

17. The original Bird and Layzell article is fascinating: J. Bird and P Layzell, "The Evolved Radio and Its Implications for Modelling the Evolution of Novel Sensors," *Proceedings of Congress on Evolutionary Computation* (2002): 1836–41; available online at Duke University, accessed May 9, 2016, https://people.duke.edu/~ng46/topics /evolved-radio.pdf; as is Steven Johnson's analysis of its implications: Steven Johnson, "Superintelligence Now," *How We Get to Next*, October 28, 2015, accessed May 9, 2016, https://howwegettonext.com/superintelligence-now-eb824f57f487#.4jjiiql4e; and if you want more discussion look at: Nicholas Bostrom, *Superintelligence, Paths, Dangers, Strategies* (Oxford: Oxford University Press, 2014).

18. Juan Enriquez outlined these in a 2016 TED talk, "We Can Reprogram Life. How to Do It Wisely," TED November 2015, accessed May 9, 2016, http://www.ted.com /talks/juan_enriquez_we_can_reprogram_life_how_to_do_it_wisely.

19. Stewart Brand is quoted in Steven Johnson's wonderful essay entitled "Superintelligence Now," *How We Get to Next*, October 28, 2015, accessed May 9, 2016, https: //howwegettonext.com/superintelligence-now-eb824f57f487#.4jjiiql4e.

Glossary

1. Article 2 (Use of Terms) of the Convention on Biological Diversity, accessed October 31, 2014, http://www.cbd.int/convention/articles/default.shtml?a=cbd-02.

2. There are many scientific definitions of evolution and they have changed over time. We provide a minimal definition which is agnostic regarding mechanism and hence can encompass all four genomes of the hologenome. Most evolutionary biologists appear to focus only on DNA mutations when defining mechanisms of evolution. For a broader discussion of this topic see the article by Professor Moran: Laurence A. Moran, "What Is Evolution?," accessed May 9, 2016, http://bioinfo.med.utoronto.ca/Evolution_by_Accident/What_Is_Evolution.html.

3. I. Zilber-Rosenberg and E. Rosenberg, "Role of Microorganisms in the Evolution of Animals and Plants: The Hologenome Theory of Evolution," *FEMS Microbiology Reviews* 32, no. 5 (2008): 723–35.

4. John Wilkins, "Species Concepts in Modern Literature," National Center for Science Education, September 17, 2008, accessed November 4, 2014, http://ncse.com/evolution/science/species-concepts-modern-literature.

5. Doug Linder, *Exploring Constitutional Law*, "Exploring Constitutional Conflicts: Regulation of Obscenity," University of Missouri Kansas City, accessed August 29, 2014, http://law2.umkc.edu/faculty/projects/ftrials/conlaw/obscenity.htm.

INDEX

:✳:

genetic engineering in, 120
head transplant in, 182, 183
life span of, 155, 156
obesity in, 40
microbes, 66, 73–93, 96, 243, 276
ancestry of, 87–88
antiseptics and, 74–75, 79
antivirals and, 77–79
effects of altering of, 90
oceanic, 82–84, 98
recolonization of, 88–89
vaccines and, 74, 78, 79
See also bacteria; parasites; viruses
microbiome, 80, 81, 86, 88, 97, 98, 99, 112,
    114, 116, 167, 174, 249, 273, 274–75,
    276, 312n
of baby, 103–4, 109–11
defined, 276
use of term, 300n
Micronesia, 338n
micropenises, 92
Middle Ages, 32
Middleton, Kate, 27
migration patterns, changes in, 37
military, 73–74, 194–95
milk, 29, 131, 230, 288n, 310n
breast, 109
unpasteurized, 111
Milky Way, 223–24, 243, 342n
MIT, 120–23, 140, 149, 150, 181, 185, 262
mitochondrial DNA (mtDNA), 137–38,
    316–17n, 338–39n
mitochondrial Eve, 137
Modafinil, 185
modern synthesis, 272, 281n
Modesta, Viktoria, 262
molecules, 54, 56, 172
Monet, Claude, 127
monkeys, 126, 171, 182, 185
monoculture, 33
monogamy, 45, 160, 163–64
Monterey, Calif., 35
mood, 150–54
Moorfields Eye Hospital, 126
morals, 209, 211, 222
Mormons, 24
morning sickness, 106, 295–96n
Mortality and Morbidity Weekly Report
    (MMWR), 9, 10
mosquitoes, 132–34, 261, 315n
mothers, 77, 136–37, 229, 296n
age of, 39, 44, 46, 48, 93–94
autism and, 92, 93, 94

breastfeeding by, 109
cesarean sections and, 42, 85, 103, 106
child obesity and, 39, 41, 42
Dutch, 65–66
male homosexuality and, 165
stress of, 41, 104, 107–8
surrogate, 47, 48
traditional Turkish, 85
mouth, 31, 98, 104
mouthwashes, 75, 114
muggings, 20
multiculturalism, 24, 204
multiple sclerosis, 89, 198
mummies, 159
Muñoz, Marlise, 206–7
murder, 20
muscles, 170, 171, 172, 180
music, 193, 238
Musk, Elon, 246, 263
mutation, 59, 95, 98–99, 112–16, 201, 231
athletics and, 172
beneficial, 26, 113, 172, 230, 231, 280n
cystic fibrosis and, 295n
defined, 276
disease and, 55, 58
gene therapies and, 125
in mtDNA, 137, 316–17n
nonrandom (*see* nonrandom mutation)
pregnancy and, 102–3
rapid, 17, 78
recessive, 113
sperm age and, 48, 94
use of term, 281n
Myers-Briggs personality test, 170, 325n
myocardial infarction (heart attack),
    59–60, 71, 199–200, 213

nanobots, 122
NASA, 245, 246, 250–51, 337n
National Human Genome Research
    Project, 197
National Institute of Medicine, 317n
National Institute of Mental Health, 319n
National Institute on Drug Abuse, 153
natural selection, 2–5, 15, 21, 26, 114, 124,
    179, 187, 225, 243, 256, 265, 272
complexity and messiness of, 15
defined, 276
famine, disease, and war in, 25
microbes and, 74
premature babies and, 203, 204
sex and, 43